T0184836

Navigating the Engineering Organization

Transitioning new engineers into professionals who can immediately blend in and contribute to the technical organization is, at best, doubtful. Trained in the "nuts and bolts" of a technical subject, new engineers have little to no training in the "soft" skills of how to work within an organization. This robust guide shows new engineers how to quickly operate and succeed within their new engineering organization.

Navigating the Engineering Organization: A New Engineer's Guide focuses on the group behaviors of technical organizations. It provides a rigorous organizational framework to operate from and delivers guidance using a dual approach of academic insight and professional experience. Through numerous case studies, the book presents actual experiential guidance and offers a method on how to extend the insights covered in the book and turn them into a valuable personal model, valid throughout the engineer's career. It helps readers understand quickly the unique values and expectations within their new engineering organization and guides them in discovering the proper ways to respond to these expectations. They can then act on these insights to deliver successful results, now and throughout their careers.

The approach and goals found in this book provide a building block to help all new engineers cross the "Great Divide" from student to professional and succeed in their new engineering organization.

Navigating the Engineering Organization

A New Engineer's Guide

Robert M. Santer, Ph.D.

CRC Press is an imprint of the
Taylor & Francis Group, an informa business

First edition published 2023
by CRC Press
6000 Broken Sound Parkway NW, Suite 300, Boca Raton, FL 33487-2742

and by CRC Press
4 Park Square, Milton Park, Abingdon, Oxon, OX14 4RN
CRC Press is an imprint of Taylor & Francis Group, LLC

ISBN: 978-1-032-10252-8 (hbk)
ISBN: 978-1-032-10251-1 (pbk)
ISBN: 978-1-003-21439-7 (ebk)

DOI: 10.1201/9781003214397

Typeset in Times
by Deanta Global Publishing Services, Chennai, India

Access the Support Material: www.routledge.com/9781032102511

To Beth

Contents

PART ONE Preparing for Change

PART FOUR *Creating a Personal Roadmap*

Foreword

Having taught undergraduate and graduate engineers for 35 years at the University of Michigan, I was always impressed by their high energy, general smarts, and unbridled optimism that if they can think it, they can do it. When companies hire them, it is a bet on the future. Engineering graduates are a prized resource, and companies often treat them in a special way: paying well, offering interesting assignments, and even providing pathways to rapid promotion. View it as a honeymoon period. Unfortunately, all honeymoons end.

At some point, young engineers must carry their own weight and deal with the vagaries of organizations. Organizations are great at mobilizing resources toward big goals but also filled with infighting and strong personalities. Some of those people will have a strong desire for control, and that could include the boss who wants to prove the young graduate is not as smart as they think.

Navigating through the good, bad, and ugly takes more than high grades in courses and enthusiasm. It takes savvy. By savvy I mean the ability to understand what is going on and overcome obstacles toward your goals. Ultimately this will come for some with experience, while others might struggle throughout their careers, complaining to their family about the "politics."

The book you have by Dr. Bob Santer is intended to accelerate the process of developing savvy. Bob became my Ph.D. student at a mature age after years of experience rising through the ranks of the Ford Motor Company. He was an aerospace engineer, a manager of engineers, someone who hired engineers, and ultimately was recognized as having a special ability to connect with people of various persuasions. That included people in advanced engineering, in different stages of product development, and in production.

Ford had a problem. Some of their best ideas for new technologies were not getting into vehicle design. They would ultimately get rejected in the product development stage as too expensive or unnecessary. Bob was tasked with using his savvy to improve this process. In fact, his dissertation was about this very technology transfer process.

After Bob earned his Ph.D., I asked him to teach my popular course on *Work Organizations*, which was about organizational design and behavior. As I expected, he excelled and spent years helping students understand the human dynamics in the real world. He gave them a view of what to expect in the transition from student to professional. Ultimately that led to this book that is filled with practical advice but also draws on the best research by psychologists and sociologists about what makes organizations work.

You will learn about getting into a position to develop yourself, to connect with people with a wide variety of backgrounds, to work effectively in a team, and hopefully to ultimately become a leader developing future generations of people. The transition from doing the required work to becoming a leader is challenging and at the same time exhilarating. You will feel your power to positively influence people and society will grow. I envy you on your journey ahead.

Jeffrey K. Liker, Ph.D.
Professor Emeritus, Industrial and Operations Engineering, University of Michigan Author, The Toyota Way

Preface

Navigating the Engineering Organization stems from my own journey as a freshly graduated engineer into the complex and mysterious world of the large engineering organization.

My first professional position was as an aerospace engineer at a major U.S.-based aerospace firm, hired to help design the latest, next-generation navy fighter jet. A major "clean sheet of paper" project, I was excited and not a little prideful over my selection to be part of this leading-edge, high-technology aircraft program.

Arriving on my first day, my world instantly changed. Where was the private office I had seen in that eight-color brochure the recruiter had given me? Where were the clear-eyed, dynamic co-workers I was expecting to work with? Where was the smiling, fatherly boss with the time and interest to gently guide me on what I should do? And where was my arrival party, you know, the "we're glad you're here" gathering? Instead, I found 1950s office furniture, a table in a large "bullpen" area with drawings and folders stacked high that was to become my desk, and a work group made up of middle-aged engineers who seemed to not know a new guy was starting that day. After a cursory round of introductions accompanied by my new supervisor ("welcome aboard" was the operative phrase), I was told to make myself a spot at the refuse table and "settle in." Looking for a bucket to wash my new desk, I discovered I didn't even know where the men's room was.

I'm still trying to forget that day.

Thank goodness, times have changed. I'm certain your first day will be a pleasant and enjoyable experience. But here's the commonality between then and now. Coming in, I sincerely had no idea what was happening within the organization around me, and no one to help guide my way through. Every action I took seemed to result in alarming surprises and cringeworthy moments. And I believe most current graduates will have the same feelings.

I distinctly remember my first project meeting with a medium-sized group of engineers, financial people, manufacturing specialists, marketers, and other assorted business office representatives. Obviously, at that first meeting I didn't speak; I just listened. The same with the second meeting, and the third, fourth, fifth, and sixth. In fact, I didn't speak up for literally months. Only when I realized the people talking didn't really know as much as I thought they did was when I finally started to contribute.

By writing this book, I am wagering countless new engineers who have (or will) faced these same kinds of challenges, looking for some way to improve their chances of success while reducing the natural mistakes and missteps everyone experiences. Helping newly graduated engineers is the aim of this book, those who are looking for some way to improve their transition experience and quickly learn just what is going on in their new organizations. In short, helping new engineers progress from students to professionals is the goal.

As with many books, the material within is a result of my own professional journey of over 40 years. Added to these experiences are the corresponding organizational theory and research concepts, providing a more complete picture of underlying causes and effects. This means the points I share reflect my take on the common situations found in engineering and its management. This book is created through my "lens" and reflects the experiences I witnessed over time, those experiences that are as important today as they are in the recent past. To that end, please assume all persons and company names in this book are pseudonyms, except for my own personal experiences.

It's important to note that not all the situations and advice, guidance and counsel in these pages will be totally relevant to you as an individual at all times. To say that not all people (or engineers) are the same is obvious. The points made here may only be applicable to you in certain instances or with certain people. That's fine, and I depend on you to "take the best and leave the rest."

Finally, you may become (at times) a bit uncomfortable with one or two points in this book. If this happens, that's great news. This book is about change, and change can be inherently discomforting. Embrace the discomfort, and you'll get the most out of this book.

Enough of the preliminaries: let's begin.

A modifiable copy of Table 12.1 can be downloaded from the Routledge Website: https://www.routledge.com/9781032102511

Acknowledgments

Writing a book is a team sport. While the author's name may appear on the front cover, an entire roster of players contribute to making the final product a win for everyone involved. I am fortunate to have many of these players.

First, I'd like to thank my Executive Editor at CRC Press, Cindy Carelli, and her Senior Editorial Assistants Erin Harris and Christina Graben. Without their faith in this project (and well-timed suggestions), *Navigating* would have been no more than an orphaned idea.

My thanks also go to Dr. Brian Denton and Tina Sroka of the University of Michigan College of Engineering's Industrial and Operations Engineering Department. Their support in providing access to key research material was essential, and I am grateful for their help.

Interviews are a key part of any book of this type, and several contributors made a significant difference in the direction it took. Additionally, several individuals were gracious enough to read portions of *Navigating* and offer their sage advice. My thanks go to Peter Bejin, Bob Allan, Steve Santer, Chuck Nagy, Marcus Sprow, and Steve Karbownik. I also thank Katy Santer for reviewing and suggesting much-needed improvements to the graphics and figures, and especially to Matt Karbownik for his special support at Schoolcraft College.

Nothing beats experience when searching for insights within any engineering organization. Among the many who contributed their encounters and learnings, Ed Krause deserves special attention.

Remarkables are individuals who have special talents and experiences that, over a career, make them unique and valuable contributors to any project such as this. I am fortunate to have three. Paul Mascarenas taught me an entire new set of skills when we worked together at the Ford Motor Company. Colonel Jack Lousma gave me a new perspective on professionalism and what really matters in his many years as a guest lecturer in my capstone course at the University of Michigan. And as the 74th Secretary of the United States Navy, Dr. Donald Winter not only contributed valuable insights into engineering leadership but also continuously asked me those difficult questions as he reviewed the book's contents. I thank all three for their support.

Special thanks go to Professor Jeffrey K. Liker at the University of Michigan, who has continually given me his support and guidance throughout my doctorate and the writing of this book. I appreciate his understanding and insights during the long journey of bringing this book to light. His suggestions have always been helpful and unfailingly "spot on," and this work has greatly benefited from his effort.

Another individual deserves special mention. I wish to thank Dr. G. Fredric Bolling for challenging me to think beyond my normal boundaries and sending me on this path in the early 1990s.

It's no stretch to realize that without my family, this book would not have been possible. I want them to know that their kind words and encouragement have meant a lot to me. Special recognition goes to my mother, Mrs. Betty Santer, for her support and encouragement throughout. To my children Stephen, Steven, Nicholas, and Matthew, I extend my love and appreciation.

Most importantly, I want to thank my wife, Elizabeth, for her love, an endlessly giving nature and her continuous support as I sat at my desk and attempted to complete this work. Without her amazing ability to care for both me and our family, and her willingness to take on so many added responsibilities, I know this day would not have come. She makes it very easy to love her. She is a very special person.

Author

Robert M. Santer, Ph.D., has over 40 years of domestic and international industry experience in the aerospace and automotive fields as an engineering management professional and product analyst. His specific experience covers aerospace engineering, automotive design, product engineering, research, technology planning, innovation methodologies, organizational development, and technology communication strategy. He has international automotive design and production experience in both Europe and Asia.

Dr. Santer holds a Ph.D. in Industrial and Operations Engineering (Engineering Management) from the University of Michigan, Ann Arbor. He also holds a bachelor's in Aerospace Engineering and a master's in Engineering Management, also from the University of Michigan.

In addition to his industry role, Dr. Santer has taught the U-M College of Engineering graduate/senior-level course in engineering organizations, covering the foundations of technological organization structures, operations, and analysis.

1 Introduction
The Territory Ahead

1.1 ACROSS THE GREAT DIVIDE: THE NEED FOR CHANGE

The engineering organization is a mysterious place.

Full of surprises, unsolved riddles, and hidden forces, this territory is governed by invisible rules and unwritten expectations. And surrounding it is a Great Divide, a barrier separating it from the world of university studies. New engineering graduates entering this place could probably use some help, perhaps a good map to navigate its many paths and byways.

Fortunately, a new engineer does not need help with technical knowledge. Not with the calculus. Not with the statics and dynamics of non-linear mechanical systems. And certainly not with differential equations or C++ programming. I'm confident you have those subjects down cold. But I'm willing to bet that you and your fellow graduates might need some insight into how to *actually operate* within a professional engineering or technical organization.

When it comes to entering your first engineering position, it may feel as if you're standing on one side of that Great Divide, an intimidating mountain range preventing you from reaching the other side, that side representing the confidence, understanding and professional capability you wish for as a newly minted engineer. And the options on how to cross this Divide are limited.

Today, it's fair to say your ability to quickly blend in and contribute to your new technical organization may be a bit precarious. New engineering hires who misinterpret, miscalculate, and misstep early in their first jobs can create a rough (and potentially short-lived) start to their careers. And, most importantly, there is little published information to help engineers like yourself navigate this new and unknown territory.

But here's some good news. This book aims to reduce the mistakes, blunders, confusion, and catastrophes caused by organizational onboarding, allowing you to better focus on your immediate technical performance and contributions to your new firm.

This is a book about change and transformation. It is about preparing you to quickly transform from a university student into an effective and respected professional. Through this book, I hope to assist you in understanding your newfound profession differently than you might have before. In short, this book aims to provide you a robust guide on how to operate within your new engineering organization. It shares the universal underlying framework of engineering organizations. It provides an approach to navigate within this framework and the surrounding environment. It aims to help you create a personal model of how to work throughout your initial professional engineering career. And along the way, it provides an opportunity to

DOI: 10.1201/9781003214397-1

make sense of the many disparate pieces of the engineering organization. It gets you across that Great Divide.

Navigating the Engineering Organization provides missing knowledge and new insights into how you can become more effective during your early career. Most importantly, it provides support in solving the mystery of working within your new professional home.

As a reader of *Navigating*, I want you to benefit in four ways:

1) Quickly understand the unique values, expectations, and interrelationships present within your new (or future) engineering organization.
2) Help discover new insights into how to correctly respond to the unique organizational expectations of engineering.
3) Act on these insights to quickly deliver successful results to your management, colleagues, and the all-important external customer.
4) Rapidly build your new management's confidence in you, both through the results of your technical work and through your ability to deliver it within the expectations of your new professional home.

Together, we will do this by learning the fundamental engineering organization model that all engineering firms use and then adding to it the common transition problems you will face and the ways to deal with them directly and successfully.

This book delivers the benefits I just mentioned through three principles:

1) First, it focuses exclusively on engineering and technical organizations, addressing their unique characteristics, culture, and working environment. Engineering organizations are unique and distinctive creatures that must be treated as such.
2) Next, it provides a rigorous framework for understanding the technical organization, the key to navigating the special factors in how engineering organizations work as opposed to general business firms.
3) Finally, it employs a "dual source" approach to providing guidance. One source provides the academic understandings and research results underpinning each topic being discussed. The other source modifies and enriches the theory through targeted real-world practice, culled from over four decades of engineering organizational experience from the author's and colleagues' careers in the technology sector.

To realize these benefits, *Navigating* is structured into four main parts:

Part One: Preparing for Change

Examines our currently held assumptions regarding how engineers should approach problems, what information we should use in their solution, and what is the validity of our methods and technical belief systems.

Part Two: Guiding the Way

A simple yet strong organizational framework from which to operate, no matter the technical product, service, or other activity needing to be done.

Part Three: Operating within the Inner Core

Guidance in working within the center of this organizational framework, where the real action takes place.

Part Four: Creating a Personal Roadmap

This fashions a complete, personal construct of how to work within your organization, a foundation of understanding that can be used for the rest of your professional career.

A good question to ask is whether you believe this kind of insight is necessary or does a Great Divide even exist at all? Some graduates believe what engineering firms care about most are the new hire's technical skills, and any organizational skills will be provided by the company's "onboarding" process or are already in place through their university education.

This may be true, but the data say otherwise. Employers want college graduates with "soft skills," such as being a good listener or talented at collaboration. A 2020 survey by the American Society for Engineering Education found that the most in-demand talents for all college students were listening skills (74%), attention to detail (70%), and effective communication (69%).[1]

Unfortunately, few college students feel confident they have these skills and knowledge to succeed in a workplace. A 2017 report from Gallup, a survey firm, and the Strada Education Network, an education social impact organization, provides a comprehensive assessment of student views on this topic.

The survey size was big: more than 32,500 students from 43 four-year public and private universities participated in the study. Only 34% of those students indicated they were confident that they would graduate with the expertise to succeed in the job market, and just 36% said they believed they had the skills and knowledge to be successful in their careers.

Similar results extend to engineering students specifically. According to the same study, some 59% of new engineering graduates say they feel unprepared having the professional skills (defined as project management, business proficiency, communications, teamwork, etc.) necessary to transition successfully.[2]

This is borne out in survey after survey over many years. And I'm wagering that at least *some* of you will need at least *some* guidance in *some* areas of your professional engineering life. That's why it's critical to read each chapter carefully and actively consider the points of view presented. This approach will maximize the benefit to you, which is the purpose of this book.

With our general purpose described, let's now dispose of several small but important matters.

1.2 DEFINING A COMMON LANGUAGE

Before we can begin a meaningful discussion about how to cross this Great Divide, we must first develop a common language with well-defined and commonly understood terms. This is true not only for engineering organizations but for any other group activity, be it work, family, significant others, or any other person or group we wish to understand. And in a distressingly high number of cases, this doesn't occur.

In this book, we will be using some special terms and phrases which have very specific meanings. Here's a tip: you will soon find that the language and expressions commonly used in engineering organizations are alarmingly imprecise, contradictory, and generally all over the place. So many terms have multiple meanings and definitions that getting our common language settled is essential before we can even begin to think about understanding an organization. We'll define a few common terms now and introduce the rest as we progress through our work.

1.2.1 Management Ranks

We will spend a lot of time discussing "management," not as an action but as a group of individuals who each day control your workplace and, therefore, you. We should have some common and consistent names for these people. Simply put:

The Engineer is you.
The Supervisor is the one individual you directly report to.
The Manager has the Supervisor reporting to them.
The Manager reports to the Director.
The Director reports to the Executive Director.
The Executive Director reports to the Vice President.
Anyone above the Vice President really doesn't matter to us.

1.2.2 Organizations

We should also decide exactly what is an "organization." As with most of the terms and concepts in this book, the definition of an organization is not written on a stone tablet somewhere. Instead, it is soft and fluid, changing depending upon the context or situation being experienced and who's doing the defining. For example, ten minutes online will easily yield a dozen or more definitions of an organization, and almost all are correct. For this book, we'll choose our definition of an organization as:

> An organization is a social entity that is goal oriented, designed as a deliberately structured and coordinated activity system and linked to the external environment.[3]

Note that this definition is a combination of multiple characteristics. An organization is a social entity. It is a goal-oriented, deliberately structured, and coordinated system. And, critically, it is linked to an ever-changing external environment. It is made up of people, attempting to achieve a goal, with attention given to the relationships between those people and how they communicate. And it is not isolated; it is part of a larger, dynamic communications network.

No matter the precise wording of the definition, an organization does three important things. First, it designates formal reporting relationships, including the number of levels in its hierarchy and the span of control (i.e., the number of direct employees reporting to a particular superior). Second, it identifies the grouping together of individuals into departments and departments into a total organization. Finally, it includes the design of systems to ensure effective communication, coordination, and integration of effort across those departments.

For us, that's probably a good enough description at this point.

1.2.3 Common Organizational Levels

Another set of terms that will be useful are definitions of the common levels of an organization's hierarchy. Here, we will define three fundamental levels. Each level has a distinctly different breadth and depth of responsibility as well as its role within the enterprise.

As shown in Figure 1.1, the lowest level is *operational*, encompassing all workers like yourself, your colleagues, and your supervisor, who together perform the day-to-day operations of the organization. These are specific tasks done by individuals

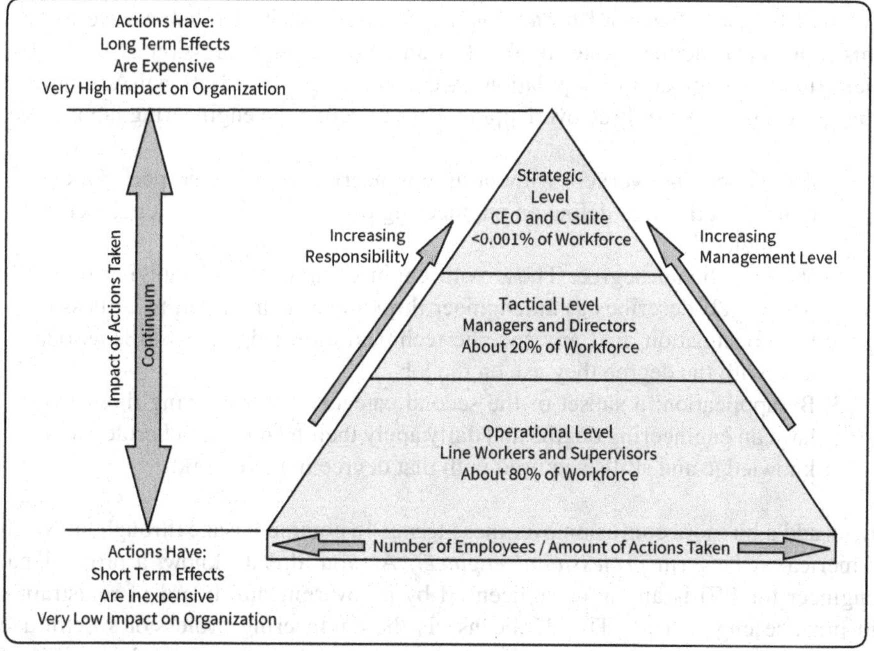

FIGURE 1.1 Fundamental organizational levels.

like yourself and very small groups. It easily encompasses the majority of the firm's employees, perhaps 80% or more. This day-to-day, singular work fulfills very specific objectives. Good descriptors of operational work include repetitive, detailed, daily, weekly, short-term, reoccurring, individualistic, and, yes, sometimes tedious activity. It can also include small problems to be solved immediately and so on. All these and more define the world of operations.

The next level up is *tactical*. This middle level is populated by managers and directors whose job is to combine and coordinate individual efforts from the operational level. This is a middle ground of not very long-term or very short-term activity. This is where subsystems are born. Here, medium-impact decisions are made and executed, such as developing schedules for perhaps two to three years hence or planning for midterm hiring needs, or where new subsystems are designed. Not ten-year thinking nor one-year work, but perhaps focusing on projects in the two to four-year timeframe.

The top level is *strategic*, where combinations of tactical subsystems are brought together to support the strategic intent of the organization. As you would suspect, the strategic level defines the big picture, where long-term goals are set, encompassing a macro view of the business. This is where visionary thinking occurs. Planning for the future, identifying, and preparing for goals and visions of the five- to ten-year variety happen here. The most senior members of your engineering organization spend their time and energy primarily at this level.

1.2.4 Professionals, Identified

We will be using the word *professional* quite a bit, so it's probably wise to take this moment to define the term. What is an engineering professional? Let's consider the U.S. engineering population. According to the U.S. National Academy of Engineering, there are three overlapping ways to define the engineering labor force:

1) By occupation: workers without an engineering degree who perform certain job duties that define an engineering occupation, i.e., an "engineering technician."
2) By educational degree: Those with engineering degrees, implying that a person self-describes as an engineer throughout their career, regardless of their occupation or if any specific technical knowledge and skills associated with the degree they use on the job.
3) By application: a subset of the second category encompassing those who have an engineering degree and daily apply their technical and professional knowledge and skills acquired with that degree in their work.[4]

Let's add a bit more confusion over these terms. In common usage throughout North America is the term *professional engineer*. As you already know, a professional engineer (or PE) is an engineer licensed by a governmental board of registration to practice engineering. The PE license is the engineering profession's definitive standard of ability, competence, achievement, and quality of work. Only 18–25% of engineers obtain their PE certification at any one time.

TABLE 1.1
Common Disciplines within Engineering

Engineering Discipline	Definition
Research	Pure investigation and creation of new inventions and enabling technologies
Development	Application of invention and enabling technologies to convert invention into innovation through a transformation
Design	Creation of a sellable product or service
Test	Verification of the function, quality, and reliability of the product or service
Production	Physical creation of the product or service, aka manufacturing
Construction	Building of enabling facilities and hardware and software, normally physical objects but may include software
Operations	Analysis and design of systems supporting all the above

On their face, none of these existing definitions alone will work so we'll create our own. Let's define an engineering professional a bit more broadly; as a combination of several of the statements above with several related characteristics:

> An engineering professional refers to anyone who earns their living from performing engineering-related activity that requires a certain level of education, skill, or training plus appropriate experience. There is typically a required standard of competency, knowledge, or education that must be demonstrated, often in the form of a university degree or professional credential. In short, an engineering professional is correctly educated, experienced, ethically strong, trustworthy, and receives compensation for services rendered.

Note our definition means no professional engineering license is required, but the underlying characteristics of a PE are expected.

1.2.5 Engineering Disciplines

It may be obvious, but we still need to define the main engineering categories or disciplines applicable to this book. We'll define seven different disciplines, based on their primary function as shown in Table 1.1.

That's it, simple and straightforward. While there are many other definitions we will run across during our time together, this is a good start to getting our language straight and creating common understandings.

1.3 SOME BOUNDARIES AND GROUND RULES

At this point it's important to discuss some fundamental boundaries and ground rules employed in this book. A *boundary* sets a limit on what we will consider in our discussions and what is out of scope. For instance, we will discuss many cause-and-effect relationships that are common in organizations, but we will not develop any mathematical expressions defining those correlations.

A *ground rule* is a general process we will or will not follow when discussing a given topic. For example, I view this book as a pathway to learning, to attain knowledge and insight in lieu of strictly checklist-style "training." Training implies only "how" to do something, while knowledge implies understanding both the "how" and "why" of the subject. Thus, our ground rules follow a pathway to knowledge and insight as opposed to simple training.

Both boundaries and ground rules are important, as we need to focus our energies on the main themes and ideas and stay away from the many rabbit holes in organizational studies (and there are a lot). That's what we're after.

With that understanding, let's look at some important boundaries and ground rules that underlie our discussions.

1.3.1 From "Great Man" Theory to Schools of Thought

Our first grounding principle has to do with the "schools of thought" approach to knowledge, as opposed to what is quaintly called the "Great Man" theory (with apologies to our female readers). The Great Man theory, by itself, can mess you up.

The *Great Man*, first described by Thomas Carlyle in 1840, is a common approach to education and training, especially in engineering and the sciences. Here, the educator selects a well-known individual as the single, end-all, and be-all expert for a given subject. We all know the list. For general relativity it's Albert Einstein; in anthropology it's Margret Mead; for sociology and psychology it's Malcolm Gladwell, and in evolutionary biology it's Richard Dawkins. Of course, we can't forget Stephen Hawking for theoretical physics. Closer to home, industrial engineering has Fredrick Taylor, the "Father of Industrial Engineering," and so on. This is a convenient approach to learning; just follow their precepts and you'll be fine. Unfortunately, this singular approach provides just a narrow fraction of what you might need to know, especially considering the importance of contingency theory (which we will discuss shortly). Because for every Edison there is a Tesla, for every Darwin there is an Owen, for every Alexander Graham Bell there is an Elisha Gray.[5]

That said, this theory isn't dead. The Great Man idea has been around for centuries before Carlyle. The idea survives and still influences all of us today.

In short, the Great Man theory is emotionally compelling. It is simple, easy to remember, and immensely comforting.

And extremely narrow, overly simplistic, slightly misogynistic, and totally inadequate when applied to engineering organizations.

The *school of thought* approach is the modern alternative to the Great Man. The idea is simple: there is no one single expert on any topic that contains any complexity whatsoever. Truth, especially in complicated systems like organizations, is not contained in one single idea or person but is an amalgam of many viewpoints, understandings, and insights. Organizational truth is not derived from straight-line analysis but is uncovered as the result of a triangulation of insights, merely getting us closer to reality but never fully attaining it. It is an approximation whose level of accuracy and precision is continuously improving but never definitive.

Currently, there are seven main Schools of Thought dealing with management theory. These schools have been created from fundamental assumptions and boundaries believed valid by their developers. These different approaches have been developed over time by specialists from a wide range of diverse yet related fields.

A summary of the principal schools of management thought can be found in Table 1.2.[6]

In this book, we will use the Contingency School as our foundational approach, while enhancing and enriching it with a collection of well-known management thinkers from psychology to sociology to anthropology, all contributing their thoughts to our discussion as appropriate. This is what we want: a holistic view of organizational knowledge today, drawn from many sources and points of the compass.

1.3.2 Which Approach: Scientific Management or Human Relations?

How an organization is managed is one of the critical factors of an engineering company's success, no matter what product you make or technical service you supply. In an area as complex as running an organization, there are many, many ways to manage.

I want to share just two basic approaches to running technical businesses. These are the *scientific management* method and the *human relations* approach. All other management techniques and tools tend to be some flavor of these two basic approaches. Let's take a minute to describe each one.

The earliest method of managing organizations was scientific management, a technique first developed in the late 19th and early 20th centuries. Following the (then) new and popular insights from the scientific revolution, proponents strictly applied scientific principles to emerging factory work. Here, systems were broken down into individual components; each was studied independently and then singularly optimized. Only after this optimizing task was completed was the system reassembled.

This method encompassed the main beliefs that:

1) Efficiency is everything. The greatest output per unit of input was the singular goal of the firm, in other words, maximum efficiency. Whether the product satisfied the marketplace need (i.e., effectiveness) was another matter entirely, leading to the possibility that the firm could produce the wrong product in a highly efficient manner.
2) Workers were seen as extensions of machines. Labor (i.e., humans) were perceived as expendable, replaceable parts to be popped in and out as the needs of the machine (the factory) demanded.
3) Maximum task breakdown, allowing for maximum simplicity of the task. Unfortunately, it also created "dumbing down" of work to produce mind-numbing jobs. Thinking is left to management, which meant that "workers work while managers manage."
4) The organization's goals and objectives were the focus of all efforts, to the detriment of the workers.

TABLE 1.2
Principal Schools of Management Thought

School of Thought	Definition	Key Beliefs
Management Process School (aka Operations or Traditional School)	A process of getting things done through people operating in organized groups.	Process management principles are universally true and applicable in all situations.
Empirical School	Management is the study of experience and decision-making. By analyzing the experience of good or bad managers or engineers the most positive methods can be applied.	Managerial experiences can be taught to practitioners and engineers. Techniques used in successful cases can be directly applied by future managers. Theoretical research can be combined with practical experiences for better management.
Behavioral Science School	Improve the organization by understanding the psychological factors driving work behavior. Based on the reality that getting work done is through people, and thus management must be focused on interpersonal relations.	Heavily oriented toward psychology and sociology. Includes the study of human relations, the role of manager as a leader, and the study of group dynamics and interpersonal relationships. Stresses the people part of management and understanding.
Social Systems School	Very closely related to the behavioral science school. This approach to management is heavily sociological in nature.	Sees management as a social system of cultural interrelationships, either as formal organizational relationships or various informal human relationships. Identical to the study of sociology.
Decision Theory	Concentrates on the selection of a suitable course of action from various possible alternatives. May deal with the decisions itself, with the person or organizational group making the decision or analysis of the decision process.	By expanding beyond evaluating alternatives, can apply theory to examine organizational structure, psychological and social reactions of individuals and groups and analysis of value considerations for goals and incentives.
Systems Approach	An integrated approach that considers management in its totality based on empirical data. The systems approach brings out the complexity of real-life management problems much more sharply than any of the other approaches. Can be utilized by any other school of management thought.	According to this approach, attention must be paid to the overall enterprise. A system has several subsystems, parts, and subparts. All subsystems, parts, and subparts are mutually related to each other. A change in one part will impact most others. This approach emphasizes study of the interrelationships of subsystems rather than in isolation.

(*Continued*)

TABLE 1.2 (CONTINUED)
Principal Schools of Management Thought

School of Thought	Definition	Key Beliefs
Contingency School	Emphasizes management as a highly practice-oriented and action-based discipline. Managerial decisions and actions are more a matter of pragmatism and less of theoretical principles. The choice of approaches and their underlying management theory and principles tend to be deterministic, while the pace, pattern, and behavior of events defy deterministic approaches. What is valid and good in a particular situation need not be the same in some other environment.	The environment of organizations and managers is very complex, uncertain, ever-changing, and diverse. It is the basic function of managers to analyze and understand the environments in which they function before applying their techniques, processes, and practices.

The belief in science underlying this method was appealing, as it was clean, simple, and for many years seen as the one and only correct management system. Yet while the scientific management method initially resulted in tremendous productivity gains, it also caused abuses to the human component of the system.

Please don't think that today scientific management is a vestige of some ancient production system. It is stunning to see how widespread the strict scientific mindset exists today, especially within small firms in second and third world locations.

Due to these issues with scientific management, a second major method emerged in the middle of the 20th century. The human relations approach reacted to the scientific method by focusing only on the social needs of the workers and ignoring the core technical (i.e., the input/output transformation) system. It attempted to focus on the entire production system, not just its individual parts. Significantly, it considered how informal groups inside the organization could actually control people's work habits and attitudes.

A famous example of this impact is the Hawthorne Effect, first identified by George Elton Mayo and his colleagues as part of an efficiency study at the Western Electric Company's Hawthorne plant. Western Electric wanted to know how much productivity would increase when better illumination was provided in their facility. The result was surprising: productivity increased as illumination was increased. Productivity also increased when illumination was *decreased.* Mayo discovered that productivity increased, not as a result of better lighting but as the result of the attention received by the workers under study.

A further analysis of the Hawthorne Effect found that, overall, people are social creatures and work is primarily a social activity. This implies that the need for recognition, security, and sense of belonging is more important in determining morale and productivity than physical working conditions alone.[7]

These (and other) findings spurred further studies in the human relations model. In 1960, social psychologist Douglas McGregor developed two contrasting theories explaining how managers' beliefs about what motivates their people can directly affect their management style. He labeled these *Theory X* (authoritarian) and *Theory Y* (participative).

Theory X managers tend to take a pessimistic view of their people. This theory assumes that workers dislike their work, avoid responsibility, need constant direction, must be controlled, be threatened to deliver work, need to be closely supervised, and have no incentive to work or have any ambition. Authority is rarely delegated, and control remains firmly centralized. Managers are more authoritarian and actively intervene to get things done.

In contrast, Theory Y managers have a more optimistic, positive view of their people and use a decentralized, participative management style. They encourage a more collaborative, trust-based relationship between management and workers.

Theory Y assumes that workers wish to work on their own initiative, be involved in decision-making, are self-motivated, take ownership of their work, accept responsibility for results, and need little direction. They see work as fulfilling and challenging, and Theory Y organizations also provide employees with frequent opportunities for promotion.

Obviously, Theory Y is viewed as superior to Theory X, reflecting the desire for more meaningful work and providing staff with more than just money.[8]

These theories continue to be important today, providing the basis for the ongoing experimental work methods being developed in startup and entrepreneurial firms, especially digitally-based organizations.

So what approach are we going to take in this book: scientific, humanistic, or something else? Frankly, compared to the length of human history, we really don't know much about organizations and management. And yes, many of today's organizations are still heavily imprinted with hierarchy, bureaucracy, and formalization. But another reality has emerged. Engineering organizations, both large and complex or small and entrepreneurial, are relentlessly experimenting with highly innovative approaches to the workplace that defy easy categorization. The orderly and predictable assumptions regarding these organizations, the role of managers, and their tools are furiously being replaced. What leaves us is an entire spectrum of work methods and approaches that are advancing without pause. Geographic location, age of the technology, first adopter or legacy, technology production or development, sophistication of the corporate culture, and worker age all heavily impact the firm's management style.

It's obvious the proper methodology is highly contingent on these and many other factors. So, our approach is neither: we must address it as we proceed.

1.3.3 Which Perspective: Engineering, Business, or Other?

Several of you might be pondering if this book is more of a business tome than an engineering book. Let me assure you that this book is about engineering, but from the point of view of who we are, what we do, how we think, and how we get things done, not just engineering as a mechanical function we perform each day. It is about engineering organizations, their operation, and your role within them.

There is a long-standing mental model about the relative knowledge of engineers vs. businesspeople. Knowledge has both depth and breath. The engineer tends to have a depth of knowledge: a specialist who has a precise and complete understanding of a narrower discipline. Businesspeople may gravitate toward the general: thinking in broad ideas and trends, but perhaps not in depth. Obviously a stereotype, but it does highlight the basic difference between engineering and business: engineering is precise; business is approximate. Engineering looks to the objective for guidance; business tends to be more subjective. Engineering is quantitative, and business is more qualitative. The differences between the two are not merely about topics but also about their fundamental viewpoint about knowledge. While initially distressing, there is actually an advantage when studying both: the advantage of balance in pursuit of knowledge.

This brings us to the role of today's management consultant. Philip Coggan, a business columnist for *The Economist*, has written about the massive number of words devoted to and about business management, lamenting the number of books piled high on his desk that he will never read. Coggan contends the sheer variety of business advice published suggests that management gurus don't yet have a handle on the enterprise nor organizations: if they did, the pile of books on his desk would disappear.[9]

This means there is risk in blindly applying business advice from a single source, as some advice can actually be harmful to the firm. The good news is that these once fashionable but damaging management ideas do (thankfully) get thrown on the scrapheap. *Shareholder value* had its time in the sun: it led to poor or unsustainable business practices such as, in some cases, businesses falsifying financial information to boost shareholder returns. *Business process reengineering* was identified as one of the most important solutions for organizational improvements until high failure rates of 70% or more were reported.[10] The list of these ideas has led some to refer to these notions as the "fad of the month."

What all this means to answering our question of a business vs. engineering approach is a mixture of both: the more certain and precise aspects of engineering applied against the generalist and highly contingent facets of business. For you this implies a struggle: the uncertainty of balancing the science-based training you've experienced against the contingency-based foundation of organizations. At this point, just know it's coming, and let's move on.

1.3.4 What We Won't Address

As important as what we will cover in this book, of equal importance is what we won't cover. There is absolutely no way this book can cover every single topic or subject regarding organizations, especially engineering organizations. And that's okay: we don't intend to. Instead, I'd like you to use this book is a first step in understanding the situations you might find yourself and then developing effective solutions based on those situations. In this way, this book provides the tools to arrive at a first understanding, and then help develop a holistic path to resolution. Remember, we are about knowledge and insight here, the how and why, not checklist-style training.

For the easy questions, yes, this book hopefully provides some simple and straightforward options, but for more difficult questions we intend to provide a method or approach and ultimately answer those questions, what our friends studying philosophy would call a "path to Nirvana."

1.4 REALIZATIONS

We've started this book by talking about the idea of transforming students and recent graduates into engineering professionals. The stakes can be quite large, and while some may find this transformation easy, most others will discover it to be quite difficult without some guidance. The simile here is that of crossing a large barrier, a Great Divide, and we need to arm ourselves with knowledge, insight, and a few tools before setting out to make that crossing.

The basic idea is, as you transition from an engineering student to a professional, you realize there is indeed an important change going on, and the success of this change depends on you, and only you, to make this happen.

This book aims to help you understand the unique values and expectations present within your new engineering organization. It will help you discover insights into the unique organizational expectations of engineering. It will assist you in acting on these insights to deliver successful results to your management, thus rapidly building your management's confidence in you as a professional.

It's important we set out some fundamental boundaries and ground rules employed in this book, ring-fencing the specific topics and what (and what not) will be covered. Of particular importance is the school of thought approach, represented by the contingency model that we will follow. Related to the contingency model are two important concepts: scientific management vs. human relations, and the engineering vs. business perspective. Both are discussed in detail.

So that's our introduction. We are on our way to crossing a Great Divide to a new destination, gaining some helpful ideas about the strange land you now entering, or perhaps just avoiding some skinned knees. If either of these outcomes occurs, we can count it as a win.

Let's get on with it.

NOTES

1. Sorby, Sheryl and Smith, Dora. 2020. *2020 Survey for Skills Gaps in Recent Engineering Graduates*. Washington, DC: American Society for Engineering Education.
2. Strada Higher Education Network and Gallup. 2017. *2017 College Student Survey: A Nationally Representative Survey of Currently Enrolled Students*. Washington, D.C..
3. Weick, Karl. 1993. The Collapse of Sensemaking in Organizations: The Mann Gulch Disaster. *Administrative Science Quarterly* 38: 628–652.
4. National Academy of Engineering 2018. Understanding the Educational and Career Pathways of Engineers. Washington, DC: The National Academies Press. https://doi.org/10.17226/25284

5. Mouton, Nico. 2019. A Literary Perspective on the Limits of Leadership: Tolstoy's Critique of the Great Man Theory. *Leadership Journal* 15(1): 81–102.
6. Kukreja, Sonia. 2021. Major Schools of Management Thought. *Management Study HQ*. https://www.managementstudyhq.com/schools-of-management-thought.html.
7. Daft, Richard. 2010. *Organizational Theory and Design*. Mason, OH: Cengage Learning.
8. McGregor, Douglas. 1960. *The Human Side of Enterprise*. New York: McGraw-Hill.
9. Coggan, Philip. 2020. Lessons from 100 Columns: Management in Theory and Practice is Found Wanting. *The Economist*. https://www.economist.com/business/2020/06/11/lessons-from-100-columns.
10. AbdEllatif, Mahmoud, Farhan, Marwa Salah, Shehata, Naglaa Saeed et al. 2018. Overcoming Business Process Reengineering Obstacles Using Ontology-based Knowledge Map Methodology. *Future Computing and Informatics Journal* 3(1): 7–28.

Part One

Preparing for Change

2 The Impending Problem
From Student to Engineering Professional

2.1 WELCOME TO YOUR BRAVE NEW WORLD

Let's see if you agree with this. On graduation, you believe the hard part is over. No more exams, no more all-nighters, no more Ramen noodles at 2:00 am. Instead, you begin your professional life as an engineer. Think of it: just a mere eight hours a day, working on an exciting, life-altering product, basking in the praise of your management and peers as your work is celebrated across the industry. And don't forget the free snacks.

Of course, you already know these comforting thoughts aren't necessarily true. But just what *is* going on? What will happen as you begin your new job? Will you fit in or be a pariah? Will anyone help you? Great questions, and this chapter will discuss the idea that yes, something big is lurking out there but you just can't see it yet: the reality of your first professional job. The pre-season practice is over; it's time for the game to count.

This moment describes the concept of "crossing the Great Divide," where graduation and winning a new job immediately makes your current engineering education, well, less relevant than you might have thought. When you finally arrive at work, you will probably experience confusion, uncertainty, and worry while immediately making mistakes of omission and commission, leaving your new boss shaking their head as to why they hired this minnow. And there will be a cost to you. You will quickly be assessed using the concept of confidence and comfort, that is, their confidence in your work result plus comfort in your fundamental social ability and behavior, both of which are necessary to integrate into your new home.

Indeed, welcome to your brave new world.

2.2 THE BASIC TRANSITION: STUDENT TO PROFESSIONAL

The essential problem you face is simple yet also complex. On starting your new job, your singular and immediate challenge is to successfully cross the barrier that stands before you. It's a problem you probably have not faced before. While the goal is clear, the method of how to get there is not. You probably do not have the necessary tools to achieve it. Even if you do know how to transition and have the necessary

DOI: 10.1201/9781003214397-3

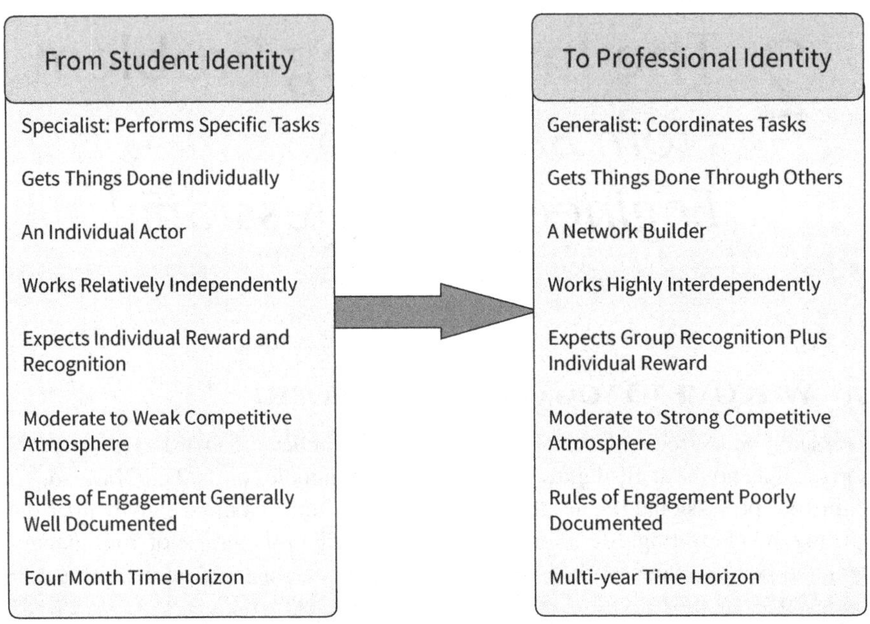

From Student Identity
Specialist: Performs Specific Tasks
Gets Things Done Individually
An Individual Actor
Works Relatively Independently
Expects Individual Reward and Recognition
Moderate to Weak Competitive Atmosphere
Rules of Engagement Generally Well Documented
Four Month Time Horizon
To Professional Identity
Generalist: Coordinates Tasks
Gets Things Done Through Others
A Network Builder
Works Highly Interdependently
Expects Group Recognition Plus Individual Reward
Moderate to Strong Competitive Atmosphere
Rules of Engagement Poorly Documented
Multi-year Time Horizon

FIGURE 2.1 Transformation characteristics from student to professional identity.

tools, do you have the skill to use them? And how fast do you need to perform this metamorphosis?

At least the goal is simple. What we're looking for here is a transition, a transformation. A movement from one state of being to another. And what are you transforming into? Simply stated, a professional identity. Figure 2.1 sums it up.

2.3 EASY OR NOT? YOUR CHOICE

Up to now, as an engineering student you have dealt with the immense world of science, mathematics, and technology. The calculus, differential equations, statics and dynamics, exotic materials and digital design; all were expected to be mastered within four short years. Simultaneously, the amount of technical information available to you during that time has snowballed, forcing you to cram ever more technical facts and concepts into that same four-year period. You learned the detailed "nuts and bolts" of technical subjects, but scant time was spent on the overarching skills of understanding and operating (never mind succeeding) in the actual day-to-day work of your new organization. This duty is left to you alone, perhaps (if you are fortunate) supplemented with a sketchy onboarding program run by your company's human resources organization. Sadly, the reality is you will probably have little knowledge or understanding of your new professional home and how to operate within it.

Of course, the good news is that everyone experiences this phase, and I doubt anyone avoids it, but that is small comfort at this moment. The question being posed to you now is this: Will you transition from student to professional with an easy laissez-faire approach, letting things take their own course, confident in your perceptions and knowledge of people and the workplace? Or will you proceed with a more difficult proactive mindset, acknowledging that this is new territory for you, yet wanting to guide the trajectory of your transformation as you see fit?

Let's consider some examples.

CASE EXAMPLE 2.1 BUILDING THE F-18 JET FIGHTER

One of my first experiences in transitioning had to do with an early assignment as an engineer. I had just entered the McDonnell Douglas Corporation (now Boeing), helping to design and build the F/A-18 fighter jet for the U.S. Navy. At that time the F-18 was the world's most advanced fighter, and I was very pleased (if not somewhat prideful) of my assignment as a design engineer and my part in this important project. As the fighter progressed from "drawing board" to first prototype build, I followed it onto the factory floor as a liaison engineer, responsible for solving initial build problems and correcting them for the production phase to follow.

In the factory, I became acquainted and worked with a foreman by the name of Bob Tyler. Bob was an excellent foreman, with about ten years' experience building advanced aircraft. I quickly observed that Bob had the expertise and knowledge to solve an amazing number of assembly issues, and had the communication skills to successfully direct his build crew in correcting these unique problems.

Late one afternoon, while working on the plane's outer wing, one of Bob's crew mistakenly drilled about 20 holes into a critical structural fitting. Since all flight stresses from the wing went through this key fitting, it had to be built perfectly. I was deeply concerned; there was not a spare part available so this damaged part had to be fixed correctly, otherwise the wing (and thus the entire aircraft) could suffer a major structural failure in flight. Consulting the structural engineers, we developed a repair that would work, but it had to be executed precisely with absolutely no margin for error. At the end of the day, I worriedly briefed Bob on the repair his crew must perform that evening. Bob assured me he and his crew would handle the job correctly that night.

That evening during dinner, I continued to worry about the problem. What if the crew mishandled the repair? They had already made a mistake, why wouldn't they make another? What if somebody didn't get the word and inadvertently wrecked the part? The first flight of the jet would be delayed, and I would be held responsible. Finally, standing it no longer, I returned to the factory after dinner to see what the crew was doing. As I approached the wing fixture, I saw Bob standing there. He was not happy to see me. As I approached, he asked sharply:

Tyler: "What are you doing here?"
Santer: "I was worried about the repair and wanted to see if the crew was doing it right."

I'll never forget what happened next. Looking me in the eye, he said:

Tyler: "Santer, I'm really disappointed in you. You coming in here sent me the message that you don't trust me or my crew's ability or desire to do this job right. Do you think this is all about you? It's not. My crew and I are just as interested in doing this right as you are. I suggest you go home, grab a beer, and watch the baseball game. We got this."

I was stunned and more than a little embarrassed. But he was right. I assumed no one else cared about the job but me. I exposed a bias toward Bob and his crew by thinking of them as merely factory workers, showing I lacked confidence in their professionalism to do the job right.

From that experience, I always tried to put myself in another worker's position and certainly not judge them based on a stereotype. Instead, I now wait to see the capabilities and attitude my colleagues actually demonstrate before I make any judgments. Bob did me a great service that evening, and I haven't forgotten it.

How about the situation in an office rather than the factory floor? Take Brad Collins, a new engineering hire in a Fortune 100 multinational. After a day or two of settling in, his supervisor, Craig Wilson, came to him early one morning with his first assignment. "Brad, calculate the weight difference between these two competing compressor designs." Brad, eager to show off his engineering prowess, begins a detailed analysis, calculating the weight to the third decimal point. At 8 o'clock the next morning, Craig approaches Brad and asks, "So, Brad, what did you find out?" Brad proudly reports that the first compressor weighs 31.724 lbs., and he will have the second compressor weight completed by day's end. Craig is shocked. "Brad, we're reporting out this weight study in 30 minutes, at the weekly project review. You mean you only have half an answer? Brad admits that, indeed, he only has a partial answer, saying he didn't know about the review. Craig responds "Well, we'll have to tell them something."

At the review, Craig covers for Brad by making the status report himself, stating they have not completed their analysis but will send out the complete study by the end of the day. The program manager is annoyed, but grudgingly accepts the plan.

What happened? A classic case of unspoken expectations and conflicting assumptions between an experienced supervisor and new hire. Brad thought the important deliverable was a highly accurate analysis; Craig wanted a fast, approximate answer. Brad didn't ask when the assignment was due; Craig assumed Brad knew the due date was the next day's meeting. Brad cringed when being told of the misunderstanding; Craig was annoyed having to take the blame for his new hire. Brad could have avoided this outcome if he knew enough to ask for an explicit due date and what level of accuracy was required; Craig could have prevented the incident by realizing that Brad needs explicit instructions and explanation at this point. But neither understood the other. The takeaway? Both Brad and Craig lose a little confidence in Brad.

This is what a new engineer's transformation is all about: understanding both the overt and covert workings of the technical organization. Yes, learning these lessons and benefiting from the resulting experience can be uncomfortable. And sadly, there is little help available to assist you. Relevant books and published papers from professional engineering societies or other sources are virtually non-existent. The small number of "engineering management" books available offer a series of disjointed, stand-alone technical topics in finance, production, decision-making, and the like, but either ignoring engineering organizational behavior (EOB) entirely or treat it as an afterthought. The EOB aspects are unique to engineers: their thought patterns, work methods, definitions of success, work quality, and other engineering-specific behaviors are missing. Most importantly, no unified framework exists for new hires to understand organizational structure and then successfully apply these EOB concepts to their own technical organizations.

So, we arrive back at our initial question. Do you choose to move into your first professional job as so many people have done before, leaving your transition to luck and Brownian motion? Or do you wish to maximize your transition opportunities (and minimize your cringeworthy moments) by actively applying some of the concepts shared here? By reading this far, you are probably open to the latter. I hope that assumption is correct.

2.4 CROSSING THE GREAT DIVIDE

Congratulations on choosing to move forward. Let's next talk more about the idea of the Great Divide that separates your current organizational skills from the expectations of your new organization. For-profit businesses want new engineers to contribute rapidly and effectively, with little or no incurred cost. Unfortunately, you do not yet have these skills, and as I've said, little guidance exists on how to obtain them without excessive time or embarrassment.

Obviously, this matters. If you as a new hire cannot effectively and efficiently integrate your effort into the organization, your work will certainly be relegated to the organizational shelf or, worse yet, the dustbin.

You have work to do in understanding what is really going on within your new group and with the organization in general. With that starting point, what information or guidance do you actually possess, right now, that you can use in your transition? Your new company probably did not give you any brochure on what to expect, no "how to" guide on what's next. Your past professors may not be the best source of current information, and perhaps the paid (or, more commonly, unpaid) internship you completed during your junior year was less than ideal. Maybe you have the good fortune of having friends or acquaintances already in the company. They can certainly give you a bit of a head start on what to expect, but they will probably be in a different department than you, and the differences between departments may be substantial. You will be exposed to a new language of the company, with acronyms galore. And what's important to your boss and your new organization is something you've probably never considered. So based on what you actually have in hand, you may be a bit unprepared for what's to come.

Yet you must proceed now to cross the Divide. You may feel woefully unprepared. And that's okay.

One thing you have going for you is this: once hired, all companies (including yours) want you to succeed. After all, they have devoted substantial money and time to your recruitment and onboarding. Managers are held to objective metrics on how well their talent acquisition systems are operating. As a result, they care deeply about understanding how your generational cohort (at this writing, Generation Z) thinks and believes to better recruit and apply your fresh knowledge for the firm's greater good. This brings us to the question of internships.

2.5 INTERNSHIPS ARE NOT PREPARATION

We all know the situation. The expectation is that you, a recent graduate engineer, should have had one (if not two or even three) engineering-related internships, spending your summers at major, multinational companies on the leading edge of digital or technical innovation. You should have spent your time with fellow students sharing an awesome engineering experience, glowing with a special feeling as they prepared you for a new, exciting professional life.

That certainly is the hope, and many engineering undergraduates like yourself have enjoyed some of these eye-opening and exciting experiences. They may have even paid you, or provided free apartments, moving allowances and business trips.

For those lucky few, great. But for the vast majority of engineering students the internship experience may have been less than optimal. While the current emphasis on engineering internships can certainly help you for a few insights, the reality is that internships are generally much too short, limited in scope and very, very uneven in quality to be of much help. Let's face it, some internships can result in the old joke: "If you pretend to train us, we'll pretend to learn."

You may wonder: why wouldn't your internship adequately prepare you for your Brave New World? After all, isn't that the purpose of an internship, a two-way street where you see if you might like the work, and the company gets an early look at your capabilities, talents, and technical prowess? Of course, that's the general idea. But internships are not professional jobs.

There is no doubt, the intent of internships is well-meaning. Internships are meant to sharpen classroom knowledge against the stone of real-world experience. You may have read about the benefits of gaining experience, increasing your marketability and competitive advantage for jobs while building your professional network. Yet there are common issues with internships that can negate the preparation aspects they are meant to address. A significant issue is underutilizing interns, where some students are ignored or perform menial tasks for the entire duration of their tenure. And no one is fooled; everyone is fully aware that what they do is not real-life experience. Another issue is being neglected by your supervisor or other person assigned to guide you as your supervisor or leader is much too busy to expend any energy on someone who will be gone in three month's time. Still another is the lack of performance feedback. For those who have had at least one internship, this doesn't qualify as news.[1]

TABLE 2.1
Differences between Internship and Entry-Level Professional Positions

Internship Position	Professional Position
Unnoticed Work	Highly Noticed Work
Uncooperative Supervisor	Fully Engaged Supervisor
Struggle with Time/Self-Management Issues	Imperative to Quickly Resolve Time/Self-Management Issues
Allotment of Trivial Work	Allotment of Meaningful Work
Inadequate Compensation	Negotiated Compensation
Competitive Co-interns	Competitive Collogues
Underwhelmed with Work	Overwhelmed with Work
Hesitant to Ask Questions	Better Ask Questions

It's clear that, while internships hold so much potential, they fall short in giving you the critically needed insight and skills to help you make a successful transition.

So, what does this transition from student to professional look like? Table 2.1 is a good start in summarizing this metamorphosis:

Each of the eight points in the left-hand column can help describe your internship experience. More importantly, the right-hand column can give you a hint into what your professional work life might look like.

I'm certain we all wish the internship experience could be better. But the internship experience is generally not a good preparation. So, what about other resources to help you conquer the Great Divide?

2.6 THE BOSS IS NOT THE PROFESSOR

Let's start with a single basic fact: the boss is not the professor. I repeat, the boss is not the professor. A boss and a professor are two fundamentally different management positions designed to accomplish two very different tasks. You cannot substitute one for the other.

Your relationship with each is based on an implied social contract, which is an implicit agreement among the members of an organization or society to cooperate for mutual social benefits. However, in this discussion those social benefits are quite different for each relationship.

A professor is a facilitator charged with assisting you in receiving your education. While they are an authority figure, they are nominally your guide, enabler, and judge who is obligated to provide training or knowledge (or both) to ensure you have reached the academic standards imposed by their institution, your profession, and overall society. Their role is to provide a credential that certifies you as "educated" in your profession. After you receive that credential, the relationship normally ends.

A boss is also an authority figure, a person who exercises control or power over workers in a business setting. Remember, an organization is about converting time, talent, money, and other resources into an output that is of more value than its inputs. Your boss, be they a supervisor, manager, or higher-level director, is there to ensure

TABLE 2.2
Differences between Professor and Professional Manager Roles

Characteristic	Professor Provides	Supervisor/Manager Provides	Comment
Benefit Provided	Professional training, technical skills preparation and academic understanding	Employment and opportunity for future professional growth	Both can individually counsel you but are not required
Exchange of Value	You pay them	They pay you	One is an investment to gain access to the other
Judgment Type	Judges you in a narrow band of technical, academic-based skills only. Provides certification for external organizations	Judges you in a wider band of both corporate behavior and performance	Severity of their judgment varies widely by who is the judge
Relationship Duration	Relationship normally extends for approximately four months	Relationship typically extends for a few years, longer if mentorship involved	Rarely extends for much time after either party moves on

your efforts translate into achieving the organization's goals, normally (but certainly not exclusively) financial.

The differences can be seen in Table 2.2. But here's a somewhat cynical insight: there are actually a pair of similarities between the two. The first similarity is that you provide resources to both. You pay universities, and you invest your talent and skill into firms. The second similarity is more impactful. Both institutions (universities and engineering firms) openly promote the notion that their management (professors and managers) energetically guide and counsel their charges (students and new hires), devoting plentiful resources to your training and development. Instead, you will find they will devote plentiful resources on their own annual objectives and goals, which normally have nothing to do with a student's or new engineer's development.

No matter what, the simple and obvious truth is your immediate future is now determined by the boss, not the professor. The totality of your attention should be directed at those supervisors, managers, and directors who hold the path of your future in their care.

Your professors are now obsolete.

2.7 CREATING AN ENGINEERING ARCHETYPE

As you and thousands of your new brother and sister engineers stand ready to cross the Divide and arrive at the promised land of the professional, it's instructive to try to categorize this group of new graduates into some broad categories. Certainly not at the individual level (that would be flat wrong and, frankly, a bit insulting) but as a collective. Who is this group of new engineers? What are they about? And why do we care?

That last question is a simple one. As you will see repeatedly in this book, to comprehend today's technical work world means to understand not only yourself, but especially those who will be your colleagues and, yes, your competitors. You must attempt to grasp their thoughts, recognize their values, and know their limitations and strengths. This understanding is both necessary and required in preparing yourself for what is to come. And creating an archetype is a first step.

There is a difference between an archetype and stereotype. A stereotype is a widely held, fixed, and oversimplified image or idea of a particular type of person or thing. It's a preconceived idea of the characteristics which typify a person or situation, or an attitude based on such a preconception. It's a person who appears to conform closely to the idea of a type, a prerequisite belief about an individual or individuals.

An archetype is different. It's a statement, pattern of behavior, or model which others can replicate or emulate going forward. It is the original pattern or model from which copies are made. It's a prototype of an assumed ideal form. In other words, it's the original of something, the original from which a model is developed.[2]

Archetypes apply to people as well as objects. Unsurprisingly, there are data available that outline the average characteristics of who and what a new engineer might be. Yes, this can be a slightly dangerous exercise (no one likes to be profiled), but in the aggregate it's instructive in understanding the professional society you are about to enter. (As a preview, in Chapter 3 we will ask you to evaluate yourself individually using some very specific tools, but right now we're attempting to outline and flesh out the engineering *group* archetype.)

Luckily, we have lots of data to work with. Groups such as the American Society for Engineering Education (ASEE) are awash in demographic and quantitative data regarding each year's engineering graduate population. For example, ASEE reports very complete statistical data on undergraduate and graduate graduation numbers, gender, international vs. domestic origins, and so on. From this we can begin to construct an amalgam of these graduates.[3]

So let's go ahead and create an archetype of a prototypical new engineer. We'll use three separate evaluation tools to perform this analysis: Demographics, the Myers-Briggs Type Indicator (MBTI), and the Five-Factor Personality Inventory (FFPI). Why multiple tools? These and other personality assessments merely provide clues into the engineering temperament, not definitive conclusions or predictions. Trying to create an archetype strictly from any single tool will be inaccurate and highly arguable, so we must use all three tools to triangulate in on a judgment of the population under study. In short, maximum benefit comes from creating an understanding of others through a combination of these tools.

Let's now step through our analysis.

2.7.1 Demographics

Demographics are statistics that describe populations and their characteristics, and is the easiest analysis to perform. In this case, we will limit our archetype to a North American engineering graduate. We'll call this engineer Jacob (the most popular American boy's name in 2000), a 23-year-old male engineering graduate (77.5% of bachelor-level engineering graduates are men) from the Georgia Institute of Technology (the highest percentage of all national bachelor engineering graduates come from there). Jacob is a graduate in mechanical engineering (surprisingly, 24.2% of all undergraduate degrees awarded are mechanical, higher than electrical or computer) and had good grades at university with a 3.4 out of 4.0 GPA (the top 25% of the average of U.S. schools). Born in the United States, Jacob comes from a solidly middle-class family and was raised in a two-parent household that valued education and achievement. He landed an entry-level position in a large multinational product development firm in his mechanical engineering specialty.

Pretty unremarkable information, only covers an individual. What about the demographics of the related members of his generation? What insights can be gathered from the leading edge of the generational cohort of Generation Z?

As you probably already know, Generation Z is a demographic defined as individuals born after 1996, making their oldest members approximately 26 years old as of this writing. They are part of a highly racially and ethnically diverse group and extremely comfortable with digital communications and technology. Very highly educated, they are politically progressive and pro-government, and see diversity in racial, ethnic, and gender areas as positive. Currently, non-engineers are vulnerable to job loss (as they make up the majority of employees in high-risk work sectors), and they look to government, rather than businesses and individuals, to solve problems.[4]

Our graduate archetype becomes a little more interesting when we evaluate him using two additional tools: the MBTI and the FFPI.

2.7.2 The Myers-Briggs Type Indicator

Probably the most well-known personality instrument is the Myers-Briggs Type Indicator (MBTI), also known as the 16-Personality Assessment. The MBTI is a personal survey designed to assess individual personality based on Jung's theory of types. According to this model, different personality characteristics drive different career choices and vocations. Most of you are already familiar with this, and many (if not most) of you probably took the MBTI in secondary school or university.[5]

The MBTI model has four distinct traits that represent contrasts in the way individuals think and behave, with a total of sixteen subcategories. The four main traits are:

1) Introversion (I) vs. Extraversion (E)
2) Intuition (N) vs. Sensing (S)
3) Feeling (F) vs. Thinking (T)
4) Perception (P) vs. Judging (J)

Literally millions of these tests have been given to all segments of global society, providing an extremely large data set from which to pull.

Significantly, the MBTI is often used by employers to help gain a general understanding of their employees' strengths and weaknesses.[6] Engineers have not been ignored. Specific research on the characteristics of engineer's belief systems and resulting actions have been studied for some time now.

So, what can we say about the aggregate traits of new engineers compared with an average adult population? Table 2.3 defines the 16 MBTI subcategories, descriptions

TABLE 2.3
Myers-Briggs Type Indicator Definitions and Distributions

MBTI Code	Descriptor	Characteristics	U.S. General Population	Engineering Undergraduates
ISTJ	The Inspector	ISTJs are responsible organizers, driven to create and enforce order within systems and institutions. They are neat and orderly, inside and out, and tend to have a procedure for everything they do.	11.5%	19.4%
ISFJ	The Protector	ISFJs are quiet, friendly, responsible, and conscientious. Committed and steady in meeting their obligations. Thorough, painstaking, and accurate. Loyal, considerate, notice and remember specifics about people who are important to them, and concerned with how others feel. Strive to create an orderly and harmonious environment at work and at home.	13.7%	3.3%
INFJ	The Counselor	INFJs are creative nurturers with a strong sense of personal integrity and a drive to help others realize their potential. Creative and dedicated, they have a talent for helping others with original solutions to their personal challenges.	1.5%	3.0%
INTJ	The Mastermind	INTJs are analytical problem-solvers, eager to improve systems and processes with their innovative ideas. They have a talent for seeing possibilities for improvement, whether at work, at home, or in themselves.	2.1%	10.1%
ISTP	The Craftsman	ISTPs are observant artisans with an understanding of mechanics and an interest in troubleshooting. They approach their environments with a flexible logic, looking for practical solutions to the problems at hand.	5.4%	8.2%
ISFP	The Composer	ISFPs are gentle caretakers who live in the present moment and enjoy their surroundings with cheerful, low-key enthusiasm. They are flexible and spontaneous and like to go with the flow to enjoy what life has to offer.	8.8%	2.9%
INFP	The Healer	INFPs are imaginative idealists, guided by their own core values and beliefs. To a Healer, possibilities are paramount; the reality of the moment is only of passing concern. They see potential for a better future and pursue truth and meaning with their own flair.	4.4%	4.3%
INTP	The Architect	INTPs are philosophical innovators, fascinated by logical analysis, systems, and design. They are preoccupied with theory, and search for the universal law behind everything they see. They want to understand the unifying themes of life, in all their complexity.	3.3%	9.9%

(Continued)

TABLE 2.3 (CONTINUED)
Myers-Briggs Type Indicator Definitions and Distributions

MBTI Code	Descriptor	Characteristics	U.S. General Population	Engineering Undergraduates
ESTP	The Dynamo	ESTPs are energetic thrill seekers who are at their best when putting out fires, whether literal or metaphorical. They bring a sense of dynamic energy to their interactions with others.	4.3%	5.4%
ESFP	The Performer	ESFPs are vivacious entertainers who charm and engage those around them. They are spontaneous, energetic, and fun-loving, and take pleasure in the things around them: food, clothes, nature, animals, and especially people.	8.5%	2.4%
ENFP	The Champion	ENFPs are people-centered creators with a focus on possibilities and a contagious enthusiasm for new ideas, people, and activities. Energetic, warm, and passionate, ENFPs love to help other people explore their creative potential.	8.1%	3.6%
ENTP	The Visionary	ENTPs are inspired innovators, motivated to find new solutions to intellectually challenging problems. They are curious and clever, and seek to comprehend the people, systems, and principles that surround them.	3.2%	6.8%
ESTJ	The Supervisor	ESTJs are hardworking traditionalists, eager to take charge in organizing projects and people. Orderly, rule-abiding, and conscientious, ESTJs like to get things done, and tend to go about projects in a systematic, methodical way.	8.7%	10.9%
ESFJ	The Provider	ESFJs are conscientious helpers, sensitive to the needs of others and energetically dedicated to their responsibilities. They are highly attuned to their emotional environment and attentive to both the feelings of others and the perception others have of them.	12.2%	2.5%
ENFJ	The Teacher	ENFJs are idealist organizers, driven to implement their vision of what is best for humanity. They often act as catalysts for human growth because of their ability to see potential in other people and their charisma in persuading others to their ideas.	2.5%	2.3%
ENTJ	The Commander	ENTJs are strategic leaders, motivated to organize change. They are quick to see inefficiency, conceptualize new solutions, and enjoy developing long-range plans to accomplish their vision. They excel at logical reasoning and are usually articulate and quick-witted.	1.8%	5.0%

of each, and the MBTI percentage distributions of two groups: the average U.S. adult population and North American undergraduate engineering students.[7]

Looking to identify the widest differences between the adults and students in each subcategory, and selecting a + / – 5% difference as an error band results in 7 of the 16 subcategories being notably different.

These data show three major personality types (INTP, ISTJ, and INTJ) compose 39% of the sample engineers, being significantly overrepresented vs. just 17% of the general population. On the other hand, the four major types of ISFJ, ESFJ, ISFP, and ESFP make up just 11% of the engineering population vs. 43% for the population as a whole. This is shown graphically in Figure 2.2.

Now, here's where things get controversial: attempting to generalize engineer's traits. Some common characteristics emerge when examining these descriptors and attributes. Based on the MBTI data shown in Table 2.4, our archetypical engineer Jacob:

1) Seeks to develop logical explanations for everything that interests him. Tends to see patterns in external events and develop long-range explanations of those events. Focuses on the theoretical and abstract and can be perceived as being more interested in ideas than social interaction.
2) Is naturally skeptical and occasionally critical, but always analytical and independent. Demonstrates high standards of competence and performance for himself and others which some may see as perfectionistic.

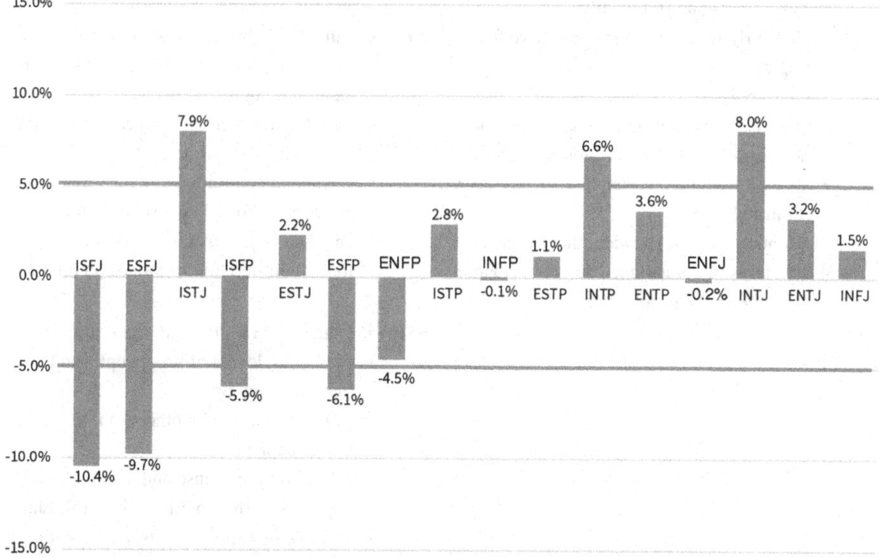

FIGURE 2.2 MBTI subcategory differences, general population vs. undergraduate engineers.

TABLE 2.4
Predominant Undergraduate Engineer Personality Characteristics

Undergraduate Engineers Scoring Well Above General Population		Undergraduate Engineers Scoring Well Below General Population	
INTP	• Seek to develop logical explanations for everything that interests them. • Theoretical and abstract, interested more in ideas than in social interaction. • Quiet, contained, flexible, and adaptable. Have unusual ability to focus in-depth to solve problems in their area of interest. • Skeptical, sometimes critical, always analytical.	ISFJ	• Quiet, friendly, responsible, and conscientious. • Committed and steady in meeting their obligations. • Thorough, painstaking, and accurate. • Loyal, considerate, notice and remember specifics about people who are important to them, concerned with how others feel. • Strive to create an orderly and harmonious environment at work and at home.
ISTJ	• Quiet, serious, earn success by thoroughness and dependability. • Practical, matter-of-fact, realistic, and responsible. • Decide logically what should be done and work toward it steadily, regardless of distractions. • Take pleasure in making everything orderly and organized – their work, home, and life. • Value traditions and loyalty.	ESFJ	• Warmhearted, conscientious and cooperative. • Want harmony in their environment, work with determination to establish it. • Like to work with others to complete tasks accurately and on time. • Loyal, follow through even in small matters. Notice what others need in their day-by-day lives and try to provide it. • Want to be appreciated for who they are and for what they contribute.
INTJ	• Have original minds and great drive for implementing their ideas and achieving their goals. • Quickly see patterns in external events and develop long-range explanatory perspectives. • When committed, organize a job and carry it through. • Skeptical and independent, have high standards of competence and performance – for themselves and others.	ISFP	• Quiet, friendly, sensitive, and kind. • Enjoy the present moment, what's going on around them. • Like to have their own space and to work within their own time frame. • Loyal and committed to their values and to people who are important to them. • Dislike disagreement and conflict, do not force their opinions or values on others.
		ESFP	• Outgoing, friendly, and accepting. • Exuberant lovers of life, people, and material comforts. • Enjoy working with others to make things happen. • Bring common sense and a realistic approach to their work, make work fun. • Flexible and spontaneous, adapt readily to new people and environments. • Learn best by trying a new skill with other people.

3) Tends to be quiet, serious, contained yet adaptable; earns success by thoroughness and dependability. Has an unusual ability to focus in-depth to solve problems in his area of interest.
4) Is practical, realistic, and responsible. Takes pleasure in making everything orderly and organized in his work, home, and life, and values traditions and loyalty.
5) Decides logically what should be done and work toward it steadily, regardless of distractions. When committed, shows great drive for implementing his ideas and sees the job through to achieve his goals.[8]

Compared to Jacob, the general adult population tends to mainly show more "people centered" behaviors and actions. Common descriptors of this group include friendly, sensitive, kind, outgoing, cooperative, and considerate. Examples of these people-centric values are common: remembering specifics about people who are important to them, concern with how others feel, or striving to create a harmonious environment at work and home. They may notice what others need in their day-by-day lives and try to provide it. Enjoying life, people, and material comforts, they can readily adapt to new people and environments. Very importantly, this group is loyal and committed to their values and to the people who are important to them.[9]

It's interesting to read these findings, but even more useful when we add some additional tools to the mix.

2.7.3 The Five-Factor Personality Inventory

A completely different personality assessment is the Five-Factor Personality Inventory (FFPI), a behavioral model used in psychology to model a person's individual personality traits. Personality traits are defined as patterns of thought, feeling, and behavior that are reasonably enduring across an individual's lifespan. The FFPI is considered the most scientifically validated of the widely known evaluation methods, and has impressive substantiation across cultures, life span, and gender.

This model is based on five independent personality traits, each having its own causes and observable behaviors. The key word here is *independent*. These five characteristics have minimum interdependence with each other and can be used as separate descriptors. These descriptors are as follows:

1. Openness/Autonomy: demonstrates intellect, imagination, and independent-minded thinking; can easily link facts together.
2. Conscientiousness: referring to orderly, responsible, and dependable behavior; likes to follow a regular schedule.
3. Extraversion: someone who is talkative, assertive, and energetic; likes to converse.
4. Agreeableness: being good natured, cooperative, and trusting.
5. Emotional Stability: displaying calm, even-tempered, and secure behaviors, can take their mind off their own problems.

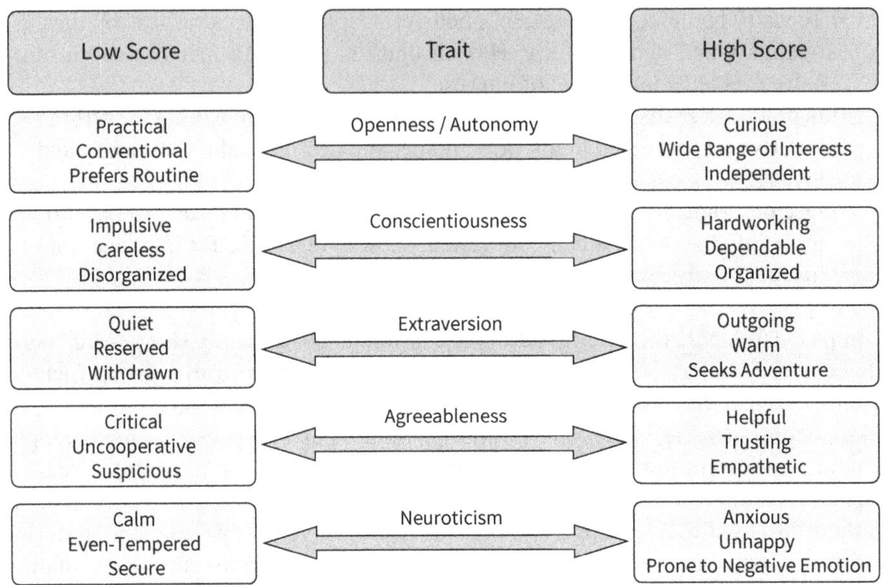

FIGURE 2.3 Five-factor model trait definitions.

The model is based on a scoring system; the meanings of a high and low score for each factor can be seen in Figure 2.3. Each trait is shown as a percentile compared to the general population. For example, a person may have higher agreeableness than 64% of the population, but higher openness than only 22% of that same population.

A number of FFPI studies have compared various groups of engineers with the general population. In one study of 419 subjects, Kichuk and Wiesner used the FFPI to examine the success and failures of engineering design teams. They found that successful teams tended to show higher levels of extraversion and agreeableness while also showing a higher level of emotional stability.[10]

Similarly, Reilly et al. showed that, in new product development teams, a higher level of team agreeableness and conscientiousness was related to more effective team performance.[11]

Meanwhile, Sayeed and Jain compared the interpersonal skills of students both with and without an engineering background. Students without an engineering background showed to be more open and understanding of others, and more focused on the feelings of others than students with an engineering education.[12]

Brown and Joslin found engineering students scored significantly lower on "need for independence" and higher on measures of responsibility, productivity, perseverance, goal orientation, and decisiveness compared with other college students.[13]

In Europe, Kline and Lapham studied students from five English universities. Engineering students scored higher on conscientiousness, tough-mindedness, and conventionality than students in other disciplines. No differences were found with respect to extraversion and emotional stability.[14]

Finally, a study by Van Der Molen and others investigated the personality characteristics of a group of engineers with a variety of years of experience. A total of 103 engineers were tested using the FFPI model. Overall, this group was shown to score lower on agreeableness, and higher on extraversion, conscientiousness, emotional stability, and autonomy, compared to the general public. Interestingly, the older the engineer, the more conscientious and autonomous they were, while engineers with lower educational degrees were more conscientious than those with higher degrees.[15]

Table 2.5 summarizes these various studies. Taken together, these FFPI-related studies strongly imply that student engineers are more conscientious and goal-driven

TABLE 2.5
Summary of Engineering-Related FFPI Studies

Study	Openness	Conscientiousness	Extraversion	Agreeableness	Emotional Stability	Remarks (Engineering Students Except as Noted)
Kichuk & Wiesner			+	+	+	Measuring success of engineering design teams.
Reilly et al		+		+		Successful performance of a new product development group.
Sayeed & Jain	(-)			(-)	(-)	Engineering vs. non-engineering students.
Harris		+				Comparison of engineers vs. nurses and psychology students.
Brown & Joslin		+				Engineers scored higher on responsibility, productivity, perseverance goal orientation, and decisiveness.
Dai		+			+	Chinese study; experienced professional engineers vs. general population.
Kline & Lapham		+		(-)		British study; engineering vs. non-engineering students at five British universities.
Van Der Molen	+	+	+	(-)	+	European study; variety of experienced engineers vs. general population.

Key: + = Engineer Trait Score Above General Population, (-) = Engineer Trait Score Below General Population than others, and show some indications toward enhanced emotional stability. Higher extraversion for engineers was slightly elevated, but only in team-based work or within experienced engineers. Interestingly, agreeableness in engineers was positive, but only in team settings. Perceptions of agreeableness in engineers in individual settings were below the general populations. However, no positive or negative trends for openness were found.

A word of caution here. With just one exception (Van Der Molen) these studies were limited to engineering students rather than experienced, seasoned engineers. It's possible that new engineering graduates demonstrate only certain FFPI characteristics due to their inexperience, and they will "grow into" more open, extraverted, agreeable, and emotionally settled professionals over time. Additionally, most of the studies lacked an adequate norm group that could be used to compare the results.

Based on the three techniques, what insights about your future collogues and co-workers can we make? What might they act like, believe in, and value? What can we say about a new engineering graduate?

Let's go ahead now and define our friend Jacob, our engineering archetype. First, I have to make the obligatory comment that the results of this little comparison are not prescriptive, nor all encompassing, nor meant to call out any individual or profession. And the qualitative data is fairly scattered. We're just trying to get a general handle on who the engineering archetype might be. Statistically speaking, this is not a bell-shaped curve shaped high and tight; it's a curve with long tails on both ends and should remain that way.

That said, who is Jacob, and what have we learned? Table 2.6 is our judgment in tabular form.

So, what can we glean from this trilogy of behavioral insight? First, very soon in your first professional position, you are going to run into some version of Jacob. Obviously, not displaying all these characteristics but certainly several of them within the same individual. Assuming your goal is to work well with these colleagues, you can use these insights as a tool to develop specific connections to different individuals, a skill that builds social capital and cements the bonds between like-minded engineers striving to achieve common goals.

Second, this is an excellent tool for designing and populating work groups. Given a particular project, and having determined the collection of personnel skills required (a highly detailed worker, a critical thinker, a creative), you can identify and build a winning team whose skills complement each other.[16]

Yet another use is to help understand your management. If you are given an assignment that, frankly, doesn't make much sense, you can use the archetype to get to the core of the difference in worldview and solve it to everyone's advantage.

One suggestion. This archetype will be of more use to you if you tuck it away in your desk and refer to it as you proceed through your early days at work. When presented with an unusual situation, this summary should be one of the first references you reach for to help assess your circumstances.

One final point. This archetype is not a profile that is stuck in time. The archetype characteristics will change over time as the toll of hard-won experiences, success,

TABLE 2.6
Summary of an Average New Graduate Engineer

Descriptor	Average Characteristic/Tendency
Name	• Jacob
Gender	• Male
Age	• 23 Years
Profession and Specialty	• Mechanical Engineer
University	• 4-Year University
Final GPA	• 3.4/4.0
Nationality	• United States/First World Country
Initial Employment	• Large Product Development, Manufacturing or Engineering Analysis Firm
Family of Origin	• Solidly middle-class family structure with two parents who value education and achievement.
Natural Preferences	• Prefers facts to beliefs. Compelled to understand and figure things out logically. Searches for clear cause and effect. • A critical thinker, can be perfectionistic and sometimes seen as arrogant. • Needs to understand everything possible about a topic or goal. Huge ability to concentrate. • Desires certainty over the tentative in all parts of life. Prefers to be in control. • Highly goal- and accomplishment-driven regardless of distractions. • Is loyal and committed to their values and to the people who are important to them. • Student engineers are highly conscientious and goal-driven than non-engineers. • Emotional stability can be read as stoicism.
Generational Characteristics	• Part of a highly racially and ethnically diverse group. • Extremely comfortable with digital communications and technology. • Currently vulnerable to job loss as they make up the majority of those in high-risk sectors • Politically progressive and pro-government, sees diversity in racial, ethnic, and gender areas as positive. • Very highly educated. • Look to government, rather than businesses and individuals to solve problems.

failures, and insight changes our beliefs and view of reality. This archetype is about new graduates starting out, not seasoned engineers near the end of their careers.

2.8 THE PAST AS PROLOGUE

Let's change gears now. This is probably a good time to talk about the role of history in technology and engineering. I generally want to stay away from strictly historical conversations in this book. But knowing what has come before is important, as an old system can be put in place many decades ago yet continue to be used even when better systems have been proven and could be used instead. In other words, inertia is common.

Engineering is no different. As mentioned earlier, in the late 1800s Frederick Taylor became known for his "scientific" approach to engineering-related work. His famous experiment of measuring pig iron moved versus shovel size established the idea of scientific management, which is applying quantifiable metrics to work output. That idea from over 100 years ago is surprisingly still in use today. Scientific management is still one of many choices when working in an organization as an engineer. The past is still with us, and let's keep this in mind as we move forward.

As a general understanding, there are three points about engineering and its story. First, when compared to length of human history, we really don't know much about technical organizations and management compared to the length of time engineering has been a discipline. Second, today's organizations are still heavily imprinted with hierarchy, bureaucracy, and formalization. Finally, as organizations become large and complex, the orderly and predictable assumptions regarding these organizations, the role of managers, and the tools first developed during the industrial age should change, but don't necessarily do.

Another factor that makes the past relevant today has to do with your new specific organization. Your organization will implicitly select an overall view of the relationship of your work to the firm based on the past. That is, without even knowing it, the company will select a general approach to their employee's work. Some companies today still apply scientific management. Others apply McGregor's Theory A or Theory B. Still others look to blend the two for guidance in how to run the organization, or maybe something totally different. The key here is to be aware and use this inertia as a starting point in assessing the approach your new company might be using.

2.9 REALIZATIONS

We've examined the impending problem of transitioning from your world as a student to that of an engineering professional. We've asked, how large is the challenge? What does this new world look like? And what does it mean to me?

We've identified some obvious (and not so obvious) points to consider and then accept or decline their validity. The level of difficulty for you to make this transformation can be easy or hard, depending on how much effort you invest in it. Your prior internships aren't going to help you much. The relationship you might have had with a favorite professor or instructor isn't the same as with a boss. Those bets are off.

We've explored ways in which broad social factors and shared experiences can influence how a group of people close in age and sharing important experiences can interact with a society and develop values. We've spent some major time and effort attempting to define what an engineering archetype might look like and how they may or may not interact with you, and how the history of engineering and education may get in the way of your transformation into a new professional.

There is an old cliché that "to know where you're going, you first have to know where you are." You are at an inflection point, and it's important to understand where you stand, not as an individual engineer or technical specialist, but who you are as part of the engineering profession. Begin to think in terms of the group. What is your group's way of working? How do you in the collective solve problems? How do you work on technical issues with both new and experienced engineers, technical specialists, and other professionals?

With all that said, let's go on to the first major step in navigating this maze called the engineering organization. Change can be hard, but many have discovered that the person undergoing change will have a much easier time if they first prepare themselves through adopting a new mindset. In the next chapter we'll discuss this idea of mindset and how it can be used to your benefit.

NOTES

1. Loretto, P. 2019. The Common Challenges New Interns Face. Thebalancecareers. https://www.thebalancecareers.com/common-challenges-new-interns-face-1986624.
2. Fisher, A. 2022. Archetype vs. Stereotype. Know Your Archetypes. https://knowyourarchetypes.com/archetype-vs-stereotype/.
3. American Society for Engineering Education. 2021. Engineering and Engineering Technology by the Numbers. https://ira.asee.org/by-the-numbers/.
4. Parker, K. and Igielnik, R. 2020. On the Cusp of Adulthood and Facing an Uncertain Future: What We Know About Gen Z So Far. Pew Research Center. https://www.pewresearch.org/social-trends/2020/05/14/on-the-cusp-of-adulthood-and-facing-an-uncertain-future-what-we-know-about-gen-z-so-far-2/.
5. Myers, I. 2022. The 16 MBTI Types. The Myers & Briggs Foundation. https://www.myersbriggs.org/my-mbti-personality-type/mbti-basics/the-16-mbti-types.htm.
6. Truity Psychometrics LLC. 2022. Myers & Briggs' 16 Personality Types. https://www.truity.com/page/16-personality-types-myers-briggs.
7. Capretz, L. 2002. Is There an Engineering Type? *World Transactions on Engineering and Technology Education* 1(2): 170–171.
8. Carayannis, E. 2020. Creativity Assessments. In *Encyclopedia of Creativity, Invention, Innovation and Entrepreneurship*, ed. E. G. Carayannis, 1589–1594. New York: Springer International Publishing.
9. Stidham, H., Summers, J. and Shuffler, M. 2018. Using the Five Factor Model to Study Personality Convergence on Student Engineering Design Teams. In *Proceedings of the DESIGN 2018 15th International Design Conference*, eds. D. Marjanović, M. Štorga, S. Škec, N. Bojčetić and N. Pavković, 2145–2154.
10. Kichuk, S. and Wiesner, W. 1997. The Big Five Personality Factors and Team Performance: Implications for Selecting Successful Product Design Teams. *Journal of Engineering and Technology Management* 14(3): 195–221.
11. Reilly, R., Lynn, G. and Aronson, Z. 2002. The Role of Personality in New Product Development Team Performance. *Journal of Engineering and Technology Management* 19(1): 39–58.
12. Sayeed, O. B. and Jain, R. 2000. Assessing Interpersonal Competency of Career-Oriented Individuals. *Abhigyan* 18: 9–17.
13. Brown, N. W. and Joslin, M. 1995. Comparison of Female and Male Engineering Students. *Psychological Reports* 77(1): 35–41.
14. Kline, P. and Lapham, S. 1992. Personality and Faculty in British Universities. *Personality and Individual Differences* 13(7): 855–857.
15. Van Der Molen, H., Schmidt, H. and Kruisman, G. 2007. Personality Characteristics of Engineers. *European Journal of Engineering Education* 32(5): 495–501.
16. Shen, S.-T., et al. 2007. Using Personality Type Differences to Form Engineering Design Teams. *Engineering Education* 2(2): 54–66.

3 Embracing a New Mindset

3.1 A NEW PERSPECTIVE

The first step in a successful transformation begins not with your arrival on your first day, nor with your first assignment or your initial meeting with your new boss. It begins within your own mind. This is a simple, but critically important, point. At this moment in time, if you do not consider the possibility that a new mindset is a necessary prerequisite for any change, then there is a reduced hope of a successful transition. This means you need to first examine your own professional belief systems (as it's been developed so far) and methods of thinking as an engineer.

This chapter offers an alternative to some current perceptions and convictions as acquired over 16 or so years of formal education, preparing you to consider and (hopefully) act on some new ideas presented in this book. It maintains that the first step to change is to weigh some contrary concepts that may be fundamentally in opposition to the outlook you currently hold. This is important, as these contrary ideas stretch and create new boundaries for understanding your pending transformation. In short, the goal here is to explore your current perceptions of engineering, study your existing assessments and beliefs, and condition you to better consider and absorb the ideas presented as we encounter them.

As I just said, this chapter argues that the first step to change is to ponder professional concepts that may be new to you and may be fundamentally in opposition to your previously learned beliefs. These contrary ideas will then constitute the new boundaries and ground rules for the rest of the book. These ground rules cover contingency theory, data as imprecise approximations needing qualitative insight, the acceptance of the troublesome paradox and paradigm shifts, of self-understanding, the need for investigation, and the radical concept of "nobody knows anything." Each of these concepts, however foreign at this moment, must be considered as a necessary first step in preparing to tackle the Divide.

After so many years of studying the "engineering way," how do you accomplish this small feat? Through a series of insights and ideas that, step by step, build from on initial understanding into acceptance and ultimately, transformation. The initial step in a successful transformation does not begin with a checklist or a YouTube "how to" video.

The first step begins with evaluating a common engineering mindset.

We all know that worldview. Engineers in training are taught that all problems, no matter their complexity have a single, well-defined, and precise answer. That clearly defined systems can and should be designed and optimized to provide supreme efficiency. That truth is exclusively data-driven, and all problems can be solved if only

DOI: 10.1201/9781003214397-4

enough prior planning and high-quality information are brought to bear. Reality is clean, clear, and relentlessly rational. And that perfectionism in all engineering work is not only desired but expected.

Unfortunately, upon entering your first professional employment, you (and your newly hired colleagues) must immediately deal with a critical reality. This stereotypical realm doesn't exist. This tidy, formulaic world as learned at university is a messy, confused, chaotic, and nonsensical place. Problems aren't only poorly defined; most can't even be stated clearly. Goals are cloudy, project timing is impossible to estimate (let alone meet), work quality is uneven, and the necessary information is often just not available.

Your first reaction may be to deny this reality and instead attempt to impose your rational will upon this strange world, thinking, "Rationality has to be around here someplace and I'm going to find it." Failure immediately results. You make mistakes; missteps abound, and your head starts hurting. And after a painful initiation, you are left to ponder, "What happened?"

The path to understanding what happened is first examining and then adjusting your values and the beliefs you've brought into the job. Each of the new concepts we're about to consider must be absorbed and deliberated upon as a necessary step in preparing to cross this Divide. Let's take a look at a few of them.

3.2 IS WHAT YOU KNOW REALLY TRUE? CONTINGENCY THEORY AS REALITY

A key concept throughout this book, the idea of contingency theory can be expressed in just two words: "everything depends." It means that reality is conditioned on ever-changing and uncertain facts that are only valid for a limited time and in a limited space. When stating obvious facts or truths, those actualities are only true if relevant to the surrounding conditions and environment at that time. This is in direct opposition to the learnings of engineering at university, i.e., there is one best, constant answer to any problem, that science develops "laws" that are unchanging, and that one can depend on facts as defined strictly by some numerical measurement. Contingency theory is fundamental and foundational to understanding a new reality.

Contingency theory is a subset of behavioral theory that contends there is no single, best way to organize a corporation, lead a company, design a problem-solving method, or make decisions. Instead, the optimal course of action is contingent (i.e., dependent) upon the internal capabilities of the organization and the condition of the external environment. This is not a new idea: several contingency approaches to organizations were developed in the late 1960s and are still used today.

The organizational theorist Jay Galbraith states clearly that, in contingency theory, there is no one "best" way to organize and that the effectiveness of different ways of organizing depends upon the external environment and internal conditions present. Our definition of contingency theory will paraphrase Galbraith:

> "In contingency theory the best way to work depends on the nature of the environment to which the work relates."[1]

What's important is that contingency theory comes from the fundamental conflict between scientific management and human relations models discussed earlier. We said scientific management holds that any organization is a purely scientific and rational animal that can be codified and controlled, treating an organization as a machine. The human relations perspective focuses only on social needs while ignoring the core technical portion of organization. Contingency theory addresses these multiple issues by handling the dynamic external environment while melding these two opposing views into a flexible whole, where both viewpoints can be accommodated while dealing with external environmental changes.

3.2.1 The Critical Importance of Contingency Theory

Contingency theory impacts this entire book. One of the most important and ubiquitous concepts repeatedly stated here is the idea there is not one best or optimum answer when it comes to engineering organizations and your role in them. There are many ways to analyze and operate within a company, with many correct answers conditional on the environment. And yes, some answers may be better than others. But this idea of many answers may be a bit uncomfortable. After all, you have been taught for years that there is one correct answer to a given problem. But unlike mathematics or other branches of the hard sciences, you will need to look at various points of view, and it will be your job to pick the most appropriate view consistent with your own education, experience, work life, values, and beliefs.

It's time for some examples highlighting the importance of contingency. Let's consider the number of planets in our solar system. For anyone over the age of 40, we all "know" how many planets there are – nine, right? Actually, it depends on what year the question is asked. Nine is true, but only from 1930 until 2006, when a few members of the International Astronomical Union (IAU) downgraded the status of Pluto to that of a "dwarf planet." Pluto is still out there floating around; it has neither exploded nor disappeared. But with a simple voice vote, our entire solar system went from nine planets to eight. This is contingency theory in action.[2]

Another example deals with simple mathematics. As good engineers we all know that 1+1 is equal to 2. It's been that way throughout our lives and centuries before us. Yet we also know that the contingency here (in this case, the unspoken assumption) is that we are considering numbers in the base 10 system. For instance, in the base 2 system 1+1 equals 10. The correctness of the answer is totally dependent on the implied use of the binary or base 10 systems.

At this point, you might be looking for a scientific fact that is inviolate, solid, unchanging, and does not adhere to contingency theory. Let's take the measurement of time. We all know that on Earth, the time of day is based on a 24-hour clock equivalent to approximately one rotation around the Earth's axis. But what about the day of the week or the day in the year? Consider the Julian calendar as opposed to the Gregorian version. In 1582 the Julian calendar gained almost two additional weeks overnight to compensate for more accurate celestial measurement. That year, the head of the Roman Catholic Church, Pope Gregory XIII, decreed that Monday, October 4, 1582, would become Friday, October 15, 1582, gaining 11 days in the process. The interesting point with this new calendar was its implementation. It was not adopted in the (then) USSR until 1922, Greece in 1923, and Turkey in 1926. Until that time two calendars were in use worldwide, creating parallel answers to the question of what day it is, to some consternation and puzzlement.

These examples can be a lot of fun. Let's look at one more time-related "fact" that sometimes is not a fact. As noted earlier, we all know that a day constitutes roughly 24 hours, no matter where you are in the world. Until 2000 this was true. But starting on November 2, 2000, this "fact" as a universal standard became obsolete when the first crew of the International Space Station began the continuous manned occupation of near-earth orbit. From then on, a day was almost exactly 24 hours, except for an average of six earthlings, whose day changed to about 90 minutes. From then on there have always been a handful of people whose day is not 24 hours, but one and a half.

The message here is that contingency theory is universal, undeniable, and with us in every aspect of our daily life. As science and technology advances, more and more facts will become less certain and more contingent on the assumptions and environment where and when those facts are being measured. Contingency theory impacts, and can sometimes even control, nearly everything we experience, and the workplace is no exception.

3.3 THE DATA WILL SET YOU FREE ... OR NOT

An unspoken tenet of engineering is this: data are our stock in trade, our coin of the realm. It has immense value: it points us to the truth, it comforts us in times of uncertainty. It can have an enormous impact on our lives. And it can steer us in the correct direction to solve our technical problems, bringing about personal and organizational success. For all engineers, data are keenly important. And when correctly analyzed, it will set us free. Yes, data are our friend.

Until it's not.

3.3.1 Statistics and the Scientific Method

Statistics (that branch of mathematics dealing with the collection, analysis, interpretation, and presentation of masses of numerical data) is a wonderful tool, a constant companion who gently and earnestly shows us the way. No agenda, no hidden advantage to be gained: statistics provide comfort and counsel when needed most. Statistics is an important and useful means to understand cause and effect relationships, significant differences, and give confidence to the user in that whatever is being measured is meaningful. Statistics has been around an awfully long time, and new techniques and ideas are constantly being incorporated into the field. For engineers, statistics may also be a satisfying topic. It's rare that one can do a calculation and come away with a 95% (or 90% or even 98%) confidence level that what you measured is the truth. How can we not like this?

Unfortunately, statistics can also be a troublesome friend.

For me, statistics has been my trustworthy indicator of reality and truthfulness. Using formulas and mathematical concepts studied intensely by mathematical wizards over the course of decades: of course I'm going to go with it. Yet as time goes on, hints of doubts about statistics have begun to appear. One early doubt came from the work of the American author Mark Twain, who famously joked:

> Figures often beguile me, particularly when I have the arranging of them myself; in which case the remark attributed to Disraeli would often apply with justice and force: "There are three kinds of lies: lies, damned lies, and statistics."[3]

Cute, and essentially harmless, especially considering its age.

Yet a growing issue of much more importance comes from today's application of statistical techniques to scientific research and its reporting, especially in medical studies. Let's consider the scientific method. Applying statistical methods to experiments normally consists of five phases:

1) Design of the experiment
2) Selection of the statistical technique to be applied
3) Data collection and application of the selected technique to the data set
4) Interpretation of the statistical result
5) Reporting of the result

A simple search in Google Scholar under "statistical errors and research" yields page after page of statistical errors in all forms of science, medicine, software engineering, public health, and essentially every facet of research that relies on numerical analysis.

There seem to be two main issues in the application of this method. The first is *misapplication*, the unintended selection and usage of incorrect statistical techniques in research papers, PhD dissertations and the like. Selection of a statistical technique is a minefield of error-producing choices. Fundamentally, in selecting which statistical method to apply, there is literally no limit to the techniques available. Without tight boundaries or ground rules surrounding the method to be used, an entire universe of answers can be generated, all of them correct per se for the individual technique, but none of them matching when compared against each other. Valid results are strictly contingent on technique.

The second issue is in the purposeful selection of certain data (*cherry-picking)* to drive the analysis toward a certain premeditated, desired result (*bias*). Sadly, an important contributing cause to bias is some scientist's desire to report new, blockbuster results representing breakthrough research, be it correct or not.

Another concern is *replicability*, both in direct reproducibility (same data, same analysis) and in robustness to different analytic approaches (same data, different analyses). Supporting case examples are common and are increasingly being reported due to the transparency created by the digital commons. Case Example 3.1 is a synopsis several examples of misapplication or bias in experimental statistical application. Medical research examples are chosen here as truth in medical research is incredibly important (people can die due to incorrect results) and the scientific method is held in very high regard.

CASE EXAMPLE 3.1 MISAPPLICATION AND BIAS WITHIN RECENT MEDICAL STATISTICAL STUDIES

There are scores of medical research studies addressing statistical misapplication and bias. This case example merely reports a brief sampling of current exemplars in the medical field.

Data Don't Lie, But They Can Lead Scientists to Opposite Conclusions

A research team led by the European School of Management and Technology asked 29 independent statistical analysts to analyze a single research question. While the data provided to each analyst was identical, each of the 29 was free to choose whatever analysis techniques they believed were valid. The research team reported that no two analysts employed exactly the same method, and none got the same results. This included statistically significant effects in opposite directions for the same research question.

The problem was not that any of the analyses were "wrong" in any objective sense. The differences arose because researchers chose different definitions of what they were studying and applied different techniques. Common factors contributing to this problem include a simplistic reliance on "statistical significance," where a statistically significant result must have odds better than 1 in 20, meaning many "meaningful" results are wrong just by chance. Other factors contributing to false results include small sample sizes, poorly designed studies, researchers' cherry-picking their data, analyst's bias, confirmation of pet theories, and financial interests. In short, general readers of scientific results (including engineers) should remember the method chosen to analyze scientific data directly influences the experimental results.[4]

Same Data, Opposite Results? A Call to Improve Surgical Database Research

According to the *Journal of the AMA*, Christopher Childers and Melinda Maggard-Gibbon contend there is a current explosion of new clinically based medical studies, driven by the easy availability of digital registry data which can feed statistical programs in seconds, answering a broad range of questions. But:

> The limitations are also well known: because the data are observational, they may be prone to bias from selection or confounding. In the absence of

> randomized data, clinicians often rely on database research to.... make patient care decisions. With increased use of database research, greater caution must be exercised in terms of how it is performed and documented.

In short, researchers use data for a statistical study because it's available, not because it's necessarily appropriate.[5]

Statistics Misapplication and Countermeasures in Medical Papers Preparation

Statistical misapplication is not limited to just North America. Researchers in Wuhan, China, determined the main causes of numerical misapplication in medical research design, material description, and analysis in medical papers mainly are:

a) Violation of randomization, control, replication and equalization principles
b) Inappropriate material description
c) Misused statistical methods, regardless of the characteristics of the data
d) Confusion in statistical and practical meanings when explaining results.

They write:

> [In] conclusion.... medical researchers should strengthen their learning on statistic theory and improve the ability of transforming practical issues into statistical issues.[6]

Statistical Errors in Medical Research – A Review of Common Pitfalls

Strasak et al. bluntly state that statistical standards in medical research are generally low. An ever-growing body of evidence points to continuing statistical errors, flaws, and deficiencies in most medical journals. Strasak argues that these statistical pitfalls occur at different stages in the scientific process, ranging from planning a study, conducting statistical analysis, documenting the statistical techniques used, presenting the data, and interpreting the study results. Strasak states:

> Statisticians should be involved early in study design, as mistakes at this point can have major repercussions, negatively affecting all subsequent stages of medical research. Consideration of issues discussed in this paper, when planning, conducting and preparing medical research manuscripts, should help further enhance statistical quality in medical journals.[7]

Poor Scientific Methods May Be Hereditary

In 1962, the scientist Jacob Cohen investigated under what conditions a scientific study would detect a real effect. He estimated most studies would have detected a meaningful effect only about 20% of the time, yet on examination nearly all studies reported significant results. Cohen concluded these scientists were not reporting their unsuccessful research. He also suggested some papers were actually reporting false positives, i.e., noise that looked like data.

Today, results don't look much better. Recently, Paul Smaldino and Richard McElreath showed that repeatability in published studies in psychology, neuroscience, and medicine is little better than it was in 1962. Their 2015 study focused on the ability to repeat another researcher's experimental test result reliability. Their experiment was to employ over 200 researchers in the field of psychology to repeat 100 published studies to see if the results of these could be reproduced. Only 36% could.[8]

Ideally, everything would be put into context and statistics would be used honestly, but until then it's wise to reflect critically on statistical studies on complex subjects such as cancer research.

So, statistics or numbers don't lie. Yet they do. As engineers, statistical testing is the one technique we put our faith into, and that faith is being shaken.

The key is to know what the numbers mean or don't mean, to know their sources, their origin, and what they really convey. This is hard and certainly not convenient, but necessary.

3.3.2 Partnering Qualitative Data with the Quantitative

The supremacy of hard, numerically based data is one of the pillars of the engineering world. Best summed up by the common saying "What is measured is real," the solid foundation of high-quality quantitative data is a reassuring and constant belief within any engineer's mind. It defines our reality and our "truth." And surprisingly, a universal source of highly valuable data is overwhelmingly ignored in the engineer's life, that of *qualitative insight.* Qualitative data is defined as information that is normally described in prose that focuses on insight, action, belief, and values. To have full understanding of any engineering problem, any technical concern, then both quantitative and qualitative data must be considered in tandem. To borrow from medicine, quantitative data is the skeleton, the "bones" of the situation, while qualitative data is the meat on the bones that fleshes out the complete picture.

Why isn't qualitative data applied more in engineering? The reasons are simple: it's hard, time-consuming, and expensive. Qualitative data is collected by ethnographic means, which are normally interview-based and observational, and collected in prose. And qualitative data are slow to arrive. Very few commercial engineering firms are willing to spend the money and wait for a substantial time to receive the results. And even if qualitative data is provided, engineers will first assume that qualitative data can indeed be converted to a quantitative representation, and second, they can apply simplifying assumptions to analyze the data as required by management timing or other desires. Remember, time and money is everything in any firm, including engineering, so ethnographic data are rarely used in our profession. The nearest thing an engineer may use is marketing or survey data converted into numerical representations. This is not qualitative data. An example of the differences between of quantitative and qualitative data is demonstrated in Case Example 3.2.

CASE EXAMPLE 3.2 RICHNESS OF QUALITATIVE DATA TO IMPROVE UNDERSTANDING

Take the example of a capstone engineering course at a major national-level engineering school. This course requires all students to provide certain biographical information to create an overall description of the class. Two types of data were obtained. The first (Type 1) lends itself to numerical analysis; the second (Type 2) is more qualified in nature. Both types were obtained from a standardized questionnaire as follows:

Case Example 3.2 Type 1 and Type 2 Data Questions

Type 1 Student Data:

- Name:
- Place of birth (Country / Region):
- Education:
- High School (Secondary) Education
- Where? (Country / Region)
- When?
- Undergraduate and Graduate Education (as applicable)
- Current University Level
- Where? (Country / Region)
- When?
- Major / Field?

Type 2 Student Data:

- Affiliations and Professional / Work Experience
- What? (Industry / Organization / Position)
- Where (Country / Region)?
- When?
- Notable Experience Related to Either Educational or Professional Areas
- Other Information the Professor Should Know

Case Example 3.2 Part 1 Student Biographical Data

The Type 1 student data produced the insights shown in Figures 3.1a through 3.1c.

Very nice data. A clean distribution, well behaved, easy-to-understand, and yielding a very precise and satisfying result. Now let's look at our Type 2 data:

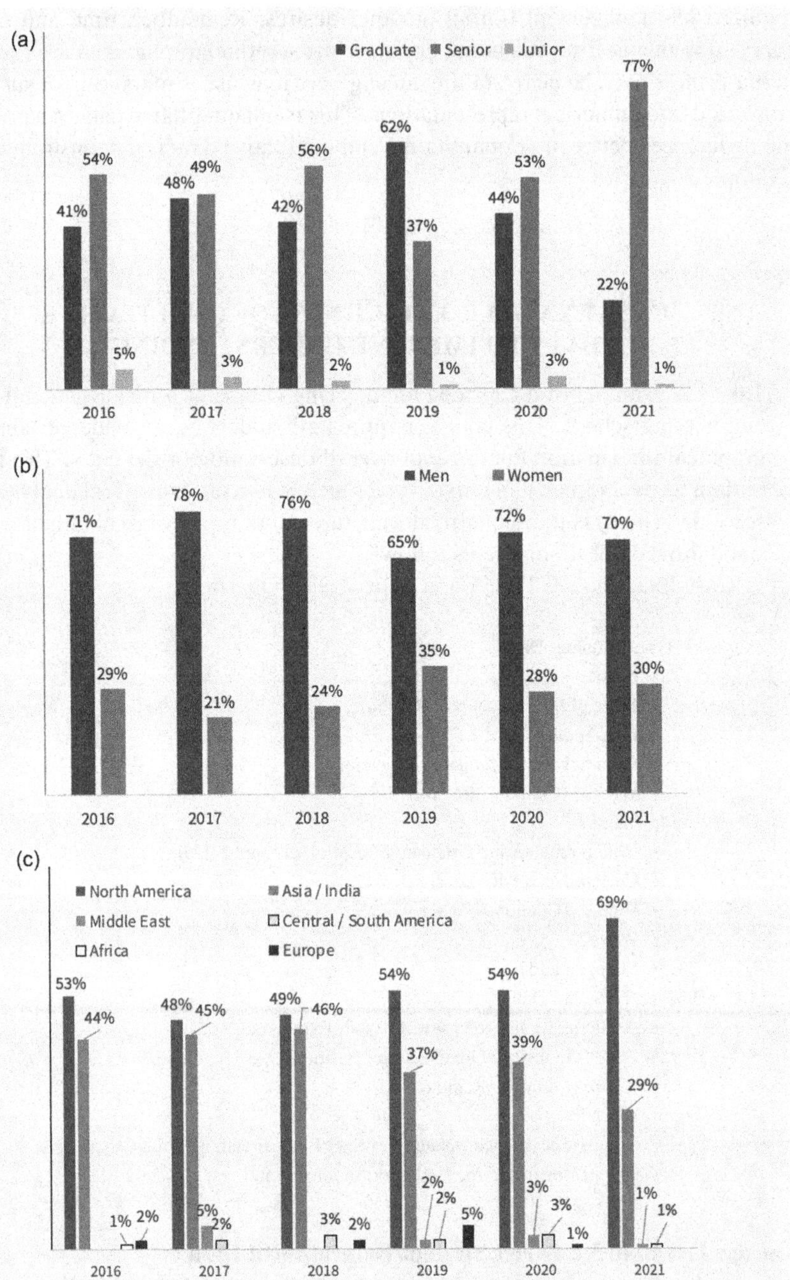

FIGURE 3.1 Graphical representation of student biographical data. There are three charts in total: (a) Distribution by Academic Class Level. (b) Distribution by Class Gender. (c) Distribution by Geographic Birth Location.

Case Example 3.2 Type 2 Student Biographical Data

Qualitative Data for University Course

- Investment Banking Summer Analyst at Goldman Sachs in NYC
- Supply Chain Intern at United Illuminating (utility) in New Haven, CT
- Founder of startup called "Connect to Cars" – automotive retail. Achieved LLC status.
- Supply chain intern at Anderson Windows
- Business Analytics intern, ZS Associates
- Supply Chain simulation intern at Infireon Technologies, (semiconductors) in Munich, Germany
- Join a large consulting company in Chicago
- Researcher at Eli Lilly and Company in Indianapolis, IN.
- Supply Chain Intern in Sintra, Portugal at Hikma Pharmaceuticals
- High School in Amman, Jordan
- High School in San Pedro Sula, Honduras
- Business Consulting in Honduras through operations cost reductions
- High School in Taiwan
- Operations Intern at L'Oreal (beauty) in Taiwan
- SHEI Communications Team Member
- Taiwanese American Student Association – Publicity Chair
- ITASA Midwest Conference – Media Chairman
- Born in Sao Paulo, Brazil, High School in Barcelona, Spain
- Distribution intern at Red Bull NA, in Santa Monica, California
- Worked at logistics company in Germany for Lufthansa Airlines
- JanSan Chemicals as Business Intelligence Analyst in Racine, Wisconsin
- High School in Dubai, United Arab Emirates
- Credit Risk Management Intern at Standard Chartered in Mumbai, India
- Business Strategy Intern at Ernst & Young
- Phi Chi Theta Professional Business and Economics Fraternity
- Co-Founder, Director of Operations and Sr. Advisor for University Entrepreneurship club
- Net-Texts, Inc. in NYC
- Supply Chain intern at Dow Chemical, accepted a full-time position at Dow starting in June 2014.
- Distribution intern at AMWAY consumer goods firm
- Conducting consulting projects through MECI consulting group
- Membership Chair for U-M Engineering Global Leadership Honors Program
- IE Intern at Nexteer Automotive in Saginaw, Michigan
- Member, Undergraduate Student Advisory Board for IOE
- Global Business Planning Intern for Estee Lauder in NYC – prediction tool success
- Internships in consumer electronics and pharmaceuticals
- Navy ROTC at U-M, spent (4) weeks attached to HSM-41 helicopter squadron for US Navy
- Will be entering the US Navy as a commissioned officer
- Lean management/6 Sigma intern at Witzenmann Corporation in Pforzheim, Germany
- Business Analyst intern at Ericsson Corporation (telecom) in Plano, Texas
- Lean Manufacturing Engineering internships at EOTech, EWIE, and ADP
- Cardinal Health Inventory Optimization intern in Dublin, Ohio
- Anderson Window manufacturing intern
- Spend time traveling and volunteering in Latin America
- Professional Intern for Business Integration and Planning at Walt Disney World
- Intern at both Dow Chemical and Microsoft, In Midland MI and Redmond, WA
- Summer Day Camp Counselor at Salvation Army in Midland, MI
- Dow Corning Process Engineering Intern
- Become a strategic operations consultant at a Big Five consultancy

(*Continued*)

Qualitative Data for University Course

- Research assistant in Materials Science and Engineering Dept. at U-M.
- Lithium-ion battery research
- Internship at Suzion Energy in Coimbatore, India – flow manufacturing
- Design Engineer/DFMEA analyst intern at Fallbrook, Inc.
- (2) Patent holder in China, (2) more pending
- Company founder in 2010, shut down due to lack of experience and business knowledge (admitted to the experience)
- High School in Auckland, New Zealand
- Held internships in supplier quality at Apple, audits at Delloitie, operations at P&G in China, Hong Kong and California
- High School in Lahore, Pakistan
- Internship at Jones the Grocer (food/beverage) in Dubai, UAE
- GTZ Chemicals Manufacturing in Calcutta, India
- Program Manager for internet hardware at Cisco Systems in Shanghai, China
- Channel Sales Department at Cisco – sales promotion
- Sourcing Department internship (quotes to purchase orders) at Grainger Corp.
- IOE 481 project at U-M Hospital this semester
- Goldman Sachs Securities – fixed income intern/analyst in Salt Lake City, Utah
- Already have a position as analyst at Goldman Sachs Operations Division in NYC
- Business Strategy Intern at GE Capital International Services
- Lean Operations Intern at Coca-Cola Company
- Summer Analyst in Government and Infrastructure Advisory Services at Grant Thornton
- Major in Electrical & Computer Engineering and Space Physics
- Aero engineer from Beihang University in Beijing, interned at ST Aerospace
- Internship at Toyota Kiroskar Motor Limited in Bangalore, India
- Intern at Cummins Power Generation in Minneapolis, Minnesota
- Currently, I do not know what I want to do after graduation
- Quality Engineering Intern at Alcoa Wheels and Transportation Product Division
- Secretary for Tau Beta Pi Engineering Honor Society
- Outrage Dance Group Member
- Engineering intern at VITEC (gas tank supplier)
- Internship at JR Technologies – travel agency website
- Taught math and science to underprivileged children in Peru
- High School FIRST Robotics Team President
- Financial Intern at Belmont Capital Group in Los Angles
- Healthcare Researcher at Alcides Carrion Hospital in Huancayo, Peru
- Volunteer Teacher for Blue Sparrow (NGO) in Huancayo, Peru
- Analyst for Chalkfly (ecommerce) in Detroit
- Chrysler Quality Intern
- Startup Career Fair Director
- Supply Chain Management Intern at GE Aviation (Avionics) in Grand Rapids
- Engineering intern at Light Corporation in Grand Haven, MI
- U-M Health System Program Operations and Analysis Department
- Worked remotely for startup called Relationship Science LLC – data mining
- Startup called Buzzn (Social Media) in Ann Arbor
- American Express in Phoenix
- Cisco Systems in India and Virginia
- Co-Op at Walt Disney World on the attractions team
- Quicken Loans Business Analyst internship, also internship in Project management
- GM Body Manufacturing Engineering in (3) plants in Midwest
- Tauber Institute for Global Operations
- VP of fraternity
- President of the Michigan Materials Society
- Newport News Shipbuilding Production Control intern

Qualitative Data for University Course

- Program Administrator for the Go Blue Box program
- Experience in Lean Manufacturing, Kaizen, 5S, Supply Chain and Quality/Safety
- Lean principles implementation project in Coimbatore, India
- Cisco Supply Chain Intern in San Jose, California
- Studied at Technical University of Berlin (International laboratory)
- Has a standing offer at Newport News Shipbuilding already

Now we're really getting someplace. Whereas Type One data is clean and lends itself nicely to graphical analysis, the qualitative data is messy but very rich in understanding. Simple inspection shows many class members had significant internships during their university career. Many had traveled a long distance for their internships, such as to Europe, South America, and Asia. Others have taken internships in non-engineering areas such as financial services in New York City, pediatric health, and even Imagineering at Disney World in Florida.

Qualitative data is unparalleled in developing insight. This means as a new engineer, you will need to develop a way to deal with this messy and undisciplined data. It takes time and a lot of effort to make sense of information like this, but you must eventually develop this skill by searching out assignments and (more rarely) ethnographic training. It takes time and effort.

In short, quantitative data alone only provides a basic framework for understanding a problem or analyzing a situation. For true insight, you must also use qualitative data in equal or greater measure. While your early assignments will be almost exclusively quantitative in nature, fairly soon you will be given tasks that begin to involve some combination of qualitative and quantitative analysis. And the time to start learning this skill is now. Having the patience, openness and proper mindset to handle softer qualitative information, melding it to the quantitative, is a necessary prerequisite to becoming a fully developed professional.

3.4 THE FALLACY OF THE OPTIMUM ANSWER

As mentioned above, we are all taught in engineering that there is one, and only one, optimum answer to every problem. That if we try hard enough, really put our backs into it, we can develop and identify not only an adequate answer to a problem but also the single, best answer possible. If we just increase the accuracy of our measurements, if we define a constant to seven decimal points as opposed to just two, if we just have one more little bit of information, we can achieve that optimum answer. And reaching that optimum answer in itself is not enough. There is a second goal: reaching the best answer in the most efficient manner possible. What we want is to follow a straight line to the optimum answer, with little or no deviation from this path of righteousness. The single best answer, obtained by the most efficient technique possible: that is our definition of success as taught to us.

Peter Bradford, a chief engineer at a major manufacturing multinational, tackles this point when coaching new engineers joining his department:

> There's a couple of slides I like to show [new] engineers on what is a plan and what is reality. The first slide shows "the plan", which is a drawing of someone on a bicycle following a straight, level path with a finish line flag dead ahead. That's the plan. The next slide shows the reality. It shows the bike as much further away from the flag. It's planted behind a pit full of rocks, followed by a river you have to cross. Then there's another concrete barricade you have to climb over, followed by a deeper river. Then another deeper pit with a ladder going up the far side. Heading uphill, you find at the finish line a fake flag, and there's a rainstorm overhead. And don't forget the guy standing at the pole holding a clipboard and saying "What took you so long?" That's reality. Oh, and don't forget when you asked the finance people for the budget, they only gave you 80% of what you need.
>
> A plan is like what military commanders say about waging war. The first causality in any war is the plan.[9]

Answers are never optimum. They only approach optimal by varying degrees. No one really has available the perfect information to achieve those optimum answers. And no path to this imperfect answer is straightforward. The path is circuitous, winding its way through the environment and can only be guessed at when beginning the solution process. If you can bake into your plan these realities (aka contingencies) you stand a much better chance of setting the correct expectations with your management and fellow colleagues. Another way to say it is "get real."

3.5 THE PARADIGM AND THE PARADOX

A key reality in engineering organizations is the continuous presence of paradox and new paradigms permeating the technical field. A *paradox* is a contradictory situation or proposition that when investigated or explained may prove to be actually true. A *paradigm* is a mental framework of understanding in which theories, hypotheses, or generalizations are created and accepted, resulting in a certain belief system. This system can sometimes be so strong as to overwhelm contrary facts that are obvious to others, but not to the belief holder. A paradigm shift is when that belief changes.

As a new engineer, both paradoxes and paradigms can significantly impact your view of reality, and your mind must be trained to both recognize a paradigm shift when it is occurring and become comfortable with resolving continuous paradoxes.

Let's talk first about paradigms and where they might be found through the use of a few examples.

3.5.1 The Paradigm

As just stated, a paradigm is a belief system, a set of ideas. Paradigms are ubiquitous and have been in existence since humans first climbed out of the primal ooze. They are powerful, and it is no exaggeration they have the capability of governing life or death. And a paradigm shift can instantaneously change from a docile entity into a powerfully different creature impacting all who come into contact with it.

As we all know, at 11:40 pm on April 14, 1912, the British ocean liner *Titanic* sank on its maiden voyage between Southampton, UK, and New York City. Over the next four hours an estimated 1,517 passengers and crew died before help arrived. Investigations proved the *Titanic* slammed into a large iceberg at 23 knots, mortally damaging a ship advertised as "unsinkable." Contributing causes to the deaths were not enough lifeboats for the number of people onboard, a goal to set a world speed record between the two cities, and a sailing route away from normal shipping lanes. But another cause of the heavy loss of life involved something else:

> We can also say another cause of the heavy fatalities was an unseen paradigm shift. The passengers and crew on the Titanic overlooked the possibility that the iceberg could have been their lifeboat. Newspapers estimated the size of the iceberg to be 50 - 100 feet high and 200 - 400 feet long. Titanic was navigable for a while and could have pulled aside the iceberg. Many people could have climbed aboard it to find flat places to stay out of the water for the four hours before help arrived. Fixated on the fact that icebergs sink ships, people overlooked the size and shape of the iceberg, plus the fact that it would not sink.[10]

The paradigm visible to the passengers and crew was simple: humans need a boat to survive in the near-32-degree water of the North Atlantic. The operating paradigm can be summarized simply as "Iceberg Bad, Ship Good." But what the passengers and crew overlooked was the possibility that a new paradigm was in play, as in Figure 3.2.

As the unfortunate souls discovered that evening, paradigms tend to be invisible to those within them, yet obvious to others outside the paradigm. People hold tightly to their paradigms, even when it is obviously against their own best interest. When presented with overwhelming evidence that a paradigm is no longer valid, people

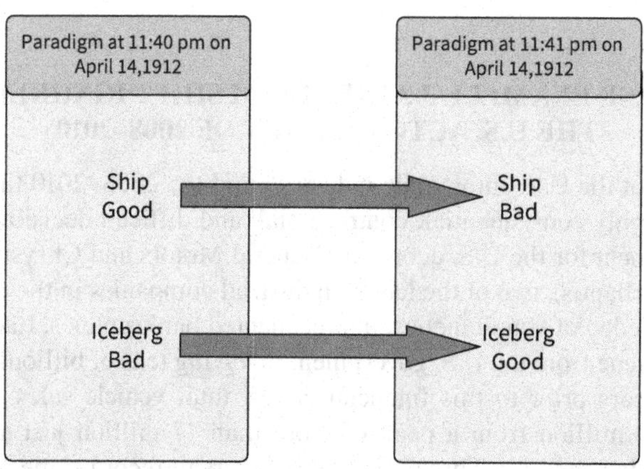

FIGURE 3.2 Comparison of *Titanic* paradigms.

(especially those with a large stake in its truthfulness) will not abandon it. Yet the instant a paradigm changes, the people supporting the previous paradigm become in the "wrong" and their paradigm becomes immediately irrelevant.

The presence of paradigms is not limited to human psychological situations only. Paradigms are a critical factor in pure and applied science. In his seminal book *The Structure of Scientific Revolutions*, author Thomas Kuhn applies the paradigm effect to the sciences and mathematics.

In his book, Kuhn argues that science does not progress via a linear accumulation of new knowledge but undergoes periodic revolutions (read "paradigm shifts") where the nature of scientific inquiry within a particular field is abruptly transformed.

Kuhn pointed out the common pitfall of scientific experimentation, where during a period of normal science, the failure of a result to conform to an existing paradigm is seen not as refuting the paradigm but as the mistake of the researcher. As these nonconforming results build up, the science reaches a crisis at which point to a new paradigm, which replaces the old framework, is accepted.

Kuhn's work has been applied extensively and is credited as a foundational force behind the sociology of scientific knowledge. In fact, an annual international award, the Thomas Kuhn Paradigm Shift Award honors those who present original views that are at odds with mainstream scientific understanding.[11]

A good question at this point is: can paradigm shifts really impact your work as an engineer? Consider this. Remember, a paradigm is a predetermined way of thinking, a framework containing the basic assumptions, methods of thinking, and methodology that are commonly accepted by members of a community, say, an engineering society. It is a cognitive framework shared by members of any discipline or group. For you it might be your company's technology development plan: the plan is a paradigm. A paradigm contains orderly and predictable assumptions regarding the holder's beliefs and understandings. It is a reassuring way of thinking.

Let's consider a recent paradigm shift, resulting in the fundamental alteration of manufacturing thinking in the United States from the 20th century into the 21st, both destroying and creating billions of dollars of company value in just a few short months.

CASE EXAMPLE 3.3 PARADIGM SHIFT IGNORED: THE U.S. AUTO BAILOUT OF 2008–2010

The rescue of the U.S. automobile industry amid the 2008–2010 financial crisis was a highly consequential, controversial, and difficult decision made at a critical moment for the U.S. economy. General Motors and Chrysler (the precursor to Stellantis), two of the largest industrial companies in the world, were both on the edge of simultaneous, unprecedented bankruptcies. Each sought a financial rescue from the U.S. government involving tens of billions of dollars.

In the years prior to this financial crises, total vehicle sales plummeted to below 10 million from a peak of more than 17 million just a few years earlier, causing the "Big Three" U.S. automakers to record some of the worst corporate performances in American history. General Motors (GM) alone lost almost $40 billion in 2007 and another $31 billion in 2008. Ford lost $3 billion and then $15 billion; Chrysler also lost comparable amounts of money.

By the fall of 2008, the automakers' financial situation was so dire that they would soon be unable to even make their payroll, let alone pay their suppliers. In November of 2008, the CEOs of Ford, General Motors, and Chrysler requested from the U.S. Congress a $25 billion "bridge loan" to make these payments and avoid bankruptcy. In the CEOs' minds, the crisis centered on major economic forces outside of their control (such as customers' lack of access to credit) as the core reason for destroying their business.

This was ostrich thinking.

The Big Three had generations-old issues of high cost, questionable quality, and other factors contributing to the industry's problems, yet the CEOs argued that they had already done all the restructuring necessary to fix those problems: the automakers themselves were no longer an issue. In reality, the Big Three's problems had built up over many years and were certainly not caused only by the downturn. For example, falling demand was a severe problem. There was a sustained downward trend in demand of over 2 percentage points per year for the combined automakers. The Big Three's market share in 1998 was 71 percent; by 2008, it was 47 percent. Hiding this problem was substantial price discounting relative to their overseas competition: by 2008 the Big Three were discounting comparable cars by $2,000 to $3,000. Other factors had also taken a toll on the new car demand, such as perennial quality and reliability issues, lower resale values, poorly received new models, and a dearth of high gas mileage vehicles.

Still another headwind was the increasing domestic U.S. production of foreign-owned companies like Honda, Toyota, Nissan, and others who were expanding production in the United States, using predominantly non-union plants in the American South. From 2000 to 2013, employment at the domestic transplant carmakers almost doubled to 163,000, while Big Three employment was cut nearly in half to 253,000. Additionally, the Big Three automakers were paying overall labor costs almost 45 percent higher than the transplants. Many concluded the problems of the Big Three automakers were particular to those firms and not fundamental to the national automotive industry.

In short, the Big Three's problems included long-term falling market share, a massive short-term drop in vehicle demand, and large fixed costs resulting in huge short-term losses. And even if they could reduce their fixed costs, when the recession ended and demand returned, the continuing decline in market share meant the automakers would soon be back in trouble, as no one knew if the prior sales levels of 17 million vehicles would ever return. If demand rebounded only partway after the recession ended, it was not clear that all the Big Three automakers could survive.

Simply speaking, the Big Three were living in the automotive paradigm of the 20th century, where failure to address systemic problems could be ignored and rely instead on yearly record sales levels to "kick the can down the road." The new 21st-century paradigm of an entirely new business model exposed the lie.

The future became even more frightening as various governmental solutions were proposed and rejected. For example, in Washington at a critical meeting of the National Economic Council, members held a straw vote on whether the Council believed Chrysler would survive for five years if a government-supported merger with Fiat went through. The Council voted no.

But the final decision-maker on what to do was not the Council but the President of the United States, Barack Obama. Obama weighed the economic arguments as well as the political, financial, and social realities. The administration determined it was essential to rescue GM to prevent an uncontrolled bankruptcy and failure of countless suppliers, potentially destroying the entire U.S. auto industry. Yet the failure of the much smaller Chrysler would probably not have dire effects for the whole industry, even though Chrysler's failure would cause considerable individual hardship. With this thinking, Obama choose to rescue GM as a stand-alone company, but also rescue Chrysler with the express intent to merge it with the automobile manufacturer Fiat, creating a new company.

If GM and Chrysler had been allowed to fail, in all likelihood the Great Recession would have been deeper and longer, and the recovery that began in mid-2009 would have been much weaker. The rescue was more successful than almost anyone predicted at the time. Some of this success resulted from actions the auto companies took; others happened because the rebound in consumer demand for autos was especially strong during the next five years, meaning the auto industry turned out to be one of the drivers of the economic recovery. A combined paradigm shift of the U.S. automakers, the federal government, and the financial markets saved the North American auto industry.

It's interesting to mention the Ford Motor Company's actions during this time. Ford decided not to take any government support like GM and Chrysler, but to "go it alone" financially. Ford had large losses in 2006 but had already borrowed a significant amount of money in advance and begun restructuring before the financial crisis struck, so the company was able to withstand the cash crunch. (An interesting side note: By not taking any government funds, Ford's reputation as a true "up by their bootstraps, all-American" company was greatly enhanced. Ford gleaned substantial, unexpected market share and unmeasured goodwill by this approach.)

So, what did the leaders of the Big Three learn about paradigm shifts and their failure to recognize them?

First, the Big Three lost their power to decide their own future. Second, they did not have the tools in hand to control the path of the crisis even if they did have the power. Third, they were lucky, at least some were: Chrysler was absorbed by Fiat soon afterward and ceased to be a stand-alone automobile manufacturer.

The U.S. government also learned. Economic analysis provided key insights into the decision to rescue and restructure GM and Chrysler. That decision was risky. While those involved gathered all the information available and attempted to place the companies on a sustainable footing, they certainly did not know if it would work. More importantly the administration, living outside the automotive bubble, was clearly able to take the necessary actions to change the paradigm. As Obama summarized in 2015:

> There was clear-eyed recognition that the auto companies couldn't sustain business as usual. That's what made this successful. If it had been just about putting more money in without restructuring these companies, we would have seen perhaps some of the bleeding slowed but we wouldn't have cured the patient.

In other words, the willingness to abandon the "normal" way of solving an economic crisis and rapidly embrace truly innovative methods without paralyzing fear was key in saving these two firms.

Finally, Ford was able to weather the economic downturn and resulting financial crisis because it had taken proactive, anticipatory steps to restructure early. Ford had seen the paradigm had changed. That GM and Chrysler did not is a lesson about the importance of management seeing and acting on these shifts. The bailout's ultimate success does not undermine the reality that the failure of seeing and dealing with a major paradigm shift can destroy entire industries.[12]

Reprinted by permission from Goolsbee and Krueger (2015).

3.5.2 The Paradox

A *paradox* is different than a paradigm. A paradox is a set of "contradictory yet interrelated elements that exist simultaneously and persist over time."[13]

Like paradigms, the paradox exists everywhere, not just in the world of engineering or technology, but in any and every portion of our lives. Paradox has existed for centuries and is common in all aspects of our being. And everyone must face the cogitative dissonance of resolving two opposing facts or conditions simultaneously: we cannot escape it.

Everyone has their favorite paradoxes; you may have already heard some of these:

The Paradox of the Pessimist

The paradox of the pessimist is well known. A pessimist sees something in its most negative light, while an optimist perceives things in its most positive condition. Invariably, things are not as bad as the pessimist sees nor as good as the optimist would like to experience. Therefore, the pessimist's poor expectations are normally exceeded, while the optimist's overreaching positive expectations are normally dashed. So, the pessimist is fundamentally happy while the optimist is eternally disappointed.

The Epimenides Paradox

There are many versions of the Epimenides Paradox, some of which going back thousands of years. Epimenides was an ancient Cretan philosopher from the island of Knossos who made the immortal statement: "All Cretans are liars." The paradox is that Epimenides himself was a Cretan: if all Cretans are liars, then Epimenides is also a liar. And if Epimenides is a liar, then the statement "all Cretans are liars" must be a lie, which would mean all Cretans tell the truth, which means Epimenides tells the truth, which means the statement "all Cretans are liars" is both true and false. Epimenides the Cretan saying all Cretans are liars is the fundamental paradox.

The Paradox of Promotion

A more current example is the Paradox of Promotion. The business literature lists a number of definitions of this paradox. For this discussion, we'll define the Paradox of Promotion as when you or another employee, through hard work and conscientiousness, finally receives what you have been working for: a promotion to the next level. To your surprise, you quickly find that the vast number of skills you have learned and honed to this point are now instantaneously obsolete. Yes, obsolete. The skills that get you to a certain level are not the skills that you immediately need to keep you there. What results is a scramble to quickly find and apply skills that many times are not engineering or technically based, but instead grounded in social or management practice.

Take the example of Derek McAllister, a five-year materials engineer who in the last several years has shown the characteristics his management values in a first-level supervisor. One day Derrick's manager calls him to a conference room and announces, "Congratulations, you are now the supervisor for the Willowcreek Advanced Materials project. Go ahead and introduce yourself to the group, find out what's going on, and begin to draft a development plan assuming a two-year product lifecycle. I need to see it in about three days." Derek is shocked. How will he approach his new group, understand their needs, and provide a realistic development plan when he doesn't even know how to deal with the variety of engineers he believes make up his group? Unfortunately, Derrick's training as a materials engineer is of no use in handling this immediate assignment, or, in Derek's eyes, this crisis. Derek is left to fumble about as he attempts to fulfill his management's direction. And he will be fumbling for at least the next six months as he attempts to train himself for this position. Welcome to the paradox of promotion.

The Sigmoid Curve

The Sigmoid Curve is an S-shaped curve which describes an organization's performance over time. In engineering, it is the story of a product's lifecycle or a corporation's rise and fall, or even the arc of your own career. First described as a concept for businesses and organizations by Charles Handy, the curve's X-axis is time while

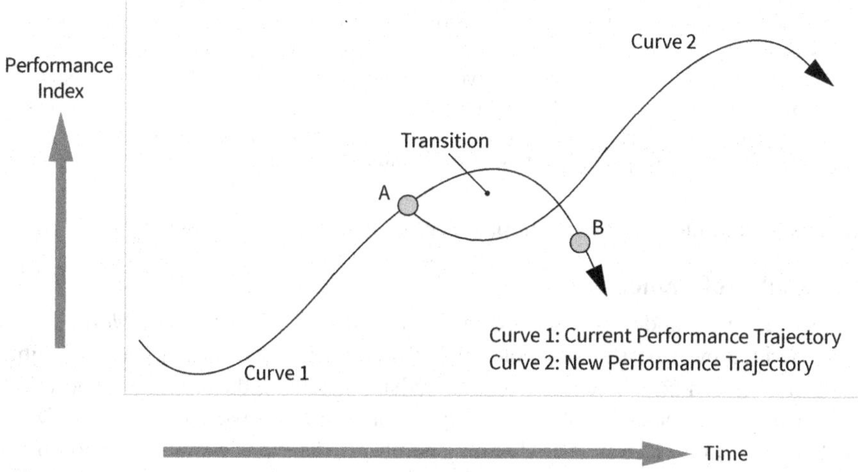

FIGURE 3.3 Sigmoid Curve representation of organizational performance.

the Y-axis is whatever performance measure chosen, say, the market share growth of an engineering design business. Whatever measure is selected, the curve starts flat, then accelerates upward until a maximum is reached, then flattens out and begins a shallow decline as shown in Figure 3.3.

As Handy shares in his book *The Empty Raincoat: Making Sense of the Future*, the impact of the Sigmoid Curve on your organization is a paradox:

> The secret of constant growth is to start a new Sigmoid Curve before the first one peters out. The right place to start that second curve is at point A, where there is the time, as well as the resources and the energy, to get the new curve through its initial explorations and flounderings before the first curve begins to dip downwards. That would seem obvious; were it not for the fact that at point A all the messages coming through to the … institution are that everything is going fine, that it would be folly to change when the current recipes are working so well. All that we know of change, be it personal change or change in organizations, tells us that the real energy for change only comes when you are looking at disaster in the face, at point B on the first curve. At this point, however, it is going to require a mighty effort to drag oneself up to where, by now, one should be on the second curve. To make it worse, the current leaders are now discredited because they are seen to have led the organization down the hill, where resources are depleted and energies are low.…
>
> Wise are they who start the second curve at point A, because that is the pathway through paradox, the way to build a new future while maintaining the present. Even then, however, the problems do not end. The second curve … is going to be noticeably different from the old. It has to be. The people also have to be different. Those who lead the second curve are not going to be the people who lead the first curve. For one thing, the continuing responsibility of those original leaders is to keep that first curve going long enough to support the early stages of the second curve. For another, they will find it temperamentally difficult to abandon their first curve while it is doing so well.… For a time, therefore, new ideas and new people have to coexist with the old

> until the second curve is established and the first begins to wane. The area beneath the peak is, therefore, a time of great confusion. Two groups of people, or more, and two sets of ideas are competing for the future. No matter how wise and benevolent they be, the leaders of the first curve must worry about their own futures when their curve begins to die. It requires great foresight, and even greater magnanimity, to foster others and plan one's own departure or demise. Those who can do it, however, will ensure the renewal and the continued growth of their organization.[14]

There is the paradox: at the height of success, change and begin again.

The Paradox of Choice

Another favorite paradox is described in *The Paradox of Choice – Why More Is Less*, a book written by American psychologist Barry Schwartz. Schwartz argues that eliminating choices can greatly increase a person's happiness by reducing their anxiety. The book analyzes the behavior of different types of people (in particular, "maximizers" and "satisfiers") who face an overwhelming number of decisions. Schwartz demonstrates how the dramatic explosion in choice challenges our ability to balance career, family, and individual needs, paradoxically becoming a problem instead of a solution. In short, our obsession with choice encourages us to seek that which makes us feel worse.

Specifically, Schwartz contends:

> Autonomy and freedom of choice are critical to our well-being, and choice is critical to freedom and autonomy. Nonetheless, though modern Americans have more choice than any group of people ever has before, and thus, presumably, more freedom and autonomy, we don't seem to be benefiting from it psychologically.
>
> Thus, the growth of options and opportunities for choice has three, related, unfortunate effects: it means that decisions require more effort, it makes mistakes more likely and makes the psychological consequences of mistakes more severe.

Of course, Schwartz could be talking about engineers. After all, our core job is to select the optimum choice from many and demonstrate why that choice is correct. With the explosion of data, information, and resulting choice options, we can easily become overwhelmed. As Schwartz argues in this paradox, less is more.[15]

The Paradox of Crisis

"Never let a good crisis go to waste" has become a popular saying in commercial companies, government circles, and other organizations ever since the Great Recession of 2008–2009. First spoken by Winston Churchill during World War Two, the quote remerged when uttered by Rahm Emanual, Chief of Staff to President Barack Obama, during his first term:

> You never want a serious crisis to go to waste. And what I mean by that is an opportunity to do things that you think you could not do before.

The paradox is, rather than avoiding or fleeing a severe crisis, embrace it as an opportunity. The underlying idea is simple: a crisis makes a company's normal operating rules irrelevant. New assumptions and guidelines must be developed immediately.

And whoever sets those rules controls what will happen next and what will be the ultimate outcome. A crisis radically changes the game, so opportunities abound.[16]

So what do all these examples tell us? These paradoxes are a preview of coming attractions. From the moment you begin your professional life, you will be continuously presented with conflicting choices. It doesn't matter if it is strictly technical, strictly personnel, or any combination: becoming at ease with the uncertainty the paradox brings when dealing with two contradictory ideas simultaneously is the goal. In a word, embrace the ambiguity.

3.6 AND JUST WHO ARE YOU, ANYWAY?

And now the trouble begins.

To be honest, this next section may cause some heartburn. The territory we are about to cover does not come naturally to anyone, never mind someone steeped in the science, mathematics and precision of engineering. Yet in many ways, this section holds the key to crossing the Great Divide. Of course, who we're talking about is the reader of this book and consumer of this information.

This section will focus on "who you are": your view of yourself as a single individual and how your perceptions are the key to fitting within your organization.

Why have a section on who you are? Why does that matter? After all, what matters is your technical prowess as the driver of success in the workplace. Isn't the ability to analyze complex technical puzzles and arrive at the correct answer in record time the goal? Perhaps, but the simple question of who you are is critical in establishing the necessary mindset on entering your new professional home. Because your mindset, as the filter that establishes your personal view of reality, must reasonably match the actual environment you perform within each day. If reality and mindset don't match, trouble results.

Thus, the simple question of "And just who are you, anyway?" becomes one of the key inquiries we need to ponder at this point of our time together.

Knowing who you are is a fundamental requirement of all new engineers (and, indeed, all successful employees). What is reality to you is actually the result of the "sifter" your mind uses to help make sense of the world. Knowing who you are helps us to figure out which filters are operating; to discover and analyze just what is going on around you. This is not an easy task for anyone, and is an even tougher job when attempted by anyone unfamiliar with the process. After all, most of us have been taught throughout our entire school life that facts are facts; there is little room left for subjectivity, perception, belief, values, or similar thoughts. Yet knowing who you are is a fundamental insight for current and future understanding, which leads to professional accomplishment and satisfaction.

Luckily, there are several methods to get us closer to understanding who we truly are. Similar to our engineering archetype exploration, a combination of methods and techniques allow us to approach this question from different directions and triangulate in on a satisfactory answer. We are going to use two different methods to help determine self-awareness and apply them to the object of interest, i.e., you. Let's look at the first one.

3.6.1 Gardner's Multiple Intelligences

Our initial task is to investigate the entire idea of intelligence and how it might apply to each of us. Howard Gardner's seminal book *Frames of Mind: The Theory of Multiple Intelligences* has had a major effect on understanding human intelligence, especially in the area of education in the United States.[17]

Gardner is a developmental psychologist, who in the 1990s developed a new way of thinking about human nature. Gardner contends that each of us are a unique combination of many "intelligences" (or what you and I might call "talents") that are not generally measured in traditional education. Here, Gardner defines intelligence as "the capacity to solve problems or to fashion products that are valued in one or more cultural settings." He posits that intelligence is not a single entity nor can it be measured solely from IQ tests. Instead, he uses eight "markers" or signs of intelligence that subsequently create a fusion of personal characteristics. These eight intelligences are defined in Table 3.1.

TABLE 3.1
Definitions of Multiple Intelligence Categories and Definitions

Intelligence Category	Definition
Linguistic intelligence	• Sensitivity to spoken/written language • Ability to learn languages • Capacity to use language to accomplish certain goals • Ability to express oneself effectively • A means to remember information
Logical – mathematical intelligence	• Capacity to analyze problems logically • Carry out mathematical operations • Investigate issues scientifically • Detect patterns • Reason deductively • Think logically.
Musical intelligence	• Skill in performance, composition, and appreciation of musical patterns. • Encompasses capacity to recognize and compose musical pitches, tones, and rhythms.
Bodily kinesthetic intelligence	• Using one's whole body or parts of the body to solve problems. • Ability to use mental abilities to coordinate bodily movements.
Naturalist intelligence	• Recognize, categorize, and draw upon certain features of the environment. • Combines a description of the core ability with a characterization of the role that many cultures value.
Spatial intelligence	• Recognize and use the patterns of wide space and more confined areas.
Interpersonal intelligence	• Capacity to understand the intentions, motivations, and desires of other people. • Allows people to work effectively with others.
Intrapersonal intelligence	• Capacity to understand oneself, and appreciate one's feelings, fears, and motivations. • Having an effective working model of ourselves and to use such information to regulate our lives.

The first two markers are the typical subjects valued in schools. The next four are linked to the arts and the last two are called "personal intelligences."

It's not much of a surprise that I'm going to ask you to download the Gardner Multiple Intelligences evaluation instrument and complete the evaluation. After all, we're on a quest for self-awareness and insight here, and this is foundational work all of us need to know. So, please take a moment and take the assessment. You can find it on your favorite search engine under keywords "Gardner," "Multiple Intelligences," or similar phrases.

These 8 categories are useful to begin to understand yourself and your particular talents. Ideally, it would also be helpful to have this assessment as a tool to help understand your colleagues and co-workers (provided your workgroup also takes the evaluation, which is not likely). Unlike our archetype engineer who is an amalgam of characteristics, the Gardner model is applied to yourself and other individuals on a more targeted level.

Experience has shown that the Gardner evaluation can uncover some interesting intelligences of coworkers, allowing us to develop some "shorthand" sketches of types of people you may encounter along the way. Case Example 3.4 shares an experience about one person's store of intrapersonal and interpersonal intelligence, or lack thereof.

CASE EXAMPLE 3.4 THE STORY OF TOM TIMID

One of the surprisingly common individual types in organizations is the "Nervous Nelly" or "Tom Timid," an insecure, anxious person who is uncertain, easily upset, and hesitant to act autonomously in work settings. These people, especially as supervisors or managers, can at best cause extra work for you and at worst create embarrassment or distain if your work is too closely tied to theirs.

My favorite example of a Tom Timid in action happened several years ago. Our advanced engineering division was about to have its annual "all hands" meeting, where all members of the organization (some 400 people) gather in person to hear the group's annual objectives, significant accomplishments, and generally get reacquainted after a year apart. (You may have gone through at least one of these all-hands meetings already). One of the rules of the all-hands (at least until COVID) was that everyone attends in person; no digital attendance is allowed, as a main purpose is to reconnect face-to-face.

The manager responsible for running the meeting was a classic Tom Timid. Every detail had to be correct, and they were in a continuous, low-grade state of terror as the meeting approached. One of the topics of the meeting was a two-minute video clip from a popular comedy movie about a Silicon Valley business internship. The clip was meant to poke fun at large technical organizations and provide a little comic relief for what could be a dry technical meeting.

All was well until the day before the conference. Someone asked Tom, "Did you get permission from the studio to show that clip in public?" Tom froze. "Oh my God, we have to get permission to show this thing? How do we do that?" In a panic, Tom calls the company's legal department. The attorney on duty gave the required answer, the only answer they could legally give: "You have to pay the rights to show it and it's going to be expensive, perhaps $10,000 or more."

On hearing this, Tom went into nervous overdrive. Where was he going to get $10,000 with less than a day to go? Keep in mind the company had no "Department of Video Clips for Major Motion Pictures" he could go to. Instead, Tom had three options: one, cancel the showing; two, pay the money within 24 hours: or three, get somebody else to make a decision. Tom was loathe to cancel the showing, as his boss had specifically requested it (even though it was a low-importance, "nice to have" request). Tom was also scared to ask anyone for $10,000 on no notice, as he would blow his entire meeting budget and didn't even know who to pay the money to anyway. So choice three is what he settled on: get somebody else to make the decision.

On talking to his team, Tom decides the only person who could make this decision was the global vice president of human resources, Kendra Jackson. At that moment, Kendra was in the corporate jet midway across the Pacific Ocean, headed for a company meeting with the China division. Tom places the call to the corporate jet.

I can't imagine what went through Kendra's mind as she receives an emergency telephone call at 35,000 feet from a manager in full crisis mode, asking her to decide whether to show a movie clip or not. Obviously, this is not a vice president's job, especially midway across the Pacific. Kendra, incredulous, said yes. (She probably also asked her Chief of Staff to not take any more of this manager's telephone calls.)

And the meeting proceeded.

This "crisis" significantly disrupted last-minute planning and execution for the meeting. I think you all know the correct answer that was never considered: just go ahead and play the damn clip. The chances of a movie studio caring about a private meeting showing two minutes with no electronic connections anywhere else is, at best, infinitesimally low. But Tom increased his own chance of having a pulmonary embolism.

So what went wrong? First, the request was in the "nice to have" category, not an essential part of the meeting. After all, it was a two-minute comic break from the main business of the meeting: it was certainly optional whether to show it or not. Second, the legal department was brought in. Of course, the company attorneys can give only one answer due to their own legal exposure: tell Tom he has to pay the royalty, no matter how impractical. Third, interrupting the work of the global vice president midway across the ocean should obviously be reserved for true emergencies. Fourth, the manager should've just made the decision as it was such a low-stakes question. And, fifth, precious preparation time was wasted chasing irrelevant apprehensions. Five errors in five hours: not a good use of the day.

An advance dose of intrapersonal intelligence analysis plus interpersonal self-understanding might have changed the outcome. Unfortunately, needed strengthening in these two areas wasn't accomplished.

A postscript. Tom never changed his ways, and one day about three years later, Tom disappeared. Quietly and without fanfare, Tom was gone. Though never spoken, everyone knew the cause. *Requiescat in pace.*

3.6.2 The Bolling Four-Step Exploration

There is an old saying that you can't figure out where you're going until you know where you are. This adage is very appropriate at this point in our journey across the Divide. The entire theme of this section is to be able to answer the question posed in the title: Just who are you, anyway? What is your makeup as an individual and as a new professional? What are your values, your understandings, and your beliefs? This is not easy, and I'm willing to wager that most new engineers have not done this type of self-reflective work before, and may be reluctant to do so now.

But I'll ask for a little slack here. Unlike Tom Timid, figuring out who you are and what makes you tick will pay great dividends in the near-term and even greater benefits in the future. Knowing yourself allows you to predict your reactions to certain work experiences and situations you may not anticipate yet. To be prepared for the right now; to provide insights into your behavior in relation to your co-workers and management, is a critical part of understanding your thought processes and emotional reactions to the wide variety of situations you are about to encounter.

To that end, there is something called the Bolling Four-Step Exploration Process. First developed in the 1990s by Dr. G. Frederic Bolling at the University of Michigan, this qualitative technique is a simple and straightforward way of getting a handle on these questions. Please note there are also a number of other self-realization instruments out there, and using those is fine also. The point here is to do this self-exploratory work, by whatever technique, as a necessary step in moving forward.[18]

Bolling outlined four distinct steps in self-discovery. Table 3.2 provides the method.

At this point, you've probably realized that the Bolling evaluation is not magical, merely a systematic listing and culling of who you are and what you believe you are, today. Don't attempt to be what you are not: honestly find out your authentic self. Because, ultimately, this is really a process to search for the pattern of how to manage yourself in the future. And when we examine who we are, we begin to understand how to change in the direction we desire.

So, what do we have at this point? Including our archetype investigation techniques from Chapter 2, you now have five assessment tools available to you: three psychological instruments (Myers-Briggs, Five-Factor, and Multiple Intelligences),

TABLE 3.2A
Bolling Four-Step Exploration

Step Number	Action
Step One	Be determined not to look at Steps Two through Four until you finish Step One. Knowing the other steps does not kill this exercise, but it can influence its results. Find a quiet room or location where you'll be comfortable, and set aside a predetermined time (perhaps one hour) with no interruptions. Provide for your needs like a cup of coffee beforehand, and arrange in advance for a way to record your responses. Most importantly, adopt a calm and quiet mindset. Your task is to think of everything you are, and write down these items in simple ways. Since it is you and "you know you," a one-word description may suffice for each item. These can be "daughter," "son," "brother," "sister," "caregiver," "volunteer," "musician," and so on. These should not be temporal such as "hot" or "hungry" nor moralistic, such as a need to be more punctual or want to exercise. There are potentially hundreds of things that you are; endeavor to list as many as possible. Be honest. This is just for yourself, and you are the only one who will see this exercise. Because you will run out of things in 10 or 12 minutes, rest a while by doing nothing and start again with a clear mind. Generally, you will find a trigger that leads you down other paths that didn't occur to you in the first place. Repeat this listing exercise until the hour is up and put away the list for a few days. It is important to spend at least a full hour on Step One.

TABLE 3.2B
Bolling Four-Step Exploration

Step Number	Action
Step Two	If you listed under 50 or over 80 items in Step One, you probably need to reconsider a little before you continue to Step Two. A few days after completing Step One, arrange another hour for yourself alone as before. Now classify your items. Think about how you can classify them, then create the groupings. Cluster everything that you can, but you can have a category of just one item if you wish. It is important not to evaluate your answers while you are doing the groupings. Look at the list you generated. For example, you might be a tennis player and ride a bicycle for fun. You may be a singer, collect rare music recordings and are an ex-smoker. Keep creating groupings until you feel that everything can be set in a grouping that makes sense. Change the groups and change what you put into the groups until you feel satisfied that overall it makes sense. If you think of new things, add them. But don't discount an item once you have thought of it. For example, it may seem strange to you, because it doesn't fit with the other things that you have entered being an ex-smoker. However, it is an example of achievement that may be an important sign for you. A grouping may have only one entry and it may be important. You should end up with a dozen or so groupings; there may be less or there may be more. I would suggest that if you have fewer than eight you haven't grouped finely enough, and if you have 15 or more you have a group more coarsely. Finish the hour and put the groupings away for a few days.

TABLE 3.2C
Bolling Four-Step Exploration

Step Number	Action
Step Three	Again, arrange some time for yourself. It doesn't have to be an hour, but you may find you want to take Step Three more than once. Force-rank your groupings from most important to the least. The groupings don't have to have titles as you know what they contain and what they are. As a result of your force ranking, number one will be more important than number two, number two more important than number three, and so on. Rearrange the positions of the groups in the ranked list until you are satisfied. If you feel uncomfortable putting one group either higher or lower than another, perhaps it is because of a particular item you included in the grouping. Rethink as necessary and revisit until you are satisfied that indeed number one is more important than number two, number two is more important than number three, and so on. Now wait a few days for Step Four.

TABLE 3.2D
Bolling Four-Step Exploration

Step Number	Action
Step Four	As before, arrange some quiet time; perhaps 30 minutes or so. Look at your force rank list. Choose no more than the top four of the groups you have created, even though there are many more. Pay attention to these top four items and don't worry about the others. This list is your assessment of the most important things defining who you are. This is your personal reference of who you are. You now have a framework of yourself at this moment in time, maintaining a benchmark against what you might change, what you want to do and how you might obtain more of what you want. Always question the impact of what you would like to do against these four important groupings. This is your personal profile and value set. Note there are probably additional reflections about yourself below these top four that are also important, but these four will probably represent more than 80% of what you are in your own life. Finally, keep this exercise in a place where you can reference it later.

a demographic analysis, and the Bolling Exploration exercise. Each points to different characteristics you may exhibit. Of course, these reflect only the present, not the future. I would suggest performing a number of these assessments, as later in this book you will make use of this information as you move through your transition. In the meantime, hopefully you do not cheat yourself out of this experience.

3.7 THE ENGINEER AS ANTHROPOLOGIST

Now let's share a little secret. It's true you've been hired as an engineer, to do "engineering stuff." Stuff like calculating, assessing, optimizing, testing, and a myriad of other tasks. One of the things you may not realize you have to do is to become a *cultural anthropologist.* For those who were not able to take anthropology at university, cultural anthropology is the study of societies: how humans organize their lives and articulate their values. More than any other way anthropology reveals what society means and why it matters.[19]

For over a century, cultural anthropologists have circled the globe, uncovering surprising facts and insights about how humans organize their lives and articulate their values. In this way, anthropologists have revealed what civilization means. By witnessing behaviors and weaving them together into systems, anthropologists develop key concepts which try to make sense of the world, from culture and values to authority and behavior. Anthropology matters because it helps us understand other points of view, and in the process reveals something about ourselves.

For new engineers, rather than studying the Chimbu Tribes of the Eastern Highlands of New Guinea, we need to study the equally exotic tribes living in the Silicon Valley of California's central coast, or the massive cauldron of technical thought brewing on the eastern tip of the Yangtze River Delta at Shanghai, or the digital society along Israel's central coastal plain at Tel Aviv. Or maybe understand the technology ideas being developed in India. Just like the anthropologist, engineers must study their surrounding society, using ethnographic techniques to understand the "civilization" around them.

Of course, if the anthropologist simile doesn't suit you, perhaps you could take the view of an investigative reporter.

Just as the need to investigate who you are is also the need to investigate and understand what is going on in the organization surrounding you. This is not news: new employees have been told this for decades. One easy way to do this is to merely follow the dictum "be curious." Another way is through the viewpoint of this "investigative reporter" concept. This is the kind of news reporter that digs into situations for clues on what has happened or what is currently happening in the environment around them. Investigators must then take these clues and assemble them like a puzzle, to understand *why* something happened, and, most importantly, what *may* happen.

Being an anthropologist or investigative reporter is a skill that is honed over a substantial amount of time. It involves observing behaviors, hypothesizing why those behaviors occur, investigating and confirming the reasons why those behaviors happened, and then judging the potential future actions resulting from those actions.

Let's take an example of this in a technical context, in this case a straight-ahead engineering calculation you are asked to perform. Say you are asked to determine the heat transfer coefficient of a new material to be placed into the motherboard of a new home alarm system. As extended exposure to heat could cause premature failure of the board and repeated false-positive alarms, your chief engineer has a keen interest in knowing how much heat the motherboard can reject over the course of a year. Thus, your calculation is a very important quality and reliability measure of the entire product.

Everything goes well with your calculation. The physics of the material is well known. It has substantial reference information available to actually calculate the heat transfer coefficient, and the calculation shows ample margin for successfully using this material. Just to be sure, you perform a double-check of your analysis using a different method, and this answer is comfortably close to your initial calculation. In other words, everything looks pretty good.

With a buoyant feeling, you and your supervisor visit the chief engineer's office to report these favorable results. Showing them the screen on your laptop, you briefly report the results and the assumptions behind the work. Instead of receiving some words of affirmation, you are instructed by the chief to recheck the result with another engineer and have that engineer report back to the chief independently. Sitting there, you are surprised, hurt, and just a touch angry that all your good work is being discounted. You feel a bit diminished.

Walking down the hall with your supervisor, you begin to feel more and more incensed and begin to believe a major injustice has been done. Back at your desk, you begin to question other group members about the chief. What's wrong with this person? Finally, another engineer clues you in.

> Two years ago on our last big project, our chief signed off on some critical code for a new voice-activated monitoring device. The chief trusted the engineer who wrote the code; after all, that engineer was very experienced and had an excellent reputation. Unfortunately, the monitor went into production with about 40 buggy lines. The thing just didn't work as advertised and the issue got lots of play online and in customer reviews. We had to pull the device after only nine weeks on the market, and Consumer's Report was all over it. It was a big deal. Nowadays. our chief is super gun-shy and has started double and triple checking everybody's work. It's changed them.

Suddenly, everything makes sense. With just a single question you now understand that your chief's reaction had nothing to do with you, but with a bad experience that occurred long before you ever arrived. That's what is meant by being an anthropologist or investigative reporter. It's about learning the "why" of any situation you might experience so you may understand at a deeper level what might be going on and react accordingly.

This approach isn't just a North American best practice. This investigative reporter idea has a long history in Japanese engineering practice. As described by Dr. Jeffrey Liker in his seminal book *The Toyota Way*, this method is part of *kaizen*, the Japanese business philosophy of applying continuous improvement to working practices and personal efficiency. This particular technique, called the *Five Whys*, became an integral part of the Toyota philosophy:

> The basis of Toyota's scientific approach is to ask why five times whenever we find a problem.... By repeating why five times, the nature of the problem as well as its solution becomes clear.[20]

Simple, yet highly effective.

3.8 NOBODY KNOWS ANYTHING

Here comes more trouble.

This section addresses the controversial contention, originally made by the screenwriter William Goldman in his book *Adventures in the Screen Trade*, that many of us find very troublesome. Goldman contends that *absolute and definitive knowledge of any area, no matter how expert and experienced the practitioner, is impossible to acquire and use.* He argues that at its core, any decision made in almost all sophisticated and complex situations still essentially amounts to nothing more than a guess. As Goldman says:

> The single most important fact, perhaps, of the entire movie industry:
> NOBODY KNOWS ANYTHING
> Again, for emphasis:
> NOBODY KNOWS ANYTHING
> Not one person in the entire motion picture field knows for a certainty what's going to work. Every time out it's a guess and, if you're lucky, an educated one.[21]

Despite its origins in the motion picture business, this pithy observation has been adopted by uncounted decision-makers over the decades, in all aspects of technical and nontechnical fields alike. It is also an alarming contention. But please be patient as we examine this important and counterintuitive idea.

Let's go back and parse the original statement: *absolute and definitive knowledge of any area, no matter how expert and experienced the practitioner, is impossible to acquire and use.*

Absolute and *definitive* means just that: someone, or some group, or some machine has in their possession total mastery of all knowledge in a given area. Since this totality of knowledge is unknown and unknowable, the practitioner is never truly a total master; they just have varying degrees of mastery (and confidence in that knowledge) at any given time. Since knowledge is never absolute, mastery is never absolute. And if knowledge is never absolute, its *acquisition* and *use* are never absolute but conditional.

Let's take Goldman's particular profession, writing Hollywood screenplays. Motion pictures are complicated, fluid creations, taking years to ideate, write, film, edit, and distribute. Hundreds upon hundreds of professionals can work on a single project (just look at the credits of a single blockbuster; if you include extras, over a thousand names can appear). The complexity of a motion picture is enormous, and that doesn't count the personalities and egos of all those involved. And don't forget the main indicator of success: how many individuals buy a ticket or stream the picture. It's safe to say, given the massive complexity and Brownian motion of the movie business, indeed nobody knows anything.

What about the engineering profession? Of course, we know millions of facts, and they are certainly true within their assumptions. But we're not talking about those kinds of simple, singular facts here. We're talking about large (even immense) projects, developments, initiatives, proposals, schemes, and enterprises that involve large numbers of systems, talented people, complex environments, unanticipated factors, and all the rest. With such a minestrone soup of interrelationships, with unknown or irrational interactions, all wrapped up with inadequate funding, short timing, and uncertain goals, it's easy to see that indeed, for us, nobody knows anything.

But what about some highly successful examples of engineering, truly amazing feats of technical accomplishment such as the James Webb telescope, the CERN Large Hadron Collider, or New Horizons spacecraft? While incredibly successful and examples of engineering brilliance, they are also loaded with backup system upon backup system to handle those "unknowns," all countermeasures to our lack of knowledge.

It bears repeating: Goldman is saying that absolute knowledge of any complex topic, by any expert, at any time, by any means is essentially impossible to possess with confidence, and that any person who believes otherwise is kidding themselves. Of course, Goldman's statement is a bit cynical, and the words "nobody" and "anything" are a writer's device to get our attention, but the underlying truth is certainly valid.

Perhaps many of you are saying right now there is nothing in common between the film industry and engineering. This is where we differ. Let's look at new product development in engineering versus major new motion picture creation.

Let's consider two organizations. The first is a major, multinational engineering firm whose business is producing and selling commercial aircraft navigation systems. New products are invented and made production-ready through the firm's separate Research and Development division. This division has a current portfolio of 435 new products undergoing development. This portfolio is costing $357 million per year with a development staff of 450 engineers, and project completion time varies between one to four years.

Each year, the R&D division evaluates about 175 brand new technology proposals for approval. Some fraction of these projects will be approved and added to the ongoing portfolio. Since budgets and staffing levels are capped, previously approved but low-potential development projects will be canceled to make room for the new.

The selection criteria are many and varied. They include total development cost, investment cost, project development timing, foreign and domestic sales potential, anticipated functional and quality performance, lifecycle, evolutionary vs. revolutionary market and technological impact, repairability, return on investment (ROI), return on sales (ROS), patent income potential, and on and on. Yet very few criteria are known with some certainty: perhaps 40% of the required criteria are not known at all, and even if known, only low quality data for the proposal is available. The Catch-22 here is the criteria to approve a project can only come from data, but that data is only available when the project is underway, yet you can't start the project without the approval. A paradox.

Let's now look at the second organization, a Hollywood movie studio, with a portfolio of 22 feature-length movie scripts in development for the summer blockbuster season in two years' time. New movies are written and produced through a studio process that is nearly 100% staffed by independent contractors. The 22-script portfolio is costing $168 million with a production staff of 842 contractors. Studio management continually evaluates new script proposals, and a small fraction of these are approved and added to the portfolio. Since budgets are capped, previously approved but lower-potential scripts will be canceled to make room for the new. Scripting, shooting, and editing completion time varies between one and four years.

New script selection criteria involve many factors. They include total development cost, investment cost, script development timing, foreign and domestic sales potential, anticipated quality of the performances, streaming potential, new market penetration, special effects feasibility, return on investment (ROI), return on sales (ROS), secondary income potential, and so on.

Once again, the Catch-22 is the data needed to approve a project is only available when the project is essentially done, yet you can't start the project without the approval.

The similarities are striking. Both firms are managing development portfolios: the script is the same as a technology project; both are governed by finance, timing, quality, and risk. And the selection of the project/script is made by management who bear a high personal risk and no real data to work with.

In both cases, nobody knows anything. Decisions are made continuously with no real, accurate, or valid data available. Yet timing of the decision (timing being the primary driver of management decisions) forces executives to make educated guesses.

But what about those business cases people talk about? A business case is a highly organized and analytical study of a future decision that provides fact-based data to create a high-confidence outcome. It is the foundation of business and engineering recommendation systems used throughout the world. But by this time, you may know the question to ask: where is the decision data and how good is it? Look behind the equations and calculations to ask where did the data come from? Sometimes, this may cause a few uncomfortable moments.

Let's take one more example of Goldman's statement in an engineering context.

Consider a much larger technology development portfolio for an international manufacturing giant who are developing nearly a thousand advanced technology projects invested at over $950 million.

The massive amount of resources being spent in R&D caused management to ask a question: How effective was the investment in their technology development effort? Should more or less funding be provided in upcoming years? What percentage of their advanced projects actually made it to sale? A multiyear, high-intensity study of their development process was commissioned and found a startling result. Their technology selection process, a highly formalized, impeccably designed system, managed by a platoon of supervisors and managers, investing millions of dollars and hundreds of people, mightily birthed a technology success rate of …… 0.6%. The causes of this disappointing result had nothing to do with formalized procedures or other quantitative measures, but with complex organizational factors firmly residing in the qualitative arena. The expert designers of the development system ignored the cultural, educational, and organizational influences of the system. Not only did they not take these factors into account, no one even thought to include them in their deliberations.[22]

Examples of other engineering disasters caused by lack of knowledge are easy to find. The 2018 collapse of the Morandi bridge in the Italian city of Genoa, the failure in

1981 of the Hyatt Regency Hotel suspended walkway in Kansas City, Missouri, and the 2003 *Columbia* space shuttle disintegration all point to the engineer's curse: that new designs often hit the wall of existing knowledge. Any desire to advance the state of the art requires moving beyond the data into the realm of judgment and risk management.

And despite the hopes of the Artificial Intelligence community, at this time reality is still too complex, too nuanced, and too situational to know with certainty exactly the outcome of any given cause and effect, or the true impact of contingencies. That any analysis, no matter how sophisticated or advanced, at some level remains suspect. While the recent triumphs of AI are promising, the examples shared by the AI community are extremely narrow, and highly selected, designed to show the promise of AI but far, far away from widespread adoption.

The point here is that sophisticated equipment or techniques can only take you so far. As a new engineer, you eventually will be forced to make critical, doubtful, and even cynical judgments counter to what may be common knowledge or belief. This is why the contingency factor approach has been used in engineering and systems design for decades. We know that after a calculation is completed a "contingency" (i.e., a safety or "fudge factor," always a nice round number) is included to give everyone involved a warm feeling that the calculation may not be perfect but still will not cause a catastrophic failure.

In a word, be confident, but remember that "nobody knows anything."

3.9 TAKING COMFORT FROM DISCOMFORT

At this point you may ask yourself a question: How comfortable are you with the idea of changing a mindset? What I mean by comfortable is a continuum, on one side perhaps enthusiastically embracing the notion, or in the middle perhaps experiencing a mild resistance, or all the way to a strong rejection of the entire idea that a mindset can or should be changed.

If you are experiencing some discomfort, congratulations.

Just like in university or other personal experience, discomfort is a key indicator of where you are at in a change process. As discussed earlier, crossing the Great Divide is nothing more than a transformation progression. That transformation can create some discomfort, which is one of the best indicators that change is truly underway and is progressing.

In a word, be comfortable with discomfort and embrace the uncertainty. You're on your way.

3.10 REALIZATIONS

This entire chapter is about change. Specifically, about changing the lens or filter of perception and understanding that we call mindset. Crossing the Great Divide will require at least some change in mindset. Yet achieving even a small change is a difficult and long-term effort. Examining and reconsidering long-held beliefs and understandings can take a substantial effort with little initial results in return. Yet an appropriate willingness to consider examining or adopting a changed mindset will pay benefits when put in the context of an engineering professional's overall judgment. Engineering judgment is a skill, and an appropriate mindset is as important a prerequisite as any training in materials, stress analysis, or the calculus. Mindset is a key gateway to engineering professionalism.

Mindset can speak volumes. It can tell you if what you believe is truly real. Are the facts as presented really facts? What portion of contingency theory is appropriate in this situation? Is the data being used to make a contention strictly quantitative, qualitative, or a combination of both? How much faith would you put into that data? And are there any paradigms or paradoxes operating in this circumstance, and if so, how is your judgment processing this as you analyze the situation? Finally, is your judgment being filtered by who you are, what you believe in, and the possibility that judgment has been hindered by previous experience.

Knowing these questions and more importantly, knowing their answer will better your chances of judging a situation correctly while gaining confidence for future action.

Some summary ideas deserve mention. The idea that hard numerical data, obtained by the most sophisticated means available, may or may not free you from the uncertainty of the analysis being conducting or the decision about to be made. Blind trust in numerical data, while anxiety-reducing, can also lead to blindness as a whole. True insight, which is the goal, is better served by combining the quantitative with the qualitative.

The desire to find the holy grail of the optimum answer, that there is only one, unmistakable, absolute, and inarguable answer to a problem, is a goal that's commendable, but ultimately unattainable. Contingency theory means there are several, if not many, near-optimum or "good enough" answers. And the good news in finding them is they will give your management options on various courses of action, which they are guaranteed to love.

Paradigms and paradoxes are fun. They can also be deadly unless you have a clear-eyed understanding of what they might really mean. For example, the merger of two mechanical engineering firms, while increasing the size and power of the newly created entity, will also create the paradigm shift of duplicate departments; who stays and who goes? The paradox of the Sigmoid Curve creates two paths for your firm: do you stay with the original legacy organization or transfer over to the new group, with all its risks and unknowns?

Decisions like this can be helped by examining just who you are at your core. For instance, in the Sigmoid example your decision rests partly on a self-examination of your own approach to risk: are you risk-averse or risk-seeking? Do you have better skills for one path or the other? These are questions that only can be answered by knowing yourself, and well.

With that understanding goes the idea of personal investigation, a predisposition to proactively discover the territory you work in, what you know, and, most importantly, what you don't know. There's a simple term for this mindset: it's called curiosity.

Finally, the idea that nobody knows anything obviously is not meant to be taken literally, but as a caution and a philosophy. We as engineers need to approach any of our responsibilities with humility and a realization for each fact we know, there are probably ten facts that we do not, and that "we don't even know what we don't know."

As engineers, we apply Cartesian thinking to a world of Brownian motion; we employ an orthogonal lens to make sense of a spherical environment. Our mindset can help us make better judgments in this confusing world.

From this point, let's move on to something a bit less esoteric: a framework for operating in the engineering organization.

NOTES

1. Galbraith, Jay. 1974. Organization Design: An Information Processing View. *INFORMS Journal of Applied Analytics* 4(3): 28–36.
2. Rinco, Paul. 2015. Why is Pluto No Longer a Planet? *BBC News Website*. https://www.bbc.com/news/science-environment-33462184.
3. Twain, Mark. 1906. *Chapters from My Autobiography*. New York: Project Gutenberg. https://www.gutenberg.org/files/19987/19987-h/19987-h.htm.
4. Schweinsberg, Martin, Feldman, Michael, Staub, Nicola et al. 2021. Same Data, Different Conclusions: Radical Dispersion in Empirical Results When Independent Analysts Operationalize and Test the Same Hypothesis. *Organizational Behavior and Human Decision Processes* 165: 228–249.
5. Childers, Christopher and Maggard-Gibbons, Melinda. 2021. Same Data, Opposite Results? A Call to Improve Surgical Database Research. *JAMA Surgery* 156(3): 219–220.
6. Li-rong, Yan, Du, Yuan-hong, Liu, Ting-ting et al. Statistics Misapplication and Countermeasures in Medical Papers Preparation. *Wuhan General Hospital of Guangzhou Command*, Wuhan Hubei, China.
7. Strasak, Alexander, Zaman, Qamruz, Pfeiffer, Karl et al. 2007. Statistical Errors in Medical Research – A Review of Common Pitfalls. *Swiss Medical Weekly* 137: 44–49.
8. Economist. 2016. Incentive Malus: Why Bad Science Persists. *The Economist*. September 24, 2016. https://www.economist.com/science-and-technology/2016/09/24/incentive-malus.
9. Bradford, Peter, Chief Engineer. Zoom Interview by Dr. Robert M. Santer, July 9, 2021. Transcript.
10. McCaffrey, Tony. 2012. Why We Can't See What's Right in Front of Us. *Harvard Business Review*. May 10, 2012. https://hbr.org/2012/05/overcoming-functional-fixednes.
11. Kuhn, Thomas. 1996. *The Structure of Scientific Revolutions*. Chicago, IL: University of Chicago Press.
12. Goolsbee, Austan and Krueger, Alan. 2015. A Retrospective Look at Rescuing and Restructuring General Motors and Chrysler. *Journal of Economic Perspectives* 29(2): 3–24. Reprinted with permission.
13. Smith, Wendy and Lewis, Marianne. 2011. Toward a Theory of Paradox: A Dynamic Equilibrium Model of Organizing. *Academy of Management Review* 36(2 April): 381–403. https://doi.org/10.5465/amr.2009.0223.
14. Handy, Charles. 1995. *The Empty Raincoat: Making Sense of the Future*. London: Arrow Books.
15. Schwartz, Barry. 2004. *The Paradox of Choice - Why More Is Less*. New York: Harper & Collins.
16. Costello, Tom and Laplante, Phillip A. 2009. Never Let a Serious Crisis Go to Waste. *IT Professional* 11(3): 72–72.
17. Gardner, Howard. 2011. Frames of Mind: The Theory of Multiple Intelligences. New York: Basic Books.

18. Bolling, G. Fredric. Professor, University of Michigan, Dearborn. Interview by Robert M. Santer, June 5, 19998. Transcript.
19. Engelke, Matthew. 2018. *How to Think Like an Anthropologist.* Princeton, NJ: Princeton University Press.
20. Liker, Jeffrey. 2004. *The Toyota Way.* New York: McGraw-Hill.
21. Goldman, William. 1983. *Adventures in the Screen Trade.* New York: Warner Books.
22. Santer, Robert. 2007. Why Technology Fails: Bringing Technology from Concept to Implementation. PhD diss. University of Michigan.

Part Two

Guiding the Way: Adopting the Essential Engineering Framework

4 Establishing the Essential Engineering Framework

4.1 MAPPING THE ENGINEERING ORGANIZATION

We all need structure in our lives. Our daily routine, the rules governing how we drive, the behaviors we exhibit when interacting with anyone from bishops to bookies – all are based on predetermined, agreed-to, and generally common behavioral structures.

These structures come directly from prior agreements of how we live together, work together, and generally interact. Its importance is that it is *predetermined.* We don't have to reinvent the structure for different situations or for each new person we meet. When presented with a given situation, we go to our predetermined structures, choose the structure closest to the situation presented, and then perhaps apply a small modification as needed to (hopefully) deal with the moment successfully. In short, a structure is a tool to help guide us through common situations without continuously reinventing the answer.

An important point emerges here. A structure is a relatively rigid construct. The ability to modify it is limited, mostly due to the assumptions we made when initially developing the structure. Let's look at the example of a traditional, routine financial transaction at your local bank (admittedly a rarer and rarer case, but it will do). Say you receive from your grandmother a personal, paper check of $50 for your birthday. You deposit the check into your checking account, which is credited to you. You receive a receipt for the deposit and continue your day, knowing that the $50 is safe and secure. These steps are the process of a typical financial transaction based on the structure of the personal banking system in the United States. But minor modifications to the system are plenty. For the deposit, you could visit the bank and have a personal interaction with a teller, or deal directly with an ATM machine, or mail in the deposit, or create an image of the check on your mobile phone and transmit the image to the bank via an app. While these options are all available to make the deposit, the core action of the transaction is to be credited $50. The structure exists to ensure you are reliably credited; the minor modification is the choice of how to receive the credit. The options to modify the structure are small and limited in number and scope.

But what about a structure with substantial flexibility? Consider the example of a local municipal courtroom. You are driving on an icy road in early morning traffic. The vehicle in front of you stops quickly, and the ice prevents you from avoiding a collision. The local police arrive, confirm no one is injured, and ticket you with a failure to stop in time. The ticket states you are required to appear in municipal court in two weeks' time to address the charge, with the fine for a guilty plea being $250.

DOI: 10.1201/9781003214397-6

In preparation for your court appearance, you attend a preliminary hearing with the local prosecutor. Based on your excellent driving record, the prosecutor offers to reduce the charge from failure to stop to impeding traffic, reducing the fine for a guilty plea from $250 to $100. You accept the offer and appear before the judge, who immediately accepts your admission of guilt. You pay the $100 fine to the cashier and continue on your way.

In this example, the court operates using a very flexible structure. For the same offense, a wide variety of outcomes can occur, all at the discretion of the prosecutor and judge. The fine can vary widely, from zero to any amount higher; the charge can become greater or lesser; the punishment can extend to something more substantial than money, such as mandatory community service or incarceration, or the case could be dismissed entirely.

For this book, a structure with such a high level of flexibility is called a *framework*. A framework's key characteristic is its expanded applicability to a greater variety of situations versus a structure. The wider the range of situations that can be successfully dealt within a single framework, the stronger the framework is.

Now to the question we face. How can we apply this framework concept to help our understanding of engineering organizations? Is there a common set of interrelationships that can be identified and assembled into a unified engineering framework, applicable to any type of engineering organization or goal, be it process, project, experimental, service, test, R&D, or any other engineering or technical situations?

Fortunately, the short answer is yes.

It's time to introduce the key framework for successfully operating within engineering organizations. Called the Essential Engineering Framework, this concept provides an integrated, simple, yet academically rigorous, model of how engineering organizations function. Developed by Dr. David A. Nadler (Harvard) and Dr. Michael L. Tushman (MIT), the model is a robust framework applicable to all those product, process, R&D, and service activities just mentioned. It shows an engineering organization as a straightforward construct yet honors the reality that these organizations are also three-dimensional, highly interdependent, and essentially organic entities. Like all good frameworks it is very simple: a system of five basic elements interacting in an external environment through a system of relationships, all remaining in balance through the forces acting on those connections.[1]

The underlying idea is simple. All engineering organizations (hopefully) create value, those products and services that make the enterprise worthwhile. Creating that value uses these five elements in an essentially linear process. These basic elements are:

1) An *External Environment*, which are all the outside organizations, laws, regulatory sources, and any other external entity that can impact the engineering firm's success or failure.
2) *Inputs*, in which raw materials (people, ideas, energy, money, skills, tangible commodities, and so on) are taken from the external environment and transformed by the organization.

3) *Outputs*, comprising objects or services exported to the environment at an enhanced value, intended to fulfill the organization's purpose.
4) A *Feedback Loop*, designed to determine if the output is what the organization initially desired, and, if not, applying corrective action.
5) A *Transformation Process*, the internal "core" workings of an engineering organization that actually performs the transformation, modifying the inputs into the desired output state. This change of state occurs in the *Inner Core*, which includes the boundary separating the external environment from the workings of the Core. The model is represented in Figure 4.1.

On first look, the model is fairly trivial. Inputs, transformation, and outputs in an environment with feedback are the fundamental model of manufacturing engineering. Only when you expand this construct to apply to *all* engineering activities, be it product development, experimental design, process improvements, testing, R&D, or technical sales and service, do we begin to see the power in this simplicity. The takeaway here is that this model is the starting point for all engineering work, independent of what that organization or enterprise produces, provides, or creates.

A note about the framework's terminology before we proceed. The Essential Engineering Framework contains two levels of classification that define the model. The upper-level classification is referred to as *elements*, the five categories just

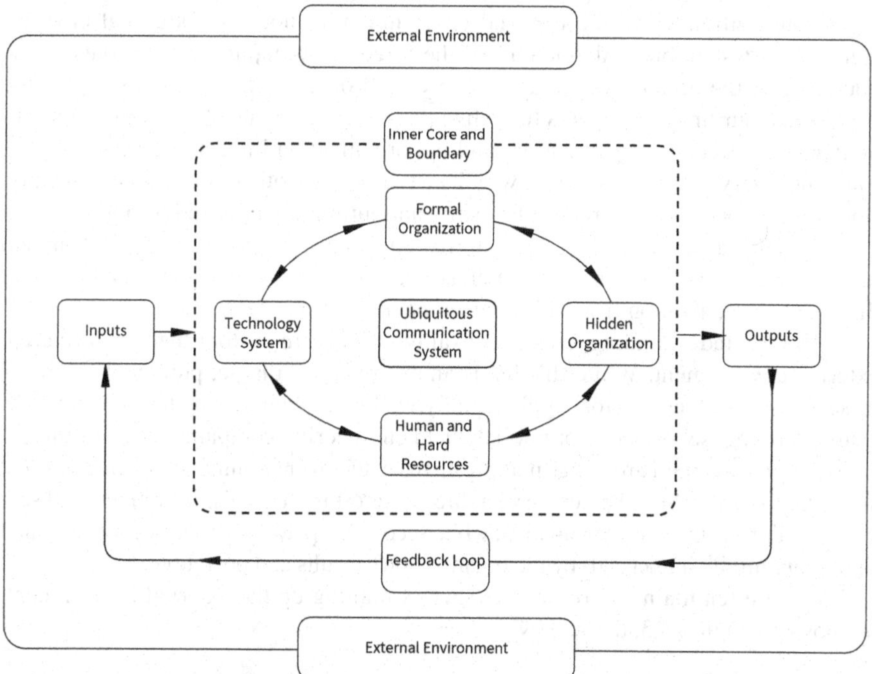

FIGURE 4.1 The Essential Engineering Framework.

introduced: Inputs, Outputs, Feedback Loop, External Environment, and Inner Core. The lower-level classification is referred to as *components* and exclusively populates the Inner Core. There are five components: the Formal Organization, Technical System, Human and Hard Resources, Hidden Organization, and Ubiquitous Communication System.

That said, let's begin to unpack the framework, beginning by defining the five individual elements of the model.

4.2 LOOKING OUTSIDE OURSELVES: THE EXTERNAL ENVIRONMENT ELEMENT

Of the five elements defining the Essential Framework, the External Environment has an outsize impact on all aspects of the functioning and ultimate product of any engineering organization.

The external environment is the world where organizations live. The obvious simile is to our own physical environment. Just like the atmosphere we breath, the water we drink, the ground we stand on, and the wind we feel, our natural environment provides the necessary conditions for us to survive and prosper. Conversely, we impact our environment through our care and stewardship of our air, water, and wildlife. In short, the natural environment impacts us and we impact the natural environment simultaneously.

The same is true with organizations. Organizations also require an environment to operate within. Here, the external environment is not the traditional environment or ecosystem but is defined as all the forces, powers, and entities outside the boundary of the organization that have any kind of interest in the workings of the firm. Thus, our first key point is to realize that everything outside the organization's boundary impacts the organization, and the organization impacts its external environment. Every firm *must* interact with its environment; otherwise, the organization slowly loses touch with its reason for being and ultimately fails and disappears.

Two major aspects describe the external environment: the external environment itself, made up of a dozen or so interest areas or sectors, and the boundary that divides the organization from this environment, as in Figure 4.2.

With that understanding, let's first examine the main sectors that constitute the external environment. While this list is an example for a large, publicly held enterprise, subsets of these sectors apply equally well to all engineering firms, from minuscule to immense, privately or publicly held engineering companies and startups.

There are two main groupings making up the external environment: *stakeholders* and *secondary players*. Stakeholders have a direct interest in the goals, operations, and success or failure of the engineering firm, while secondary players indirectly influence (and in turn are mildly influenced) by the organization's results and operation.

There are ten main environmental sectors making up the external environment, as shown in Figure 4.3, defined by:

1) Industry: direct and adjacent competitors, industry size, competitiveness, and related factors.

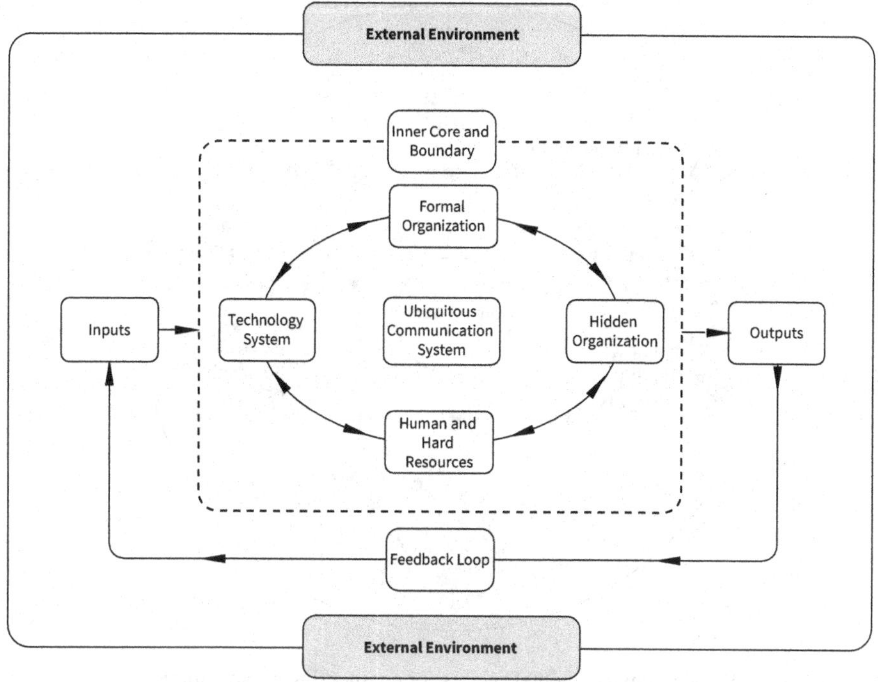

FIGURE 4.2 The external environment and boundary.

2) Raw Materials: Suppliers, first-tier manufacturers, real estate, infrastructure, etc.
3) External Human Resources: Labor market, employment agencies, national and local unions, universities, training schools, employees in first-tier supplier companies
4) External Financial Resources: Wall Street, investment community, venture capital, angel investors, banks, other private investors and higher-risk financial instruments.
5) Market: Current and future global customers, clients, and direct potential users of the firm's output. Individuals and entities that actually provide money in exchange for products and services.
6) New Technology: Techniques of production, research, product development, applied science, digital tools, and future information technology.
7) Economic Conditions: Domestic and international recession or growth, unemployment rate, inflation and investment rates, availability of investment funds, and black swan events.
8) Domestic and International Governments: Regulatory entities that examine and control the firm's products and services. City, state, and federal laws and regulations, operating taxes, infrastructure and related services, court systems, and political processes.

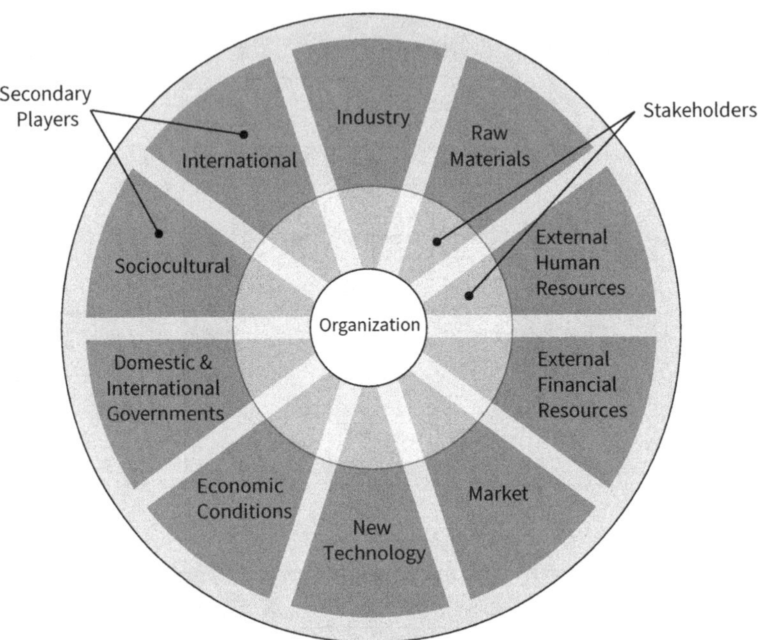

FIGURE 4.3 Primary environmental sectors for engineering organizations.

9) Sociocultural: Nonprofit and nongovernmental (NGOs) advocacy groups representing distinct societal values and beliefs. Examples include consumer, environmental, and safety protection movements.
10) International: The impact of international company operations on the proceeding nine sectors. Global competition, product homologation, potential acquisition by foreign firms, overseas markets, foreign customs, regulations, and exchange rates.[2]

We should spend a minute talking about the market segment, specifically the actual customers. There are two types of customers: in-house clients (other internal departments that use your output or product within the Inner Core as part of the firm's overall processes) and external purchase consumers (those end-product buyers who actually reach into their wallets and place money into the company's coffers).

Dealing with an internal customer is straightforward. Simply ask what they want or need, and as possible provide it to them in the form, quality, cost, reliability, and timing they require. They should be able to tell you clearly. Now I'm not saying the work of providing the internal product to them is necessarily easy, just that the internal customer requirements should be relatively well defined.

Providing for the external purchase customer is an entirely different universe. New engineers are beseeched to "understand your customer" and be subjected to the continuous drumbeat of "know this" and "know that" about who is the recipient of the firm's hard work, and exactly understand what they want and need. In the vast majority of firms, one of the most common ways to understand the external customer is through basic demographics.

As we discussed in Chapter Two, *demographics* is defined as statistical data surrounding the characteristics of a population as classified by age, gender, income, and the like for market research, sociological analysis, and governmental policy development. Demographics has been the "go to" tool for multiple business generations, as it's easy to understand, comfortable to use and the data is easily available through census data, surveys and other data sources (normally for sale).

Unfortunately, everyone tends to have the same data. Striving for a marketing advantage, some firms have moved away from strict demographic analysis and attempted to base their marketing on "softer" factors such as a customer's values, desires, aspirations, and level of self-actualization.

One advantage of these methods is their potential ability to target very specific groups of customers with individualized products and marketing programs. These methods support a "segmentation" approach to targeting customer groups based on personal characteristics and beliefs. Specifically, consumer segmentation groups are similarly minded customers binned together by any number of traits. For example, a technique called "values-based segmentation" is worth describing. Here, the customer's *personal values* are the main consideration that compel them to make a purchase.

Let's look at the values definitions for a U.S.-based automotive business in the last decade. Here, the segmentation is based on the purchase customer's attitudes and values toward vehicles and driving, personal motivations, functional needs such as product usage, emotional needs and feelings while driving, the dealer experience, and overall satisfaction.

With that background, let's compare the same U.S. market using both demographics and values-based segmentation. Tables 4.1 and 4.2 show these results.[3]

Both lists give valuable insight into the customer base; the demographic analysis defining who the customers are *today*, and the values segmentation can be interpreted as what the customer *aspires to be* going forward. Both analyses have value, but several factors impact the values-based approach. Demographic data is fast and cheap to obtain and analyze. Values-based data is expensive and slow to acquire, not only to gather the data through ethnographic means but especially conducting the analysis afterword.

Another factor is authenticity. While asking the customer directly is the apparent best way to gain that data, the reality is customers will not always tell the truth. Many times, customers will not tell you what they truly think. Instead, they will tell you *what they think you want to hear.* Customers will attempt to gauge the interviewer's views and then state whatever they think the interviewer would like to hear, not what they themselves like or dislike about the product.

An additional authenticity factor regards investment-heavy businesses such as manufacturing. These firms are constantly designing new products for sale three or four years in the future. For example, in the automotive business vehicle designers will create three levels of change (or "themes") for a proposed new car or truck.

TABLE 4.1
Demographic Segmentation for U.S. Automotive Market

Segment Name	Descriptors	
Empty Nesters	• Average age of 65 • 81% are male • 97% are married • 2% are widowed • 100% are retired	• Low average household income • 84% are two-person households • Skew to basic large and luxury sedans
Common Family	• Average age of 40 • 58% are male • 99% married or living with someone • Are large families (3 or more) • Below average income	• Average age of children is 17 • Lowest education • 75% own home with mortgage • Skew to pickups
Solo Again	• Average age of 56 • 68% are female • 66% are divorced • 34% are widowed	• Lowest household income • 68% are one-person households • Highest proportion of sedans
Single Life	• Average age of 39 • 51% female • 100% single • 61% 1-person households	• Income below average • 27% rent home (highest) • Most ethnically mixed
Middle Older Successful	• Average age of 52 • 70% male • 99% married or living with someone • 2 or 3 people in household • Has grown children, average age of 23	• Highest household income • Very high education level • 72% own home with mortgage • Skew toward luxury vehicles and SUVs
Younger Successful	• Average age of 41 • 63% are male • 99% married or living with someone • Kids at home (average age 11) • 61% are 4-person household	• High income • High education • Highest proportion of vans/minivans (more than 2 times the market)
Grown Family	• Average age of 55 • 56% are male • 100% married or living with someone • Grown children • Below average education level	• 12% are self-employed (2 times the market) • 80% are 2-person households • Highest proportion of pickups
Young Family	• Average age of 36 • 54% are male • 92% are married	• Small families • Highest proportion of living with someone (8%)

The first is an evolutionary design with minor changes to the current model. The second design is one with a moderate level of change. The third level of change is normally a "stretch" design, with highly advanced shapes, colors, materials, and the like designed to make a revolutionary statement about the vehicle and the person driving it.

TABLE 4.2
Values-based Segmentation for U.S. Automotive Market

Segment Name	Subsegment Name	Percent of Total Market
Transportation Tools	• Basic Car, Big Purchase	5.6%
	• Dependable Mainstream	4.9%
	• Commuter Solution	6.0%
	Subtotal	16.5%
Family Transport	• Family Focused, Not Vehicle Focused	6.2%
	• Vehicle as a Family Enabler	8.3%
	• Aggressive, Active Lifestyle	3.9%
	Subtotal	18.4%
Can Do	• Hard Working Truck	4.7%
	• Independent Adventurers	5.0%
	• Functional Technology	7.0%
	• Trucks with Attitude	6.1%
	Subtotal	22.8%
Comfort and Luxury	• Comfortable Car, Sound Decision	7.2%
	• Conservative and Accommodating	7.1%
	• Luxury and Elegance	4.0%
	Subtotal	18.3%
Vehicle Enthusiast	• Stylish and Smart	7.0%
	• American Heartland	5.0%
	• Fashion Statement	8.0%
	• Sports Car Driving Enthusiast	4.0%
	Subtotal	24.0%
	Total	100.0%

Typical customers of the existing vehicle are invited to discuss with market research experts what they like and don't like about the three themes. A large percentage of customers will say they like all three, keeping their real opinions to themselves. Another large percentage will prefer the evolutionary design, and less so the moderate. Yet there is also a percentage of customers who will loudly and definitively state they do not like the third, most advanced theme. These customers will invariably say, quite strongly:

"I know what I like, and *that's not it!*"

That's fine, but market researchers will tend to evaluate these strong responses with their own interpretation of the comment:

"I know what I like, and *I like what I know*."

Market researchers know that a customer's reaction to a design or product is heavily influenced by that person's prior experience, highly impacting the acceptability of the design. The important thing here is to realize that marketing evaluations of new products are more an art than science, so be aware of the softness around any customer acceptance analysis.

Let's switch gears and talk a bit about a firm's overall relationship with the external environment. When the environment is uncertain (which is nearly 100% of the time) the organization attempts to interact with that environment in three ways. The firm can adjust its own internal organization to better fit with the current or future external environment, or it can attempt to control that environment or both. This results in a continuous de facto negotiation between an engineering organization and the external environment, resulting in an unremitting change to the status quo; i.e., the organization is always in a state of flux.

Companies attempt to control the external environment through national or local political activity, supporting trade associations or initiating public relations campaigns, all of which attempt to influence outside organizations and other entities.

The second, and more difficult, organizational strategy attempts to adjust the internal organization to react to substantial environmental change. This path may be chosen when senior management judges the environmental change is indeed revolutionary and unlikely to return to its prior state. The rise of the web, ubiquitous digital products and services, and cloud computing are obvious examples of this level of change. The techniques a firm may use to effectively change are many, including absorbing uncertainty by protecting the organization's technical core, offering a high degree of specialization, zeroing-in on a specific target environment, creating formal links through ownership or joint ventures, and several others.

Of the two approaches, experience teaches that attempting to control the environment tends to be cheaper, faster, and marginally more effective than adjusting internally. Internal changes are lengthy endeavors; expensive and high risk. That said, a combination of both control and adjustment strategies are usually applied. Because of its outsize impact on the entire enterprise, managing the external environment is a primary role of the CEO and Chairman of the Board, while the Chief Operations Officer (COO) tends to handle the internal workings of the organization.

The important point here is that companies change as a result of evolving environmental factors, which constitutes the major reason for the organization's continuous metamorphous in both the short and long term.

Before moving on, we need to discuss one more topic to fully grasp the idea of the external environment, that of understanding the *Organizational Boundary*. The Organizational Boundary is defined as a semipermeable layer which surrounds the organization and through which inputs and outputs may pass, as shown in Figure 4.4.

To completely understand an organization, it's critical to precisely define the location of the boundary that separates the enterprise from its environment, that is, where to define the limits of the company. Like nearly all concepts in this book, boundary placement is governed by contingency theory. When asking, "Where exactly does my organization begin and end?" the answer will always be contingent on what level of the organization is being analyzed. The department level, the multidepartment level, the division level, or the enterprise level, all have different boundaries. This is important, as how the level is defined sets up what is an input, an output,

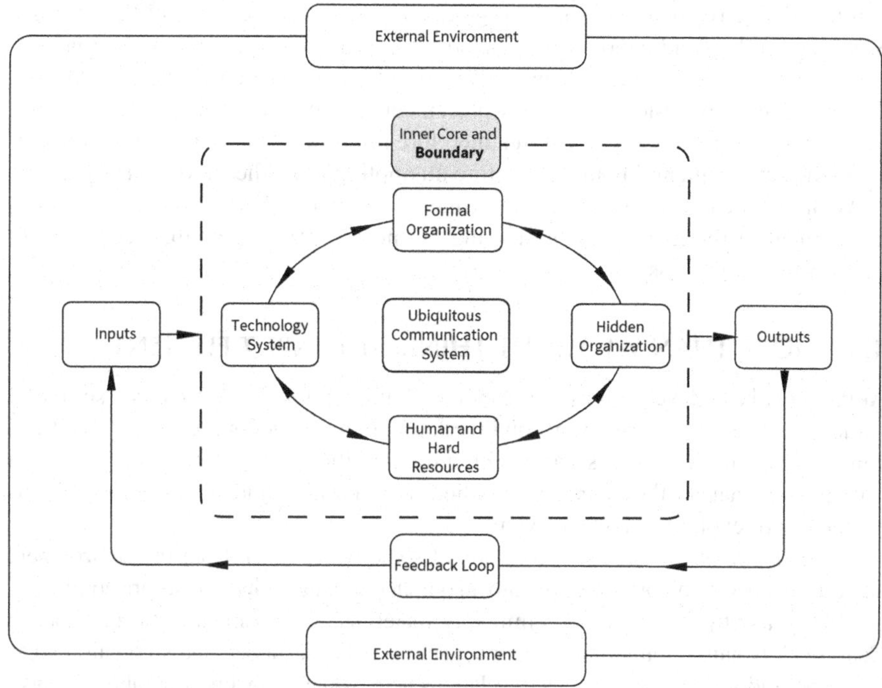

FIGURE 4.4 The organizational boundary.

and where the transformation activity and feedback loop are located. For example, when considering the enterprise level of a technical organization, an input would be investment bankers or Wall Street for finance, national NGOs for social responsibility, the federal government for regulations, and international industry players for competitors. Yet when considering a department-level analysis, inputs would be your internal finance department for funding, your firm's internal sustainability department for adherence to company "green" guidelines, local city or county governments for regulations, and other internal departments charged with controlling or limiting any technical work.

It may seem trivial that we must think about an organization's boundary. After all, the idea is simple: everything having to do with the direct workings of the organization resides inside the boundary, and everything that does not must belong in the external environment. However, a key idea about the boundary is how it is initially placed in your mind and how it relates to your understanding. Engineering organizations tend to be complex entities with imprecise language. When we talk about a technical organization, are we referring to an engineering department, a section, a directorate, a division, or an entire enterprise? Then what is considered an input, output, and external environment is completely dependent on contingency theory.

Let's make this a bit more personal. Say you are asked to prepare a plan for improving the flow-through of a continuous manufacturing process. It's critical that you first define the boundary of the organization to set up the terms of the analysis. Selecting too large a boundary that includes the company sales and marketing divisions as part of the manufacturing analysis is not likely to help improve the process. Instead, a department-level boundary placement is much more appropriate. In short, the boundary selection is critical. It's therefore important that you select and understand the environmental boundary before attempting to further develop the plan.

With this idea of the external environment fully in mind, let's now turn to a simple element of the Essential Engineering Framework, the inputs that feed a firm's transformation process.

4.3 MORE THAN WE MIGHT THINK: THE INPUT ELEMENT

At the most basic level, inputs are resources. In this book, a *resource* is a source of value that when transformed or combined, additional benefits are produced. It has value and utility. Resources can be broadly classified upon their availability and importance; that is, they can be classified as renewable and nonrenewable, or as something of explicit or intrinsic value.

Similarly, *inputs* are items or concepts of some worth taken from the environment and transformed by an organization into an output of greater value. Inputs are commonly thought of as physical items or commodity materials, as in a manufacturing process. Coal, steel, bananas, tomatoes, component subassemblies, rubber, and so on all tend to fit this definition. There are no new revelations here: you already know all about inputs.

We as engineers tend to think of inputs almost exclusively as a manufacturing or software concept. Inputs are either commodity materials used to build something or digital data applied to a computer program or simulation. The commonality here is that both these material and digital inputs are commodities, bulk items where each unit has a very low individual value. Note these are not to be confused with economic commodities (i.e., gold, silver, pork bellies, and financial contracts), which are excluded from this discussion.

Here, inputs denote a much broader concept. Inputs reflect the total number and type of resources that a group has at their disposal to create a transformation. Inputs are the raw ingredients of talent, people, ideas, energy, time, money, information, skills, tangible raw materials, and other various items of value. The key idea is still valid; just the boundary is different: organizational inputs are defined as something of value from any internal or external source and transformed by the organization into a product or service of increased value. Figure 4.5 shows the fundamental relationship to the model.

As we consider inputs to be a wide range of items the engineering organization uses to create value, they include items not normally considered traditional inputs, such as skills, energy, concepts, and ideas. Our perception of inputs needs to be something much more universal than we may have previously thought. Let's look at some examples.

4.3.1 People as an Input: Commodities vs. Talents

Today, we do not think of people's value as strictly a commodity, to be placed in the same category as iron ore or soybeans. Yet certain organizations, especially

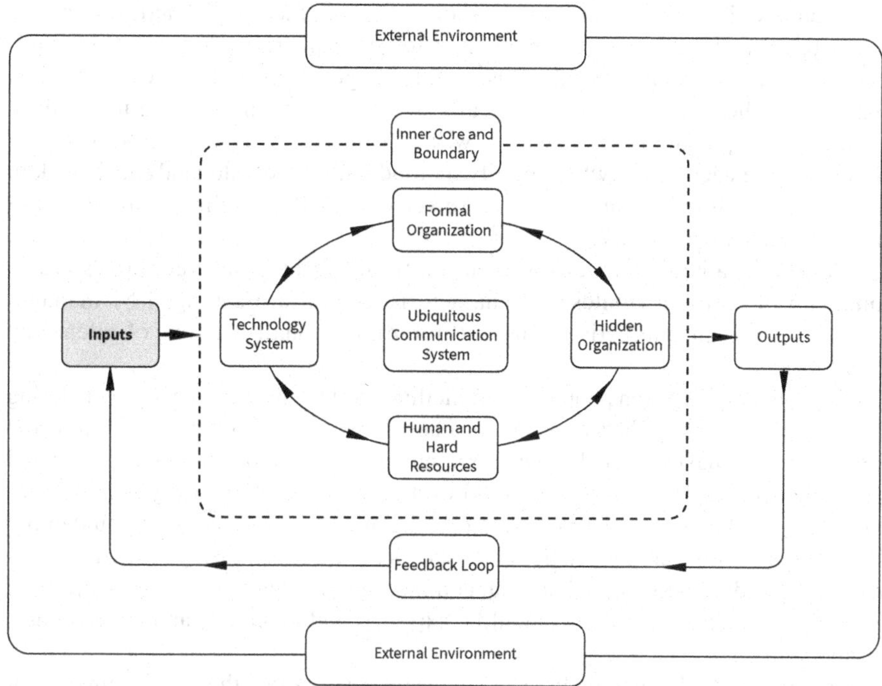

FIGURE 4.5 Inputs within the Essential Engineering Framework.

organizations in the past, saw people as raw, mobile muscle; unthinking hordes to do the will of the overseer (or, in today's lexicon, local management). The laborers who built the ancient Egyptian pyramids fit this definition nicely. Fredrick Taylor's mechanistic organizational beliefs (i.e., scientific management) are founded on workers as commodities. Vestiges of this type of input remain today: road crews and landscapers can be thought of as commodities, providing sheer numbers and muscle to the job.

The individual talent of a worker is typically not explicitly identified as an input. Yet without talent, today's technical organizations would lack the knowledge and skill to operate complex systems (in fact, the phrase "knowledge worker" was coined to refer to these talented individuals). Talent has immense value in all organizations, the amount of value varying by organization type and desired output. Consistent with Gardner's Theory of Multiple Intelligences covered in Chapter 3, talent can be thought of as a special aptitude or ability of many types, a competence in doing something uncommon. And unlike tons of green beans entering a canning plant or a thousand points of data powering a weather forecast, these talents cannot be readily measured.

Take the example of two colleagues you meet on the first day of your new job. Both are the same age, both come from big-name engineering schools, and both have the same social-economic backgrounds. As you soon discover, one is outgoing, energetic, and loquacious. The second is highly logical, stoic, and somewhat serious. Are each equally valuable to the technical firm?

Of course. But the common perception is that engineering organizations tend to see talent as limited to mathematical or logical constructs. Yet in reality, engineering managers consider talent as a wide range of characteristics, and when placed into the organization strategically can create enhanced value to the firm. Like a jigsaw puzzle, good managers will arrange talented people into positions that reinforce each other, where highly methodical workers do analytical work in association with others, and where more verbally skilled engineers are placed in more interactive positions.

The key idea here is that, as we learned from Gardner, most engineers are an amalgam of many attributes and characteristics that are arranged by management to create an interactive, almost three-dimensional, network of talent and capability.

The boundary between people as commodities and talent is frequently seen during a new engineer's early days at work. As a new hire, you are initially an unknown to the group, so it makes sense that early assignments will be more commodity-oriented than talent-based. This serves two purposes: to experiment with how you might perform, and frankly, to prevent you from causing any real damage. Once you demonstrate some positive characteristics and establish initial trust, then your assignments will begin to draw more on talent than commodity. The main point is to realize that early commodity assignments should be expected and embraced and not seen as a disappointment.

An important distinction here concerns understanding the true boundary of the input. Are the people in question (either commodity or talent) already a direct employee of the organization or contribute to it as an outside hire? Are they already internal to the firm or an external input to the organization? "Hired guns," consultants, job-shoppers, and "temps" all fall into the external category. Seeing these resources as either external inputs or internal resources has an important bearing on the accuracy of the "mental map" you are attempting to create.

4.3.2 Funding as an Input

While a very easy idea to grasp, we need to touch on some additional types of funding acting as inputs. As everyone knows, funding is the single, most critical input to any engineering project or process. During the United States vs. Soviet space race of the 1960s, the old joke was, "What fuel propels a rocket ship to the moon? The answer: "Funding." Today, this saying still holds. Is there enough funding? Is the funding of the right type? Will it arrive in time? Is the funding source reliable? Are the funds fungible, or are they locked into covering only certain expenses? The list of questions surrounding the funding input is truly large and important.

Some people suspect that new engineers tend to avoid funding questions, as they may feel that finance should be left to the finance department, upper-line management, or some unknown "they." In other words, "I'm an engineer, not an accountant." Instead, new engineers need to embrace the chance to understand, at the day-to-day level, how the funding mechanism works in any given organization.

There is an important distinction when it comes to funding. The distinction is whether the funding in question is provided internally by your existing organization or comes from sources external to it. Radically different rules and limitations apply between these two sources. This will be covered in some detail in Chapter 6, but for now be alert that a significant difference exists between the two.

4.3.3 Ideas as an Input

Ideas are a fundamental input to any engineering project or process. Without new ideas, there is no reason to do the work or strive for a successful result. Ideas are the engine that propel a product, service, or entire enterprise forward. Sometimes identified as innovations, inventions, evolutionary improvements, or revolutionary advances, ideas are the creative factor required to gain a competitive advantage or other worthwhile output.

Now for a warning. When thinking of an idea as an input, a complex and confusing situation emerges. Ideas are very nebulous creatures: they have many forms. They are described by countless definitions and terminology that is not agreed upon and are developed through an untold number of methods and techniques. They can encompass a tangle of concepts that innovators continuously argue about, never mind the near-infinite number of resulting innovations possible.

In short, ideas are a mess.

The challenge is not in identifying an idea to develop, as there is never a lack of good ideas already in your organization. A tableful of experienced engineers can generate dozens of new, worthwhile ideas in about two hours flat. Instead, the issue is the very nature of an idea. Ideas are a mess because you can't readily get your arms around them; you can't pin them down precisely. Instead, the challenge is to become comfortable with the ambiguity of ideas at the input stage and allow the natural maturity of the idea to better define itself over time. The nature of an idea will become clear, just not immediately.

For now, the takeaway is that ideas and innovations are a key input that can be nebulous, ill-defined, and very difficult to work with. And that's OK.

4.3.4 Willingness or Commitment as an Input

This may seem an unusual input for an engineering project or task. Yet a willingness or commitment to see a project through plays a surprisingly large role in the success of any undertaking. Here, commitment refers to the desire or will of the technical organization to actually achieve the goal or objective set. Organizational willingness is not a given; it varies widely depending on a number of factors, many of which are not necessarily visible. You can see indirect hints of this lack of will: low headcount assigned, funding starvation, a lack of space or proper facilities, or difficulty in getting management's attention. All point to an organization's lack of willingness to complete the project or achieve the goal. Many initiatives die, and on their tombstones is written the epitaph "Lacked the will to live."

4.3.5 Time and Information as Inputs

Time is required for any project, yet time is a flexible concept, as we will discuss in Chapter 6. As an input, time can be flexible, rigid, limited, unlimited, or any number of different forms. At this point we just need to plant the seed that time is not a single, linear commodity but a amorphic, Jell-O-like quantity capable of taking many forms and shapes.

Information is an obvious input; its availability is critical to any knowledge-based project. As an input, information can be mere data or refined intelligence or knowledge. The important considerations are obvious. Is it available? At what cost? In what form? And for what length of time is it accurate or even relevant?

All inputs have limitations, or boundaries, that directly impact their value and appropriateness to an engineering organization. Each of the input types we've discussed will have some sort of limitation set on them. Not all raw materials will be available in the quantities or quality needed. Time may be in short supply. Talent may be severely limited or just not available. While some needed inputs can be substituted (such as money for time), others cannot. Inputs are contingent and so must be considered individually and managed like any other aspect of contingency theory. In short, inputs can severely limit or enhance the success of your organization's goals, and the guarantee of their availability requires a substantial effort and investment.

4.4 IS THIS WHAT WE REALLY WANT: THE OUTPUT ELEMENT

I'm certain everyone knows what an output is. An *output* is something of greater or (sometimes) lesser value created by a transformation process. Outputs are materials, products, processes, knowledge, inventions, innovations, or a myriad of other entities exported to the outside environment intended to fulfill an organization's purpose. Of course, not all outputs are favorable. Any coder who has unintentionally created an infinite loop on a customer beta site can tell you exactly what kind of output that is.

As with inputs, the definition of an output in contingency theory is far broader than traditional engineering thinking. The point here is that outputs, like inputs, consist of an entire galaxy of objects and ideas.

Be it new products, novel process recommendations, fresh insights, or plain old widgets, outputs are the core interest because they represent the end result of your firm's considerable intelligence, planning, and investment. They receive the lion's share of an organization's attention. As with inputs, here we view outputs as broadly as possible. As the Essential Engineering Framework is meant to describe all engineering work activities, so it stands that its outputs can be new consumer products, manufacturing processes, ideas, insights, recommendations, digital applications, and a universe of advancements created at the pleasure of the firm. Just take a look at Figure 4.6: it's as simple as that.

But once an output is in hand, the question becomes this: Is this the output actually desired? Does it successfully achieve the goal? Are we where we wish to be?

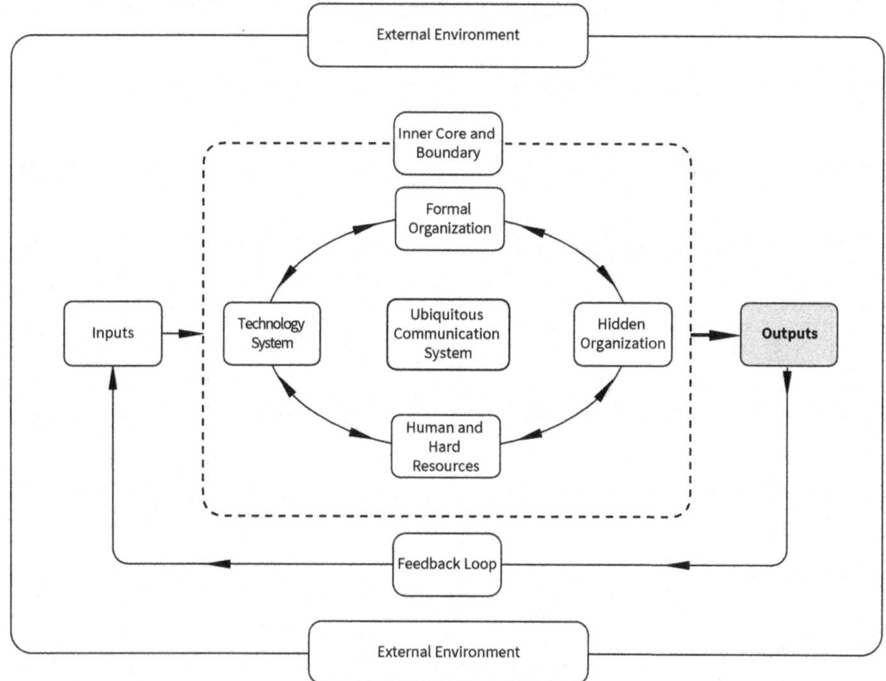

FIGURE 4.6 Outputs in the Essential Engineering Framework.

Luckily, the answers to these questions are close by. Of course, we're talking about the Feedback Element.

4.5 ARE WE DOING THE RIGHT THING: THE FEEDBACK ELEMENT

Feedback in engineering is a very familiar concept you already know. Sometimes called "closed loop control systems," these are structures and methods that automatically sense and correct for out-of-specification output conditions.

For us in engineering, a closed loop control system has a narrower definition than in general use. We see closed loop systems as a set of mechanical or electronic devices that automatically regulate a process to the desired state or set point without human interaction. Engineering closed loop methods are widely used in many industry applications, including agriculture, chemical plants, aviation, nuclear power plant cooling systems, automobile lane departure systems, and many, many other examples. Closed loop control systems contrast with open loop systems, which may sense an out-of-bounds condition but are unable to correct themselves without human assistance.

Organizational feedback systems are different in that, like the inputs and outputs we've discussed, their definition is wider, extending far beyond strictly mechanics. These feedback systems exist everywhere. All organisms in nature have closed loop

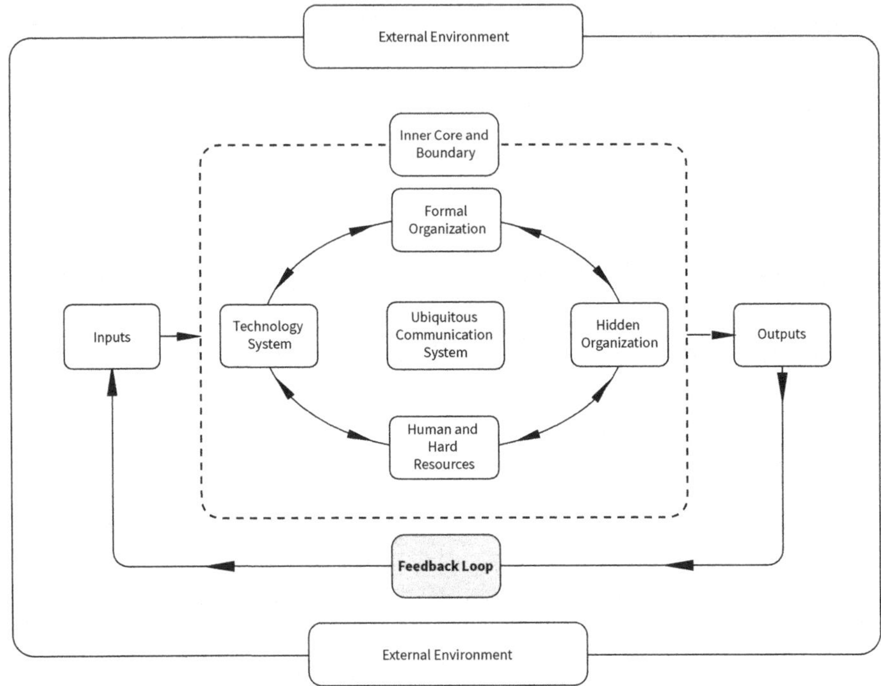

FIGURE 4.7 Feedback loop in the Essential Engineering Framework.

feedback systems. For example, pain is an example of an organic closed loop system. Pain exists to tell the body that damage is being (or has been) done to the organism and corrective action is required. In the animal world, prey rely on a variety of organic sensors to detect predators, including the visual, chemical, auditory, and tactile senses. In the legal world, a repeat offender, by the very fact they have multiple convictions, offers feedback to the defense attorney that a successful argument on a new charge is unlikely.

Obviously, organizations also have feedback mechanisms (many times highly developed) which determine if they are achieving their intended purpose. Here, feedback mechanisms are designed to measure progress toward operative goals. In the Essential Framework, shown in Figure 4.7, feedback systems are designed to answer two questions: are you effective in hitting your organizational targets, and are you achieving your target efficiently? (Of course, this assumes you are actually aiming at the right target.)

In many instances, undergraduate engineers are taught that feedback loops are relatively straightforward systems, requiring minor attention or effort. Often, just one (or perhaps two or three) indicator is measured periodically or continuously and compared to a predetermined, desired output level. Usually, these measurements are selected for ease of measurement rather than as an optimal indicator of performance. Examples here can include quality, quantity, timeliness, and cost, characteristics that are all related by the fact they are easy, quick numerical indicators that are extremely cheap to measure. This may or may not be appropriate. The key to selecting a feedback indicator is in its ability to truly measure what is important, not what is easy.

Another common misunderstanding is that many engineers view effectiveness and efficiency as the same thing. Effectiveness is not efficiency. Effectiveness is the *degree* to which an organization achieves its goals. Efficiency is the *amount* of resources (time, funds, talent, etc.) used to achieve those goals. Efficiency confused with effectiveness usually results in only efficiency being measured rather than both. Both measures must be well defined and clear; otherwise, you open the door to reaching the wrong answer in a very efficient manner.

A well-known example of efficiency being mistaken for effectiveness happens regularly in engineering decision meetings. Say you are invited to a meeting with the purpose of approving a major new design initiative within your division. Due to the importance and riskiness of the decision, in-person attendance is required. A time is set, a location reserved, and review material transmitted in advance.

As soon as the meeting begins, the majority of attendees immediately begin the ritual of bowing their heads and staring at their laps, tapping on their mobile devices under the table, sending texts, writing emails, making appointments, and perhaps booking dinner reservations. Soon, they are no longer "present" in the meeting.

The justification for this behavior is improved efficiency. Viewing "just listening" as wasteful, the attendee believes they are "making good use of their time." Unfortunately, the quality and effectiveness of the major decision suffers, as important information is commonly missed due to these digital distractions.

We already know this behavior is widespread, having seen it in lecture halls and conference rooms around the world. To combat this distraction, certain senior-level meetings require all participants, on entering the meeting room, to silence their mobile devices and physically place them on a side table. Not only does this reduce distraction, but it also sends the message that what is about to happen is important, requiring 100% attention. It also avoids the risk that the person texting is seen as disrespecting senior leadership. The message is clear: the texts and emails can wait.

As before, feedback should be an indicator of both organizational effectiveness and efficiency. Yet measuring organizational performance is a decades-old problem: it's hard to do. High-quality feedback can be expensive, slow to obtain, and unwieldy to analyze. And consistent with contingency theory, organizations have evolved many techniques for measuring these two characteristics, based on environmental conditions, business type, the product created, the structure of the organization, and so forth. Common measurement techniques or approaches include (but are not limited to) the following.

4.5.1 Goal Approach

Here, the single focus is exclusively on current output or productivity, represented by one or perhaps two simple measurements. Quarterly financial profit and loss statements or revenue are prime examples. Easiest to measure but most limited in insight, this approach is similar to a physician performing a diagnosis on you by only taking your temperature, neglecting blood pressure, body weight, blood glucose level, or any other physical measurement. While extremely fast and easy to measure, the goal approach is only a "backward looking indicator," providing no insight into future performance. In short, this approach when used alone is totally inadequate to achieving an accurate assessment of an organization's health.

4.5.2 Resource Approach

This method focuses on the ability to acquire necessary resources, such as funding, talent, time, or other resource to achieve the firm's goals. While a "forward-looking" indicator, and useful when other measures are difficult to obtain, its limitation is in being unable to track either internal processes or outputs, making it a very weak indicator.

4.5.3 Internal Process Approach

The emphasis here is on internal process efficiency only, implying that a healthy internal process guarantees a good organizational outcome. In other words, it indicates only efficiency, neglecting effectiveness. This approach is a poor stand-alone indicator of the organization's internal health.

4.5.4 Competing Values Approach

Adapts Robert Quinn's Competing Values Framework to emphasize four interlocking organizational values: human relations, internal processes, rational goals, and innovation. The point here is to provide a holistic assessment of the engineering organization's performance, considering as wide a range of indices as possible. The difficulty is in the measurability and repeatability of these four indicators, as several tend to be qualitative rather than quantitative.

4.5.5 Balanced Scorecard Approach

The natural extension of the Competing Values model, the Balanced Scorecard develops curated, reliable, and repeatable metrics drawn from the four Quinn categories. Balanced Scorecards were initially developed by Kaplan and Norton to provide a more advanced but still holistic way to consistently measure the main characteristics of an engineering enterprise. Four main performance aspects are tracked: Financial, External Customer, Internal Processes, and Employee Development. Since its introduction in the 1990s, the Balanced Scorecard (also referred to as a "dashboard") is considered a best practice and is a favorite of numerous organizations worldwide. The challenge here is that the local organization must invent their own metrics; they cannot depend on a cookbook or menu approach to establish the metrics that matter.[4]

A summary of these five methods is shown in Table 4.3.

Despite the popularity of the Balanced Scorecard, arguments are plentiful over which of two underlying philosophies is better: does success come to those firms who drive their businesses with only strict numerical targets and performance measures,

TABLE 4.3
Common Performance Measurement Systems

Method	Measures What?	Timeframe Measured	Past, Present or Future Looking Measurement	Measurement Boundaries	Ease of Application	Quality of Insight
Goal	Effectiveness	Short Term	Present	Single Measure Only ("Silver Bullet")	Very Easy	Very Low
Resource	Efficiency	Short Term	Past	Timeliness of Measurement	Moderate	Low
Internal Process	Efficiency	Short Term	Present	Limited to Process Only	Moderate	Moderate
Competing Values	Efficiency and Effectiveness	Short to Moderate Term	Present and Future	Includes Hard to Measure Aspects	Difficult	High
Balanced Scorecard	Efficiency and Effectiveness	Short to Moderate Term	Past, Present and Future	Amount of Data Available	Difficult	High

or that managing a firm through a combination of qualitative and quantitative indicators is optimum? Engineering firms can sometimes demonstrate a preset bias toward the former, as the engineering culture reinforces numerical measurement with such adages as "What is measured is real" or "In God we trust; all others bring data."

With these two approaches unresolved, many (if not most) general organizations today tend to avoid controversy and adopt the more popular and "safe" Balanced Scorecard approach.

One of the most well-known examples of Balanced Scorecards is discussed in Bryce Hoffman's *American Icon: Alan Mulally and the Fight to Save Ford Motor Company*. During 2010–2012, Ford CEO Alan Mulally and his team achieved one of the great comebacks in business history. As the global automotive industry collapsed during the Great Recession of 2008–2010 (see Case Example 3.3), Ford quickly transformed from near-certain failure into the most profitable automaker in the world. The biggest strategic driver of Ford's turnaround was a measurement process called the weekly Business Plan Review, or BPR, another name for the Balanced Scorecard. Held every Thursday, Mulally used the BPR to identify and focus on the four critical drivers of the business as defined by the firm. From this, Mulally was rapidly able to pinpoint the strengths and weaknesses of the business and apply first aid to those areas needing it while strengthening targets of opportunity.[5]

One important but rarely talked about advantage of the Balanced Scorecard is as a tool for driving transparency and cultural change. At a critical moment in Ford's transformation, Mulally used the BPR to identify an important hidden issue, forcing top executives to admit to a problem that had been previously covered up. And once Mulally demonstrated that executives owning up to this problem uncovered by the

BPR would not be punished (as was the previous practice), senior management could focus on solutions and not concealing trouble.

For a new hire, the upshot of feedback and measurement is in becoming familiar (and eventually conversant) with your company's performance measurement system, and especially its preferred method of measuring results. This is how everyone and everything is measured, not only at the enterprise and divisional levels but within departmental and local groups as well. These are the rules of the game.

4.6 THE HEART OF THE MATTER: THE TRANSFORMATION PROCESS AND INNER CORE ELEMENT

We finally arrive at the center of our framework and the subject of the rest of this book. Our previous chapters have orbited around our foundational engineering organizational model: the Essential Engineering Framework. We now focus on the center of this model, the Inner Core. The Core is where the actual transformation of inputs into outputs takes place. We can view the Core as where the actual "day-to-day" work occurs, be it in a product design house, production factory, research complex, or computing facility.

The Inner Core denotes the location where the change of state of any input into any form of output occurs, be it data, knowledge, a physical product, a new idea, an improved process, or any other transformation. The Core houses the internal workings of the engineering organization, consisting of all the systems, departments, technologies, sub-organizations, and corporate culture that provide the necessary elements to do work. The remainder of this book will deal with you as a new engineer operating within this Core.

The Core consists of five components that together perform the actual transformation process, as diagramed in Figure 4.8. It comprises:

4.6.1 The Formal Organization

Think of an organization chart: this is where the formal organization would reside. Yet the formal organization is orders of magnitude more complex and far-reaching than depicted on a scrap of paper. This is where all written policies, methods of communication, standard operating procedures, vertical and horizontal relationships, and all other prescribed activities reside. Being the most public face of a firm, this element contains the information that any outsider asking about an organization would be given.

4.6.2 Human and Hard Resources

This element addresses aspects of two different areas: *human resources*, which covers employee needs and labor relations, new hire selection, job roles, interpersonal dynamics, and formal reward systems; and *hard resources*, which is any nonhuman item of value, such as equipment, workspace, funding, and information. Note one distinction. Inputs are defined as resources that enter the transformation system. While commonly acquired from outside the organization, human and hard resources

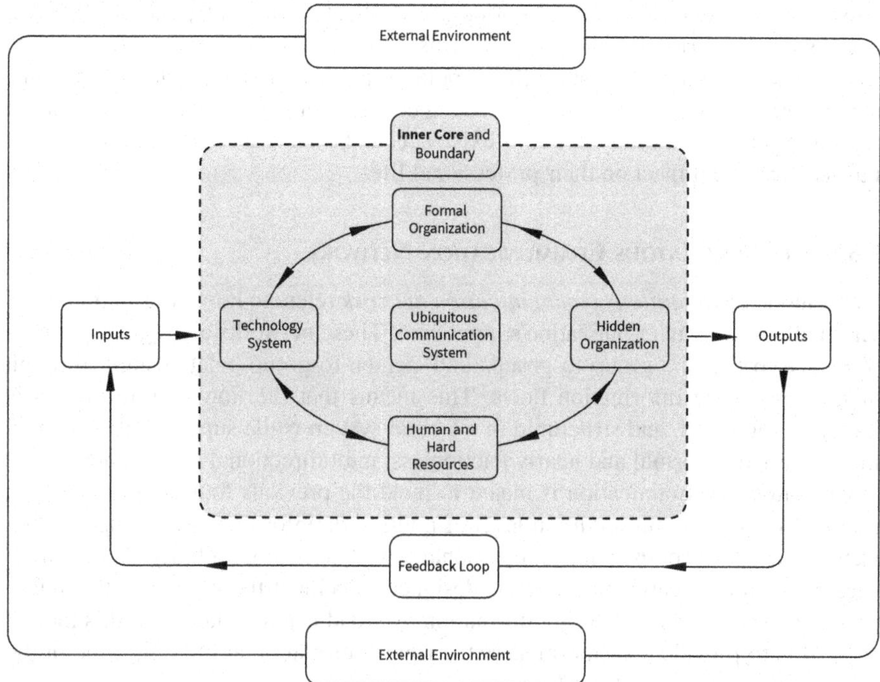

FIGURE 4.8 Inner Core within the Essential Engineering Framework.

are also sourced from inside the firm and are acted upon internally. The source of those interior inputs is ongoing budgets covering existing funds, talent, workspace, and so on. The distinction here is that all resources can come from either internal or external sources or both, and all are managed by a human and hard resource system.

4.6.3 The Technology System

The true center for converting inputs into outputs. Here, technical or technology is not the common definition of just a digital device or system; say, a smartphone or 5G network. Here, *technology* is defined by the U. S. National Research Council as the sum of techniques, skills, methods, and processes used in the production of goods or services or in the accomplishment of objectives, such as scientific investigation. These include manufacturing technologies, product and services development technologies, and technologies deployed at different organization levels such as department, individual, and information system echelons.

4.6.4 The Hidden Organization

Also called the corporate culture, this component is very important to you but largely hidden from view. The *hidden organization* is an interconnected and interrelated network of behavioral and social norms, the "collective thoughts, habits, attitudes, feelings and patterns of behavior."[6] It deals with more behavioral, "softer" topics such as leadership norms, group culture, politics, power, interdepartmental relations, and any other concerns impacting the social cohesion of a group. In some instances, new engineers may initially doubt its existence, and even if accepted, many may see it as having little impact on their professional life.

4.6.5 The Ubiquitous Communication Network

In this model, a *ubiquitous communication network* refers to how information flows within all areas of an organization's inner core. These networks are regular patterns of person-to-person, group-to-group, and person-to-group relationships through which appropriate information flows. This means that the flow of information is managed, regulated, and structured in a formal system while supplemented with an unconstrained, informal and nearly frictionless, multidirectional arrangement.

Ubiquitous communication is meant to meld the previous four components into a unified whole with an optimum level of multi-node information exchange. While each of these four components is intentionally well defined with fairly clean boundaries, the communications network is designed to cement these components into a strong and robust whole. While information overload is a real danger in this model, the goal is to provide decision-makers the right information at the right time in the right amount in the right form.[7]

4.7 REALIZATIONS

The Essential Engineering Framework is simple and easy to understand, with just five well-defined elements interacting in a basic relationship: Inputs, Outputs, Feedback Loop, Inner Core, and External Environment. Inputs are broadly defined, encompassing a much wider definition than in normal use. Equally broad is a concept of Outputs that can take a form far beyond traditional understanding. Linking inputs and outputs is a Feedback Loop, meant to adjust the system for optimum effectiveness and efficiency over time. An Inner Core provides the actual transformation function of the enterprise. Surrounding these four elements is an ever-changing external environment.

The area of our greatest attention is the Inner Core. The Inner Core accomplishes the purpose of the enterprise: convert inputs of a given value into outputs of greater value. The Core is the heart of the organization, consisting of five internal components: technology that performs the actual transformation, supported by human and hard resources, structured through a formal organization, enhanced by a hidden organization or culture, and bonded together through a multichannel communication network.

The Essential Engineering Framework is modest, unassuming, and unpretentious, yet fundamental to what all engineering organizations accomplish. That's our model.

With this introduction to the Essential Engineering Framework, we now push forward to a detailed tour of the Inner Core. This is where you will operate, where you will spend your time, where you will learn, and where your success or failure will take place. And this is where the satisfaction or regret of choices you make will play out.

The Core is where the action is.

NOTES

1. Nadler, D. A. and Tushman, M. L. 1997. A Congruence Model for Organizational Problem Solving. In *Managing Strategic Innovation and Change*, ed. M. L. Tushman and P. Anderson, 159–171. Oxford: Oxford University Press.
2. Daft, Richard. 2010. *Organizational Theory and Design*. Mason, OH: Cengage Learning.
3. Corporate Proprietary Study. 2003. *Demographic and Needs Clustering Profiles, Segmentation and Theme Definitions*. Dearborn, Michigan, USA
4. Kleiner, Art. 2002. What Are the Measures That Matter? Booz, Allen and Hamilton. *Strategy + Business* 7(26). https://www.strategy-business.com/article/11409?pg=0
5. Hoffman, Bryce. 2012. *American Icon: Alan Mulally and the Fight to Save Ford Motor Company*. Sydney: Currency.
6. Sadri, G. and Lees, B. 2001. Developing Corporate Culture as a Competitive Advantage. *Journal of Management Development* 20(1): 853–859.
7. Greenbaum, H. H. 1974. The Audit of Organizational Communication. *Academy of Management Journal* 17(4): 739–754.

5 A Tour of the Inner Core

5.1 THE FIVE COMPONENTS OF THE CORE

Yes, we're getting closer to the heart of our book, closer to learning how to cross that Great Divide. But before we get there, we need to take a deeper dive into the Inner Core of our Essential Engineering Framework. Viscerally understanding the Core and what goes into it is an important preparation step before you can begin to operate successfully within your new firm.

As introduced in the previous chapter, the Inner Core is where value is created. This is where inputs are absorbed, transformed, and released as outputs in either a continuous or periodic manner. Five basic components make up the Core, all contained within a boundary that defines what is internal or external to the organization.

The Core is where new engineers receive their initial assignments and perform their first tasks. Early reputations are made there; initial triumphs and setbacks will happen within its boundaries. Why there and not someplace else? It's very simple. Supervision will assume a new hire will know next to nothing about the business, making them an unknown quantity in a dynamic and churning environment. They will not be seen as the young, dynamic engineer as portrayed in the beautiful PDF the company recruiter sent you. At this point a new hire is considered a low-impact employee, too green to initially be given any complex, sophisticated, or high-risk task. Naturally, management will place them in a spot where "they can't cause any damage." (An engineer named Bill Freiburg once shared that his first assignment was to design a wooden base to hold a scale model of an F-15 fighter aircraft. Not something to brag about to your Mom or friends at the craft beer pub.) New hires will probably be pigeonholed into the supervisor's mental "box" for perhaps two or three simple assignments until the new engineer can build enough trust to be rewarded with their first real assignment.

To help get out of this assignment purgatory, new hires like yourself will need to know each piece of the Core, and within each piece, know who inhabits it, what they do, when they do it, how they do it, and, most importantly, why they do it. Only then comes that hall pass to receive a meaningful assignment.

In a new group, you and your fellow engineers will need to make decisions and take actions that are timely, effective, efficient, integrated, and ultimately correct for both your local organization and the overall enterprise. You do this by operating within the Core, that five-part construct introduced in Chapter 4 and summarized as:

1. The Formal Organization, that part of the organization that is easily visible to both the organization's inhabitants and those outside the firm. This is the public "face" of the enterprise. This is where the classic organization chart resides, where announced goals, strategies, plans, and policy deployment occur. This is what everyone sees as the "organization." And the degree to

DOI: 10.1201/9781003214397-7

which it has anything to do with the reality of a local engineering organization will vary.

2. Human and Hard Resources, consisting of human resources, which are the talent, skills, training, and personnel support required to accomplish the input-to-output transformation, and hard resources, those physical items that support human resources in getting the job done. These items include highly specialized equipment, fresh computer upgrades and analysis software, test stands, physical or virtual commons areas, and for those working on-site, the cappuccino machine, and snacks provided in the breakroom down the hall.
3. The Transformation System, the "engine room" where the actual transformation of inputs into desired outputs is accomplished. This is the "core of the Core," where products are created, services rendered, and value materializes.
4. The Hidden Organization, easily the most interesting and complex area of the Core. Also called the corporate culture, the Hidden Organization is a (mostly) invisible set of values, norms, guiding beliefs, and understandings that are shared by members of the organization and taught to new members as correct. This is where unwritten knowledge is shared, where deals are cut and tacit agreements are made. This is where your organization can become amazingly successful or sink without a trace. Without the Hidden Organization, a firm will be ineffective and could become irrelevant.
5. The Ubiquitous Communication System, defined as the information network linking the four previous components together. The power of the Core resides in the mutual reliance on its component parts and resulting synergy. Without a strong communications network, coordination between these four areas will break down and the creativity and performance expected of the organization will suffer. This is why each of these four components must closely interact with and support each other through this multimodal information exchange. This network can be a combination of sophisticated or simple, formal or hidden, written or verbal, electronically transmitted or performed in person, or otherwise exchanged by any means possible (ideally, all these modes are used). The important point is that a communication system must operate in a strong and frictionless manner.

The entire Inner Core is depicted in Figures 4.8 and Figure 5.1.

Let's now deconstruct the Inner Core and examine each component, their individual characteristics, and how they contribute to the totality of the product transformation. But first, we need to answer a fundamental question: just how do engineering firms organize, anyway?

5.1.1 Birthing an Engineering Organization

You may be wondering why we haven't talked about organization charts yet. After all, what's an organization without an organization chart? Doesn't an organization

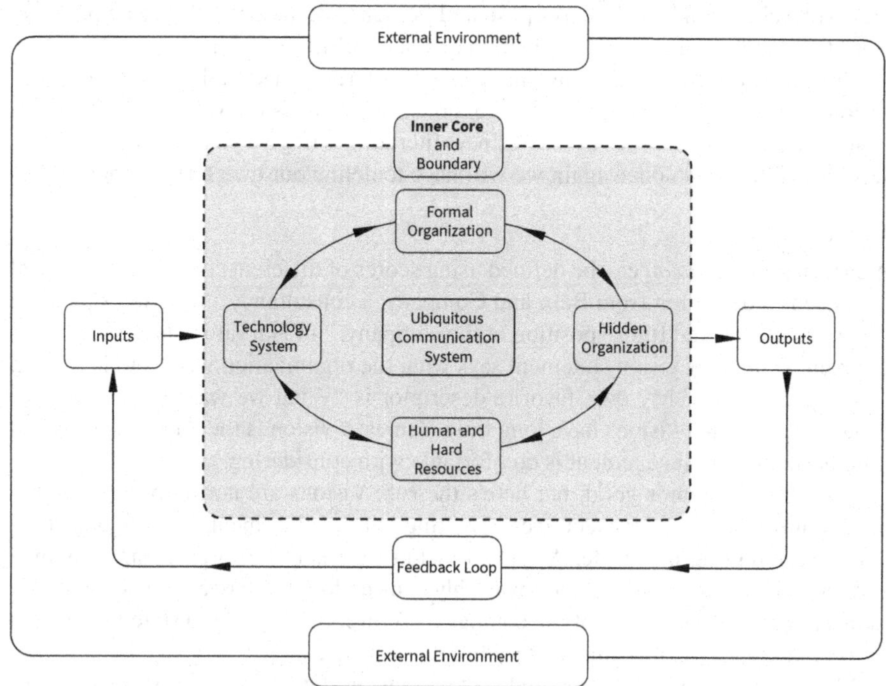

FIGURE 5.1 Inner Core of the Essential Engineering Framework.

chart map the heart of the enterprise? Isn't it the Rosetta Stone of an engineer's professional life? Well, there's a very good reason for the delay. At this point, a new engineer may believe their local organization is like an island, self-contained and independent. Yet it is actually embedded in and dependent upon a larger organization called by various names: the corporation, company, firm, or even the "mother ship." And the local organization chart is the end result of an entire process based on the larger firm's vision, mission, purpose, goals, customers, and myriad other factors. Only when these terms are identified, defined, and addressed can a local organization chart finally be created. It's the last step of what can be a long process.

We need to accomplish two things here. The first is to truly understand all these organizational terms, and the second is to outline the process that ultimately creates your group's organizational framework. Only then can we understand each of the five components.

The terms I've just mentioned have an important purpose in the world of an engineering firm: they define what the company is and is not. What it does, why it does it, who does it, when and how it is done; all these terms define the larger enterprise of which you are a part. Without these basic terms clearly defined, you and your colleagues theoretically wouldn't know what to do, and neither would anyone else.

Of course, this is oversimplistic to the point of naivete. But the fact is, any organization, big or small, high-tech or commodity-based, old or startup – all need at the highest level a definition of its purpose and guidance on how to achieve its purpose. With that in place, then the organization can help fulfill that purpose.

Unfortunately, there are many, many business terms in use and considerable confusion over the definition of these expressions such as mission, values, and purpose. Consulting any two sources in the business literature will give you two very different sets of definitions, so once again we will have to define our own terminology:

Vision

A case in point, a *vision* can be defined using scores of different constructs. A typical example is a definition from Bain and Company, a consultancy: "A vision statement describes the desired future position of the company,"[1] or Harvard Business School's Graham Kenny: "A vision statement says what the organization wishes to be like in some years' time."[2] My own favorite descriptor is "What we want to be when we grow up." Note that visions have long time frames: a vision is the longest-term planning horizon that management is comfortable with considering.

Now all this sounds good, but here's the rub. Visions are normally put together by a committee of senior executives, sweating out the placement of every adjective, verb, and dangling participle. A typical vision statement for large companies may take up to a year to complete and invariably ends up looking like a word salad rather than a clear communication of the company's future. In short, it is a statement which unfortunately may communicate very little of substance.

It's always fun to look for examples of bad vision statements. For example, Avery Dennison's mission statement begins with: "To help make every brand more inspiring, and the world more intelligent." That's a pretty heroic aspiration for an outfit that makes stick-on labels. Wikipedia has a vision more to the point: "A world in which every single person is given free access to the sum of all human knowledge." Yet neither of these statements addresses what direction the company actually wants to head.

With that said, let's define our vision from the Kenny statement: "A vision statement says what the organization wishes to be like in some years' time."

Mission

A mission statement states what we do today and what we will do tomorrow. It is strategic in scope. It describes what business the organization is in, both now and into the mid-range future, and its task is to provide a direction for its employees to work toward. One consultancy states that a mission statement "defines the company's business, its objectives and its approach to reach those objectives."[3] The mission statement is supposed to be a summary of what they do, a description of the present and a near-term expectation, not a picture of the far-field future. The mission statement should distinguish the organization from other similar organizations.

It can be hard to create a good mission statement. Barnes & Noble, a bookseller, conveys their statement as: "Our mission is to operate the best specialty retail business in America, regardless of the product we sell." Note that their mission is independent of the product they currently sell; their words and music don't match. Albertsons, a grocery chain, wants "To create a shopping experience that pleases our

TABLE 5.1
Sample of 21st-Century Digital Corporation Mission Statements

Corporation	Mission Statement
Twitter	"...to give everyone the power to create and share ideas and information instantly without barriers."
Apple	"...to bringing the best user experience to its customers through its innovative hardware, software, and services."
Meta	"...give people the power to build community and bring the world closer together." Note: As of February 2022, the company is not operating under a single mission statement.
Amazon	"We strive to offer our customers the lowest possible prices, the best available selection, and the utmost convenience."
Zoom	"Make video communications frictionless and secure."

customers; a workplace that creates opportunities and a great working environment for our associates; and a business that achieves financial success." This misses the mark, as mission statements should state what the company actually does.

You might suspect that 21st-century digital companies would have different or more sophisticated approaches to their mission statements, as these firms represent the latest corporate trends and advanced thinking. Table 5.1 lists the current missions for a number of familiar digital firms. Surprisingly, the quality of their mission statements ranges from good to poor, no different than legacy companies before them.

That said, let's settle on this definition of a mission: "In its basic form, a good mission statement states what we do to fulfil the vision. It describes what business the organization is in for the mid-term timeframe."

Purpose

A *purpose* statement declares "Why we do what we do"; that is, what is the firm's reason for existing. It's why you and your fellow engineers are there, working hard every day.

So how does purpose differ from the vision and mission? Vision and mission emphasizes how the organization views itself objectively. A purpose is more subjective. It's not just about emphasizing clear endpoints and methods to achieve them, but also what compels the organization to be in that particular business. It's motivational because it connects with the heart as well as the head. Examples of purpose include the Dutch financial services company ING ("Empowering people to stay a step ahead in life and in business"), the insurer IAG ("To help people manage risk and recover from the hardship of unexpected loss"), and Kellogg foods ("Nourishing families so they can flourish and thrive").

Two points about purpose that deserve mention. Not all organizations explicitly define their purpose; they view it as already mixed in with their mission or vision statements. And that's fine. The second may have more meaning to you in the future. To many people, the purpose of a company is to make as much profit for shareholders as quickly as possible that is legally feasible. Yet there is a substantial movement today to redefine a firm's purpose as something much broader. Called *sustainable capitalism*, this approach advocates that companies must incorporate the interests of environmentalism, worker welfare, opportunity inequality, and social justice into their reason for being.[4] The same intent is being pursued by the Business Roundtable, an association of nearly 200 chief executive officers of America's leading companies. Their *2019 Statement on the Purpose of a Corporation* committed them to modify their businesses to include these more holistic purposes.[5] The odds are better than even that your company, either now or in the near future, will begin to address this movement, meaning your work may change to satisfy these new purposes.

Organizational Goals

An *organizational goal* is a strategic target a company's senior management establishes to identify general outcomes that guide corporate efforts. These broad goals are normally qualitative, tending not to specify precise numerical values or achievement metrics. Here, desired achievements are broad-brush. Timing is given as a span of time in the mid-range future. An example of an organizational goal might be "Become the dominant player in the development of cloud-based storage through emerging quantum computing methods." Note there is no definition of "dominant," nor is any timeframe given. But employee's efforts are directed toward "emerging quantum computing."

A good definition might be this: "An organizational achievement (or achievements) delivered in the mid-range future that directly supports the firm's mission and ultimate vision."

Values and Principles

Values and Principles (also known as *Core Values*) describe a set of desired organizational behaviors. They are interchangeable terms that serve as a behavioral "compass" for all the firm's employees. Values and Principles define how the company expects people to behave and guide how employees should operate. Almost every successful company has a clear set of these principles that foster positive behaviors and deter anything that may damage the firm.

For instance, Coca-Cola's published values include "having the courage to shape a better future, leveraging collective genius, being real, and being accountable and committed."

Values are hard to express succinctly, as stating a set of principles in clear, precise wording needs a Tolkien, Capote, or Mailer more than a committee of gray-haired business types. But the idea is clear.

That said, let's define values as "the set of guiding beliefs upon which a business is based. Corporate values help people function together as one and shape the way employees (should) behave."[6]

Organizational Culture

Also called the "hidden organization," an *organizational culture* is the set of covert, shared assumptions, beliefs, and guidelines that determine how to behave and operate in a technical, product development, research and development, or technical service setting. Like values, these assumptions and beliefs have a strong influence on how people in any company act and perform their jobs. Mostly obscure and hard to see, the hidden culture exists within all companies and firms, invisibly operating at all levels and at all times. Each company's hidden organization is unique and cannot be ignored.

Organizational Objectives

Essentially, organizational objectives are a more detailed and specific version of the organizational goals discussed above. They guide employee efforts, justify a company's existence and actions, define performance and timing standards, provide barriers to pursuing unnecessary goals, and function as incentives for desired actions. Successful technical organizations need objectives that are ambitious, achievable, and clear. Most importantly, they must be measurable. They provide the basis of a detailed organizational performance tracking system which is the foundation of management.

Again, a huge number of written definitions of "objectives" exist, but a solid one we will use is:

> Objectives are measurable ends for a set process. They identify goals and take action to make them happen. Organizational objectives help in setting goals in a way that all company-wide activities move in one single direction. It is the future results that an organization wants to achieve.[7]

Strategy

Here comes a major source of confusion in business terminology. Misuse of the term "strategy" is extremely common, as it's applied to just about any and every topic having the word "business" in its name.

A *strategy* is a clear set of choices that define what the firm is going to do (or not do) to achieve an organizational objective. Unfortunately, many organizational "strategies" are in fact goals. For example, "We want to be the number one or number two in all the markets in which we operate" is a goal, not a strategy.

A strategy states what you are going to do, not what you hope to achieve. This means strategies need to be revisited often to keep them current, as the rapid change in the external environment significantly impacts strategy. Strategy is ultimately the responsibility of the CEO and their senior management group, and if done right, their engagement will align the organization more effectively.

Luckily, a strategy is easy to define. We will use: "A strategy is a clear set of choices that define what the firm is going to do to achieve organizational objectives."[8]

Business Model

Business models are very popular with MBAs and their ilk. Armies of entrepreneurs will corner you during cocktail hour and gas on about their new business model, until your eyes glaze over or you can escape to your friends at the bar. Researchers will tell you there are 16 specialized variations of four basic business models, each with specific boundary conditions and so on. That's fine, but all we need to know here is that a business model is what a company specifically does to achieve their organizational objective. It's a detailed version of how your organization transforms inputs into outputs to create value. It's a mini-version of the firm's strategy, but containing the detailed information that gets the firm where it needs to be. Our resulting definition is "A business model is a detailed transformation system that converts inputs into outputs in service of the organization's goals."[9]

Transformational Technology

This definition is refreshingly short and simple. We define a *transformational technology* as a "work processes, technique, machine, action, analysis, method, tool or any other object or knowledge used to transform organizational inputs into outputs." It is not limited to digitally based products but extends to anything that performs a transformation: it enables the creation, development, and sale of any product or service. Ideally, a transformational technology should be totally contingent on the previously defined business model, strategy, objectives, and goals.[10]

Organizational Structure

An *organizational structure* is a map of basic relationships between groups or entities who are charged with getting things done. It is a general arrangement of how groups are arranged to create value. These are the basic organizational structures such as the traditional hierarchy, matrix, process, agile, and the like that we will come to know and love. In general, no individuals are identified at this level, only functions. These are the structures shown and discussed later on in Section 5.2.2.

Organization Chart

At last, here we are. So, what is this mysterious thing called an organization chart that's been so hard to get to? An *organization chart* is a detailed diagram that conveys a company's internal structure by showing formal relationships between individuals within the firm. An organization chart graphically represents an organization's interrelationships, highlighting the different jobs and departments that connect the company's employees to each other and to management. These charts can represent all company groups (depicting the overall company) or can be department or workgroup centric.

In Chapter 6 we are going to spend some time on this subject, so for now let's define the organization chart as a construct that attempts to explain the formal authority relationship between members of an organization and the work product responsibility between those members.

Why have we spilled so much ink defining these terms? Because it's critical for the second part of this chapter: applying these terms to understand your engineering organization. We aim to use these expressions as building blocks to understand how each interacts to create a viable, successful organization.

Here's the key understanding. These separate terms interrelate with each other in a cascade of information and knowledge called *policy deployment*. Policy deployment means the deliberate and purposeful sharing of broad purposes and desired future success, downward into an organization into an ever more detailed division of effort. From the top of a multinational behemoth down to the individual engineer, all receive the same direction and coordinated into one strong and unified force with one intent in mind.

Too bad it rarely works out that way, but that's the intent.

That said, let's focus on this idea of policy deployment. The correct way to build an effective organization is to deploy the directions given from the "top of the house," consistently down the organizational levels, where ideally everyone has the same information and understands it in the same way. This means your group's formal organization chart is developed as the last step in the policy deployment process as shown in Figure 5.2.

A few noteworthy comments. The policy deployment system is based on three insights. First, the organization's structure (be it at the enterprise, departmental or group level) must follow the group's strategy, and that strategy is developed from both the group's surrounding environment and its internal capability. Structure is the direct and singular result of strategy. The result is that strategy impacts internal organization characteristics, meaning that executives must design the organization to support the firm's competitive strategy. This implies there is not one "best" organizational structure or form. Instead, structure is always contingent on the organizational strategy being pursued. In short, structure follows strategy, and strategy follows environment: environment, strategy, structure, and technology all must be aligned. This is the heart of contingency theory as applied to organizations.

So, what's our takeaway? First, it's a long journey from a company's vision and mission statements down to your individual job, and the length of that journey is a large problem. With the pressure to do more and more with less and less time to do it, most groups and individuals will merely take last year's objectives, organization chart, and performance metrics, put this year's date on them, and call them good. This obviously hurts the integrity of the process and relegates policy deployment to useless bureaucracy.

Second, however onerous, policy deployment is a critical and necessary effort to ensure you, your department, and your management are pulling together, driving for a common goal with the greatest efficiency and effectiveness possible.

So yes, it's hard.

And?

5.2 THE FORMAL ORGANIZATION

Having worked through the policy deployment idea, let's get into an investigation of the first component of the Inner Core: the formal organization. As mentioned earlier, the formal organization is where all written policies, business structures, methods of communication, standard operating procedures, official directions, specified vertical and horizontal relationships and all such reside. This component is what you would describe if any outsider asked you about your organization. By its very name, it is a collection of documented characteristics that defines a prescribed "surface level" of the company.

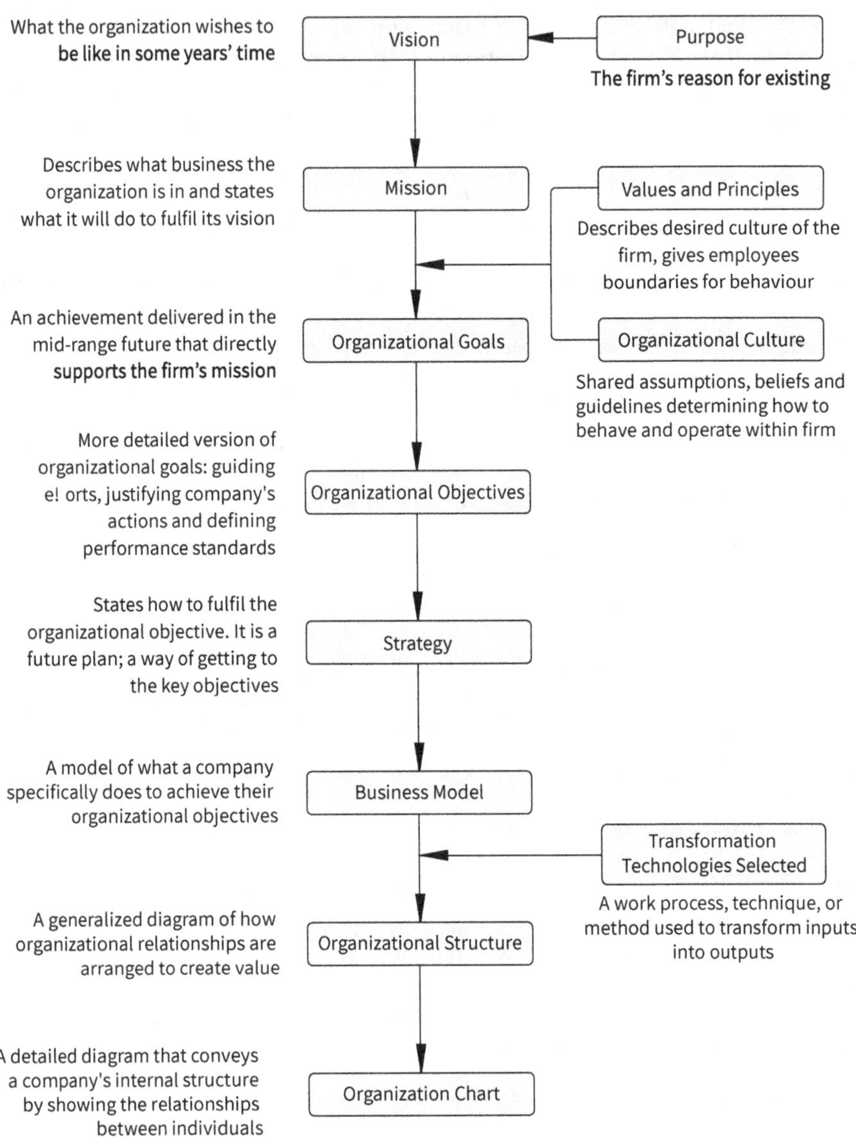

FIGURE 5.2 Policy deployment process.

5.2.1 The Key Components of Organizational Structures

We've already established that the Essential Engineering Framework is just that: a relatively flexible relationship of activities that engineering organizations use to achieve a desired result. Now we are going to consider a number of organizational *structures*, the more rigid, specified set of relationships that the engineering firm applies to achieve a narrowly defined output. It is a general map of how groups are arranged to create value.

An organizational structure does three important things. First, it designates formal reporting relationships, including the number of levels in the management hierarchy and the span of control of directors, managers, and supervisors. Second, it identifies the grouping together of individuals into departments, and departments into the total organization. Finally, it includes the design of systems to ensure effective communication, coordination, and integration of effort across departments. It is no surprise these are the three characteristics of all organizational structures.

5.2.2 Organizational Structures: Which and When

For those of you who are breathlessly awaiting to see some organization charts, your time is at hand. Remember, the rule regarding organization design is very simple: the form of the organization chart reflects the firm's goals and the method used to attain those goals. It's as simple as that; the chart's structure is uniquely and directly dependent upon the goal being pursued and strategy employed. This means that, theoretically, there are an infinite number of organizational forms and charts that could be devised. But for most companies that's too hard and slow to actually pursue. So, management experts have conveniently clustered similar structures into bins and assigned characteristics to each, creating a "menu" of organizational structures and charts to choose from. If your desired result is "A," then choose organization structure "1." If your desired result is "B," choose structure "2," and so on.

Let's take a brief look at each of the major structures and their applicability to what kind of result is desired.

Tables 5.2 and 5.3 and Figure 5.3 list eight major structures and their related characteristics.

Of course, with this universe of organizational forms available to the boss, how do they decide which one to use? After all, we just went through a long policy deployment discussion, so shouldn't that be the basis for the boss's organization chart? Well, it depends on who is doing the organizational design. Some very large engineering firms have their own dedicated organizational design departments whose job is to assist senior executives in designing efficient and effective organizations. Provided they already have a completed policy deployment analysis in hand, these groups will tend to choose from the list in Table 5.3 and then modify the selected structure to account for the unique characteristics of the group undergoing study. And this can work. However, for firms that don't have the luxury of an organizational design department, it falls to the director, manager or sometimes even the supervisor to create the organizational design of the group. And with the constant specter of "do it fast," a manager is apt to fall back to one of three basic designs. Yes, today there tends to be only three organizational structures used in engineering firms. These are the Hierarchy, the Matrix, and, just emerging, the Agile.

TABLE 5.2
Major Characteristics of Engineering Organizational Structures

Structure Name	Condition of External Environment	Interrelationship between Departments	Organization Size	Goal	Operative Goal	Planning and Budget	Formal Authority
Traditional Hierarchical (Functional) Organization	Low uncertainty, Stable	Routine, low interdependence among departments	Small to medium	Internal efficiency, quality	Functional goal emphasis	Cost basis, budget, statistical reports	Functional managers
Divisional (Product) Organization		Nonroutine interdependence among departments	Large	External effectiveness, adaptation, client satisfaction	Product line emphasis	Profit center base, cost and income	Product managers
Geographical Organization	Moderate to high uncertainty, environment changing	Nonroutine, high interdependence among departments	Large	External effectiveness, adaption, client satisfaction	Product line emphasis	Profit center basis, cost and income	Product managers
Matrix Organization	High uncertainty	Nonroutine, many interdependencies	Moderate, a few product lines	Product innovation and technology specialization	Equal product and functional emphasis	Dual system by function and product line	Joint between functional and product heads
Hybrid Organization	Moderate to high uncertainty, changing customer demands	Routine or nonroutine, with some inter dependencies between functions	Large	External effectiveness, adaption, efficiency within some functions	Product line emphasis, some functionals	Profit center for divisions, cost basis for central functions	Product manager, but with coordination responsibility resting with functional managers

(Continued)

TABLE 5.2 (CONTINUED)
Major Characteristics of Engineering Organizational Structures

Structure Name	Condition of External Environment	Interrelationship between Departments	Organization Size	Goal	Operative Goal	Planning and Budget	Formal Authority
Horizontal Organization	Moderate to high uncertainty, changing customer demands	Routine	Moderate to large	External effectiveness and adaption, efficiency of core process	Process efficiency, goal emphasis	Cost basis: budget, statistical reports	Team/process leaders and members
Modular Organization	High uncertainty	Both routine and non-routine	All	External effectiveness and adaption	Emphasis on key functional activities that establish competitive advantage; outsource all other activities to competent partners	Cost basis for contracted modules	Coordination responsibility rests with the hub module
Agile Organization	Both high and low uncertainty	Both routine and non-routine	Small to medium	Speed to market	Empower employees, reduce command and control	Cost basis assuming finance system is stable	Project leaders and group members

TABLE 5.3
Major Engineering Organizational Structures: Strengths and Weaknesses

Organization Type	Strength	Weakness
Traditional Hierarchical (Functional) Organization	• Allows economies of scale within functional departments. • Is best with only one or a few products	• Slow response time to environmental changes • Cause decisions to pile up, hierarchy overload • Restricted view of organizational goals
Divisional (Product) Organization	• Suited to fast changes in unstable environments • Leads to client satisfaction because product responsibility and contact points are clear • Allows units to adapt to differences in products, regions, clients	• Leads to poor communication across product lines • Eliminates in-depth technology competence and specialization
Geographical Organization	• Suited to fast change in unstable environments • Leads to client satisfaction as product responsibility and contact points are clear • Allows units to adapt two differences in products, regions, clients	• Leads to poor communication across product lines • Eliminates in-depth technology competence and specialization
Matrix Organization	• Achieves coordination necessary to meet dual demands from customers • Flexible sharing of human resources across products • Suited to complex decisions and frequent changes in unstable environment	• Causes participants to experience dual authority, which can be frustrating and confusing • Participants need good interpersonal skills and extensive training • Is time-consuming, involving frequent meetings and increased conflict • Will not work unless participants understand it and adopt team-based rather than vertical type relationships
Hybrid Organization	• Allows organization to quickly adapt and coordinate product divisions with efficiency in centralized functional departments • Results in better alignment between corporate-level and decision-level goals	• Leads to conflict between division and corporate organizations
Horizontal Organization	• Used when processes are to be emphasized. • Flexibility and rapid response to changes in customer need	• Determining core processes to organize around is difficult/time-consuming • Requires changes in culture, job design, management philosophy, and information/reward systems

(Continued)

TABLE 5.3 (CONTINUED)

Major Engineering Organizational Structures: Strengths and Weaknesses

Organization Type	Strength	Weakness
	• Promotes focus on teamwork and collaboration, common commitment to meeting objectives • Provides some flexibility for an environment with moderate to high uncertainty and changing customer demands. • Most appropriate for moderate to large organizations • Emphasizes process efficiency	• Traditional managers may balk when required to give up power and authority • Requires significant training to work effectively in this environment
Modular Organization	• Enables even small organizations to obtain talent and resources worldwide • Gives smaller companies immediate scale and reach without huge investments in factories, equipment, or distribution • Enables organization to be highly flexible and responsive to changing needs	• Managers do not have hands-on control over many employees and activities • Requires substantial time to manage relationships and potential conflicts with contract partners • Risk of organizational failure if a partner fails to deliver or goes out of business
Agile Organization	• Potential for speedy product creation	• Can be easily misapplied to inappropriate tasks • Product development may create customer dissatisfaction due to product being not being properly vetted • Offloading of critical development functions to inappropriate outside organizations

Your management will invariably select the organizational form in which they are most familiar. Familiarity is based on history, and the organizational history of your group is probably not based on the best organizational form but generally on what's come before. And that means it's easy and quick to implement. So, hierarchy begets hierarchy, matrix begets matrix, and the agile will soon beget the agile.

Here's the thing about the agile. The agile assumes that the external environment is not only unstable but can actually be a chaotic mass of outside factors spinning with ever-increasing velocity until it can't be confidently analyzed. This environment can send your organization into a perpetual flux, where small subgroups form, detach, and reattach themselves to other groups, complete a task, break away and

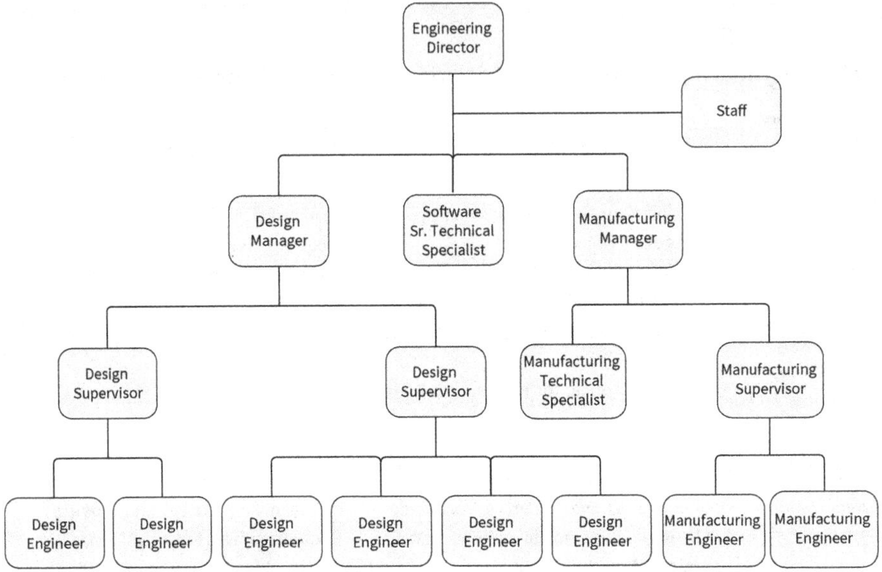

FIGURE 5.3A Traditional Hierarchical (Functional) Organization.

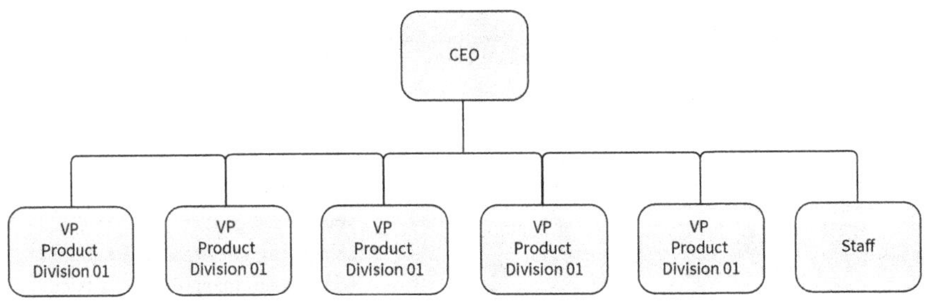

FIGURE 5.3B Divisional (Product) Organization.

move over to some other task, ad infinitum. Speed of product delivery becomes the goal in this unstable environment. Thus, the agile is very decentralized and can be remarkably difficult to centrally manage.

At the time of this writing, organizations have very limited experience with agile in the engineering workplace, as its application is early and currently limited to software development. Yet much is being written about it, including suggestions for its use in appropriate situations. A good path is to watch and wait: let the appropriate foundational experience be gained before moving forward in its application in your own situation. And don't allow your management to pressure you into using it inappropriately, purely because of its current buzz.

Finally, consider this thought. The agile represents a case of trying to respond to a chaotic environment. Yet not all suborganizations in the firm are in these unstable

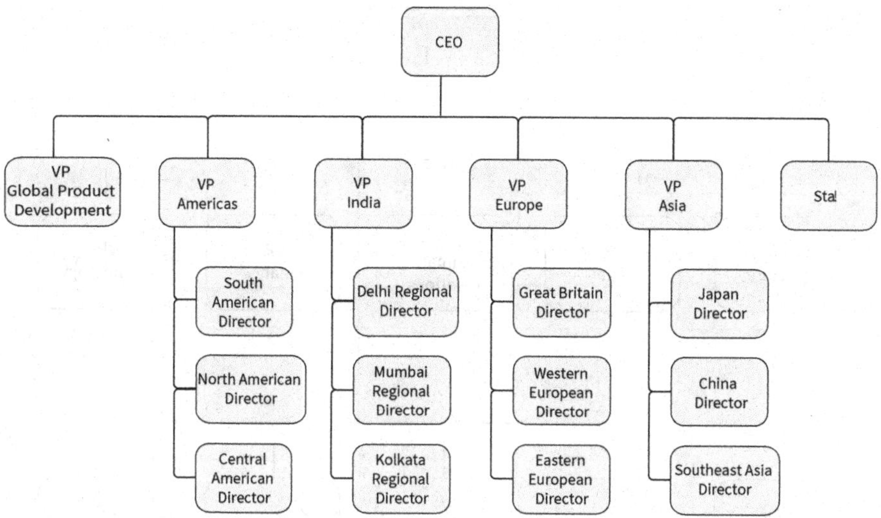

FIGURE 5.3C Geographical Organization.

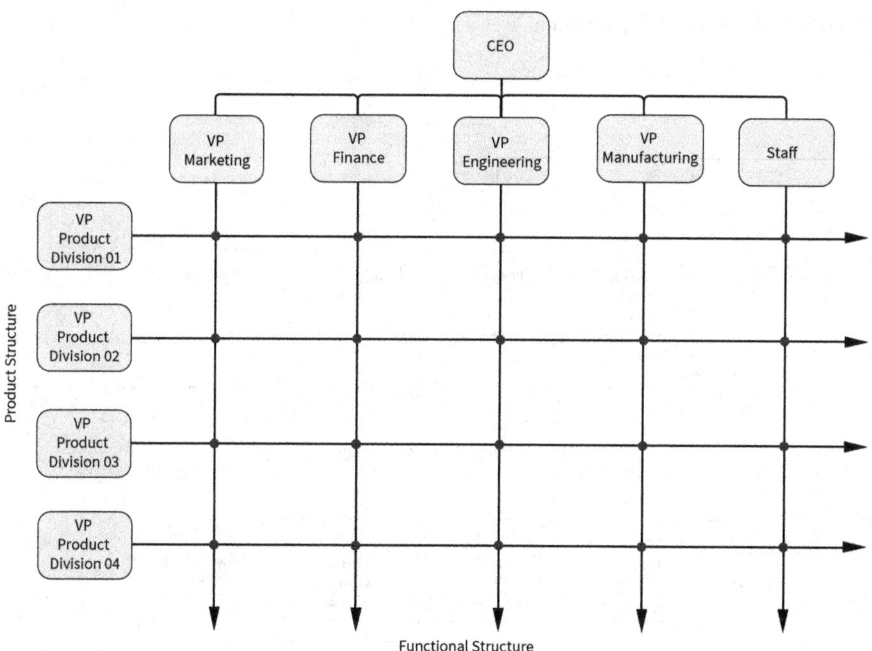

FIGURE 5.3D Matrix Organization.

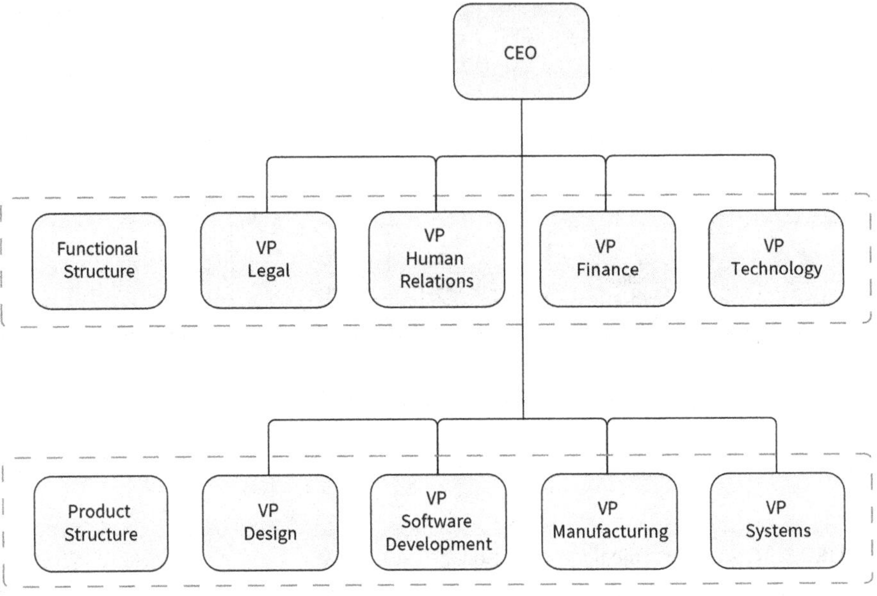

FIGURE 5.3E Hybrid Organization.

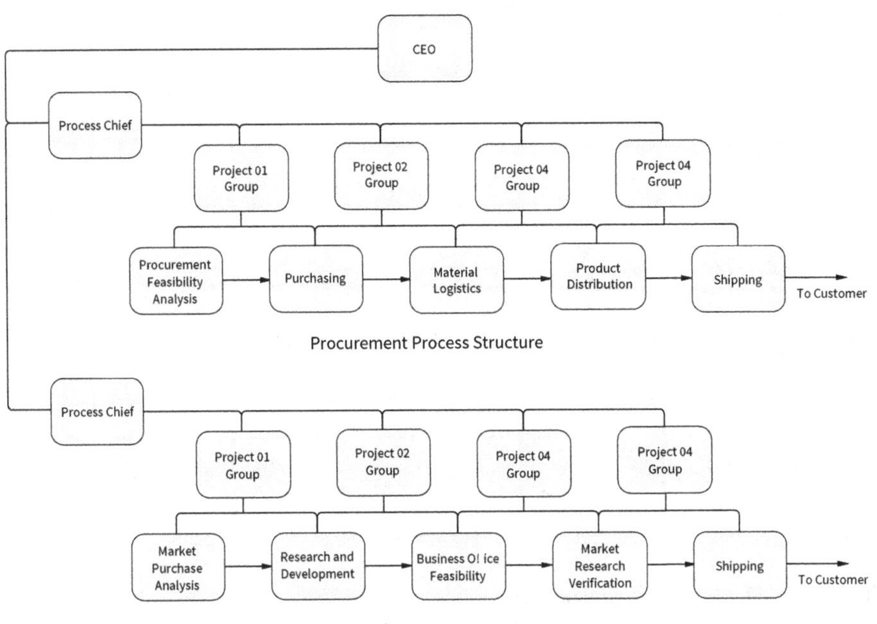

FIGURE 5.3F Horizontal Organization.

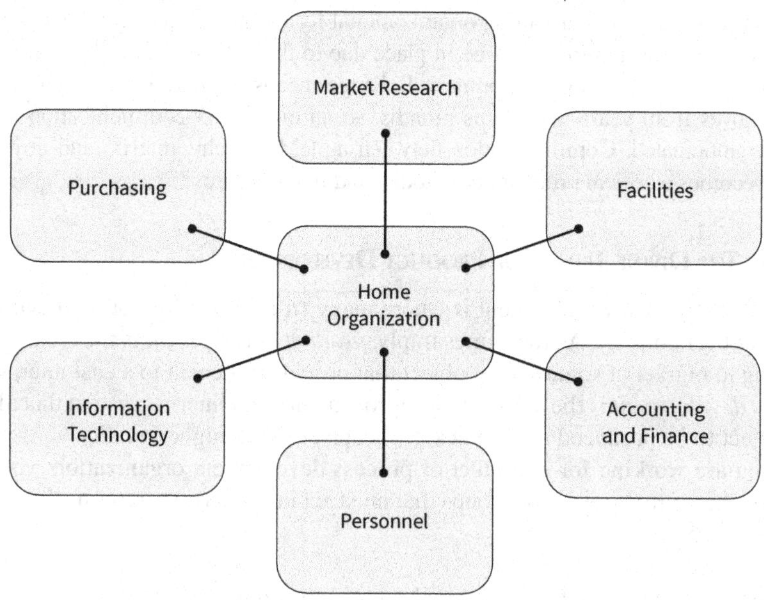

FIGURE 5.3G Modular Organization.

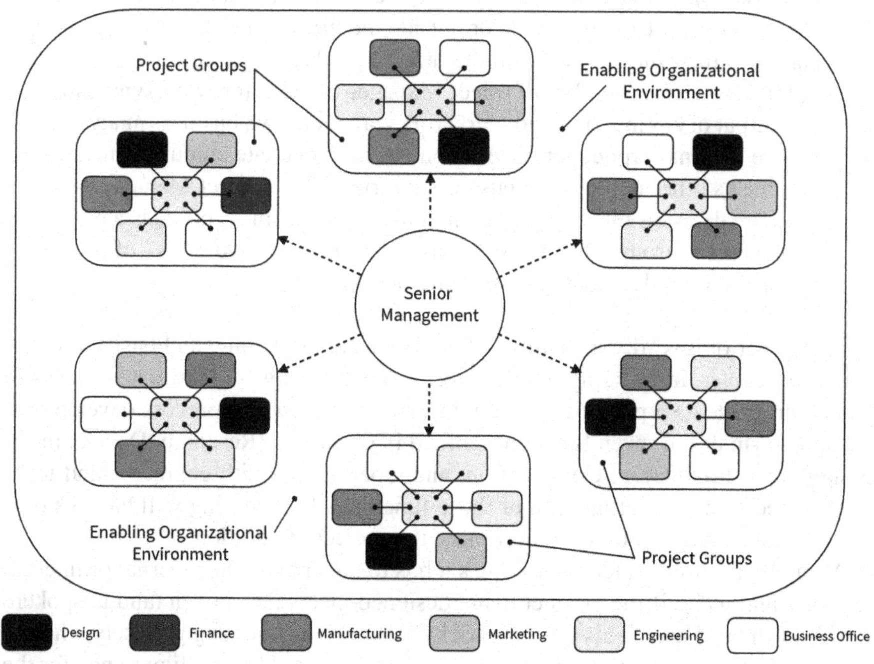

FIGURE 5.3H Agile Organization.

environments: all of the remaining organizational forms, such as a hierarchy, matrix, or any other configuration must still be in place due to their slower but still moving environments. Yet accelerating environmental change means the duration of each of these forms moves from years to perhaps months, so an ever-better communication system must be anticipated. Communication between agile, hierarchy, matrix, and any other forms becomes a critical problem, both today and in the future.

5.2.3 The Divine Trinity of Product Development

Product and process development is where many (if not most) of you will end up as you enter the company. As the names imply, *product development* is the creation and bringing to market of some sort of object that provides a benefit to a customer, while *process development* is the creation or improvement of an internal system that allows that object to be produced more quickly, cheaply, or with higher quality.

If you are working for a product or process development organization, you will soon see there are three major groups that must act in concert to successfully achieve their goal. These are:

1. An engineering activity, charged with innovating, designing, testing, and otherwise bringing the product or process into a state of *feasibility*, which is the judgment that a design will actually work within the specifications of the original idea.
2. A marketing activity, which defines and certifies the characteristic profile of the customer (be they inside or outside the firm) and if they will actually part with their money to obtain the offered product or process.
3. A business office, led by a Project Manager or Director, who synthesizes the output of engineering and marketing with their own business analysis to ensure the final product or service achieves an adequate amount of revenue and meets other appropriate business metrics. This Project Manager is the ultimate decision-maker. Also located inside the business office is a project management group, charged with ensuring the "nuts and bolts" of project timing and development gateways are met.

Each member of this "divine trinity" of engineering, marketing, and business office is a power center, influencing the final form, functionality, and ultimate success of the end product. As a power center, engineering in product or process development takes the form of the seven functions defined in Chapter 1 (Research, Development, Design, Test, Production, Construction, and Operations), with engineers and technology specialists executing each of those functions. Engineering will have its own requirements and goals for each portion of the project in question.

As another power center, marketing's job is to understand the external or internal customer and judge if the product being designed meets the spoken (and unspoken) desires of the end user. This can be a tricky business, as knowing precisely who the customer is, what are their true needs, and what they would be willing to pay for the product is typically more art than science. Complicating the problem is that customer surveys of current and potential future products are notoriously inaccurate. As mentioned in Chapter 4, customers have a way of not revealing their real thoughts about products; a reluctance to tell their true feelings to relative strangers or having the willingness to consider new ideas.

As the third center, the business office is an interesting amalgam of business school graduates, financiers, schedulers, and optimizers whose job (they feel) is to pass judgment on the information and recommendations that engineering and marketing provide. As mentioned, the Project Manager is nominally the ultimate decision-maker for the product, absorbing insights from engineering and marketing and balancing them with their own views as to what is best for the product.

These three centers will exhibit various degrees of conflict as each attempt to push the project in the direction they prefer and believe is correct. Reasonable conflict is good for the product, but there is always the specter of a dysfunctional struggle where the product suffers. Having three power centers in the trinity creates a built-in tiebreaker and makes resolving conflicts a bit easier.

What does all this mean for a new engineer like yourself? Obviously, it means learning to advocate for your engineering position to both marketing and the business office as a routine part of the job. Your arguments regarding the best engineering position will undoubtedly be based on objective data. Marketing will base their recommendations on "soft" data from market research and their prior experience, while the business office will present endless cost–benefit ratios and other analysis based on financial projections. And don't forget the decision-maker, the Project Manager, who oversees the whole muddle. How do you ensure your engineering positions are honestly considered and (hopefully) included?

The secret is surprisingly simple but challenging. In addition to engineering concerns, you must also be reasonably conversant with the everyday workings of both the marketing and business offices. When arguing any question or advocating any course of action, an advocate must know if an opponent's analysis is solid or shaky, precise or approximate, or if their technique is appropriate or their premise faulty. You need to know their financial terms (such as ROI, ROA, net revenue, capital investment, churn, bluebird sale) and so on. Marketing and business methods cannot be a black box; a new engineer must understand how their analysis was conducted (if any was actually conducted at all) and how their conclusions drawn. In a word, you eventually need to know their business nearly as well as your own or risk getting bamboozled.

This may be an alarming bit of news, but this is the essence of experience. After all, this is really what a new employee job rotation program is all about. Experienced engineers and management know this already and can help you learn this skill. Just realize that school didn't end at a graduation ceremony; you've merely changed subjects for the next few years.

5.3 HUMAN AND HARD RESOURCES

Obviously, nothing gets done without resources. Resources are a critical enabler of your work, a necessary but not sufficient requirement to achieve the organization's objectives. As mentioned earlier, it's a good bet that many engineers think of resources as inanimate objects but sometimes neglect the human portion of the resource question: the talent, creativity, and innovation people can bring to their workplace.

We also know human resources cover employee-related topics such as labor relations, new hire selection, job roles, interpersonal dynamics, conflict resolution, and the like, while hard resources are any non-organic item of value, such as equipment, workspace, funding, and information. Let's consider each.

5.3.1 The Same, But Not

But wait a minute. We already talked about this whole resource idea back in Chapter 4, where human and hard resources are part of the input element. Yes, we did. But now it's time we introduce a new twist to the resource discussion. This twist has to do with your organization's boundary.

We've learned that the boundary determines what is inside or outside an organization. A complete understanding of your engineering firm, your entire organizational worldview, revolves around the placement of that border. It's simple: what's inside the line is yours; what's outside the line is someone else's. This holds for both human resources and those hard, non-organic assets. Management defines where the border is: around the group or department, around the division or around the firm, or around some other configuration. The key is to actually define and know precisely where the line is.

This means something important when it comes to resources. The upshot is you actually have *two supplies* of human and hard resources available: those already in place within your home organization and those that are procured from outside the boundary, i.e., outside your department or firm.

The rules are simple. For any initiative or project you are assigned to, your supervisor or manager will first draw upon all the appropriate human and hard resources available from inside the organization's boundary. These tend to be a fixed, predetermined amount. Only after those internal resources are exhausted will the search begin beyond the local boundary, be it within adjoining departments or into the external environment itself. This is why the boundary is so important. A natural bias is to consider resources outside the boundary as more expensive, harder to procure, of lesser quality or tougher to manage than those internal to you. This may be true.

However, there are situations where an engineer is forced to acquire hard or human resources from outside the boundary. For example, say your department is entering into a new technology space. Being new work, the department will probably not have the required tools, facility, or talent at hand to succeed. Bringing in the required new resources can be done in scores of different ways. If the new business is an experiment, this means new resources will be temporary: leasing rather than buying space, hiring contract workers (agency workers or "job-shoppers") rather than permanent employees, and obtaining one-time funding only. If the new technology has a firm commitment from management (and we mean a *firm commitment*), permanent resources are required: training and education of new and existing employees for the new technology, long-term capital investment in hard resources like facilities and equipment, and appropriate operational funding committed for many years.

The takeaway here is that resources used by you and your group can come from a wide and disparate universe of sources, from strictly internal to company-wide, or even deep into the external environment. This goes for any resource: hard and human alike.

At this point, since resources can be human or hard, internal or external, there are four combinations that can be used to supply your needs. We'll get into this a bit more in Chapter 7. For now, let's move on to the Technology System.

5.4 THE TECHNOLOGY SYSTEM

The Essential Engineering Framework tells us the entire Inner Core constitutes the transformation system of the company, where the five components together transform inputs into outputs of enhanced value. The technology system is where the actual, tangible work gets done. It is the "Core of the Inner Core," and its product is the whole point of your and everyone else's employment. A technology system can be simple, like making popcorn at a movie theater, but within engineering it is normally exceedingly large and complex, incorporating uncounted procedures applied within intricate and interlocking subsystems. The technology system must be designed to support everything from product development to R&D, advanced manufacturing technologies to new technology processes, test methods to sociotechnical design, all connected to enabling departmental technologies at the enterprise, departmental, individual, and information system levels.

5.4.1 Technology vs. Transformation Systems

We should take a moment and specify two items that are normally confused. What is a *technology*, and what is a *transformation*? They are different.

As previously mentioned, this book defines technology differently than what is commonly thought or meant. In everyday usage (e.g., popular media or sitting at home among your friends), "technology" refers exclusively to common, digitally based products and features. Your cell phone, tablet, the safety features in your car and refrigerators that tell you when more milk is needed: these are all considered technologies in this context. These are not the technologies we will be referring to here. In organizational studies, technology is defined as the techniques, skills, methods, and processes used in the production of goods or services or in the accomplishment of objectives. These include manufacturing technologies, product and service development technologies, and work methods deployed by various organization levels such as department, individual, and information system echelons. In this construct, a technology is a tool to accomplish something.

A transformation is different from a technology. A transformation is a process where any input, when combined with all five inner core components, creates an output of greater value than the inputs and resources initially provided. A chef creates a transformation when creating a wonderful *boeuf bourguignon*: their spoon is the technology. An auto mechanic creates a transformation when repairing the transmission of your car: their ratchet is the technology.

In a nutshell, a transformation is a result, a goal achieved, a unit of work accomplished. In contrast, a technology is a tool, used in service of accomplishing a transformation.

5.5 THE HIDDEN ORGANIZATION

The Hidden Organization is easily the most interesting and complex area of the Core. Also known by other names such as *organizational culture, corporate culture, informal organization*, or *hidden culture*, it's seen as an invisible set of values, norms, guiding beliefs, and understandings that are shared by members of an organization and taught to new members as the proper way to behave and perceive. It is where unwritten knowledge is shared, where deals are cut and covert agreements are made. It is where organizations can become amazingly successful or sink without a trace. Frankly, it can be the most important component of the Core.

For those in need of a formal definition of the hidden organization, there are many to choose from, which we will discuss further in Chapter 9. In the meantime, we will use this:

> The [hidden] organization is the interlocking social structure that governs how people work together in practice. It is the aggregate of norms, personal and professional connections through which work gets done and relationships are built among people who share a common organizational affiliation or cluster of affiliations. It consists of a dynamic set of personal relationships, social networks, communities of common interest, and emotional sources of motivation. The [hidden] organization evolves, and the complex social dynamics of its members also.[11]

We can have a really good argument regarding if any organization could survive without its own culture, let alone prosper. For most experienced employees the answer would be "no." Culture has strength and depth. It has layers of subcultures, any of which can be very hard to identify. And searching for and dealing with culture can absorb a major amount of your energy and attention.

The purpose of culture is to turn the pieces of a disparate organization into an effective "whole" through integration of the group's many components. This internal integration includes using a common language, agreeing on group boundaries, the criteria for member inclusion and allocation of status, power, and authority, how to distribute rewards and punishments, and generally helping to manage the unmanageable. It is a covert adjunct to the overt structures and processes of the formal organization. It can accelerate and enhance responses to unanticipated events, foster innovation, and enable people to solve problems that require collaboration.

It can also prevent all this from happening.

As an engineer, the hidden organization is very important to you, even though it is largely invisible. And because it deals with "softer" topics such as leadership norms, group dynamics, power, interdepartmental relations, and other concerns impacting the social norms of your group, you need to know it, cold.

Let me share a personal experience. When I first started in engineering, I firmly believed anything having to do with group relationships or workplace behaviors was flat-out immaterial to me. I believed that engineering work performance was based on a strict meritocracy: correct results, delivered with high quality was all that mattered. As a newbie, not only did I initially doubt culture's existence; I believed that even if it was real, it had no business in the workplace. On a good day, I might have given it (maybe) a 5% impact on my professional life. And this view was common with the other new hires around me.

Over the next few years, my view changed. Each year I discovered a new way of how the hidden organization impacted my work. And I'm not talking about politics, which I now believe I initially confused with culture. I discovered that the formal organization, with its strong rules and boundaries, was incapable of dealing with the inherent "messiness" of real people working together. The workarounds, the invisible knowledge people might possess and share, the experiences that create learning; all come from the culture. In a word, I found the hidden organization to be my friend. By the time I reached perhaps 15 years of experience working within technical organizations, my judgment was that corporate culture impacted at least 50% of my work life. I became a believer, and I wasn't the only one. If you ask any experienced engineer or technical specialist in your company, they will tell you without hesitation the corporate culture is a large and mightily important component of their performance and professional career.

5.5.1 Cultural Strength

On first look, it may not seem that something that is hidden could be classified as being either strong or weak. Yet organizational cultures and subcultures definitely have strength. Cultural strength is the degree of agreement among members of the organization about the importance of specific values. If there is widespread agreement or consensus, that indicates a strong culture. Likewise, little agreement indicates weak cohesion. Strength (and its resulting internal consistency) depends on several factors: the stability of the group, the amount of time the group has existed, the intensity of the group's shared experiences and learnings, how the learnings take place, and the clarity of the assumptions and values held by the founders and leaders of the group. Like a championship soccer team fighting through a long season, the shared experiences and beliefs developed on the team's journey to a championship determine the overriding culture going forward. Culture is an ongoing, changing entity, not a slogan.

An interesting example of cultural strength is "the good soldier." A *good soldier* in a corporation is seen as someone who blindly (and without second thoughts) does exactly what management wants and in a way that submerges their own individuality. This behavior is a direct result of a strong culture and is commonly witnessed in new hires who want to demonstrate adherence and allegiance to their new group.

The power of this type of culture can be overwhelming. Very extreme examples are well known, such as the Jonestown massacre of 1978. The Jonestown People's

Temple in Guyana was a religious cult led by a self-proclaimed minister named Jim Jones. After the murder of officials investigating the group (including a U.S. Congressman), Jones was able to convince some 900 adherents of this cult to collectively commit suicide in the name of their leader and organization. Jonestown remains one of the most infamous examples of the good soldier culture, and in fact the phrase "Drink the Kool-Aid" came directly from the Jonestown tragedy.

5.5.2 Subcultures

As the name implies, subcultures are a subset of an overall culture and reflect the common problems, goals, and experiences that members share on a team, department, or other unit residing deeper within an organization. This makes sense, as we've learned that organizations do not share a single, monolithic organizational structure but are an amalgam of many structures based on the reality and goals of different levels. This means it's possible for a single organization to hold conflicting values and manifest inconsistent behavior while having complete consensus on the group's underlying assumptions.

We are going to spend some significant time examining cultures and subcultures in Chapter 9, sensitizing yourself to see subcultures, assessing their strength and depth, and deciding your place within them. For now, just be aware that subculture has a significant role to play in your work life, and mastery of it is not only important but also essential.

5.6 UBIQUITOUS COMMUNICATION

As you already know, a firm's communication network refers to how information flows within an organization. These networks are regular and irregular patterns of person-to-person, group-to-group, and person-to-group relationships through which information moves and is consumed. This means that information must be thoughtfully designed, managed, regulated, and structured on a continuum from an effective, formal system to an equally effective but informal, covert arrangement.

In this book, a ubiquitous communication network is meant to meld the previous four Core components into an effective "whole" with an optimum level of multidirectional and multidimensional information exchange. A ubiquitous communications network is designed to cement all these components into a strong, robust, and interdependent network, using all technological and personal means available.

This means that ubiquitous communication is not a two-dimensional construct but a three-dimensional concept. All Core areas should be connected to all others by every appropriate path, method, and mode possible. And ideally, not only should information pathways be ubiquitous; they should also be frictionless.

This is not easy. A reoccurring struggle with communications in any organization is breaking down the natural walls that each Core area erects to protect their perceived "turf." In other words, to control their domain.

If we are to create this communication paradise, several decisions are necessary. The first has to do with what should be communicated to what areas of the Core, what information does not need to be communicated, and which information needs to be communicated but is not, caused by either valid or invalid reasons. In other words, transparency.

There is one school of thought that says greater transparency is always better for the organization. Another school of thought says that, to maintain confidentiality and reduce risk, certain information should be strictly limited to those on a "need-to-know" basis.

One of the hardest tricks for any group or engineer like yourself is balancing information transparency versus information containment. Your supervisor will probably spend much of his or her time performing this balancing act. It is also a prime area that can create embarrassment or praiseworthy moments for you personally if brought before many eyes. For instance, many experienced engineers have had meetings with higher-level executives (say, a vice president plus a dozen directors from multiple departments) presenting a report of some kind. No matter how the presenter attempts to balance the information transparency vs. confidentiality, regularly one of the directors will complain "they are not receiving the information." Inevitably, the vice president will turn to the presenting engineer (i.e., you) and direct you to provide the information, even though it may not be appropriate for that director to receive it. The vice president wants to show in public they personally are a "team player" and are open, with nothing to hide. In reality, the vice president knows the director shouldn't have the information but passes the problem to you. The dilemma is usually solved through a discussion with your mid-level management, but you understand the point.

So, the whole idea of "maximum" communications vs. "optimum" communications comes into play when thinking about ubiquity. More is not always better. As you've surely experienced, information overload is a real danger that must be purposely designed out of the system, leaving us to communicate the right information at the right time in the right amount in the right form to the right people. This is no small trick to pull off, and we'll address this and other communications conundrums relating specifically to you in Chapter 10.[12]

5.7 REALIZATIONS

Hopefully, our tour through the Inner Core has convinced you this is a place to really care about. As an aspiring professional, mastery of the Core is the first and most important task as you enter this new territory of formal structures, resources, technologies, culture, and communications. The chances of receiving meaningful assignments are limited until you demonstrate some understanding of these components.

The Core defines the fundamental interrelationship between these five areas. It is the map of how you can get things done and successfully compete in both your local organization and within the greater enterprise. It's the guide to understanding just what is going on inside your engineering home.

A foundational idea within organizations is the concept of policy deployment; the cascading of the firm's vision, mission, purpose, goals (and all the rest) downward until it arrives at your workstation or desk. Following this deployment process results in an extremely well-defined and clear picture of the desired direction for your department and your company. And it is rarely (if ever) fully employed in the everyday life of an engineering company. Only bits and pieces are used, and only when they take minimum time and effort to accomplish. In short, management views policy deployment as more of a task with little upside and substantial waste. This is a mistake.

The formal organization is what you would expect, where all visible policies, structures, work methods, standard procedures, relationships, and all such reside. Combining the formal organization, policy deployment, and contingency theory together creates the many, many organizational structures engineering firms use every day. And an engineering department is not an island: it continually requires the cooperation of both the marketing and business office in an honest and competitive relationship, best described as "Trust; but verify."

Resources enable you to do your work and accomplish things. Defined as either hard or human, they come to you from various places and in various forms. Two main sources of these resources are available: those already in place within your home organization, and those that are procured from outside the department or firm. They extend from inorganic objects to talented individuals. Sometimes they are easy to obtain, sometimes they are impossible to acquire. Taken alone, they are a necessary, but insufficient, condition for success.

Technology is a tool to help transform an input into a desired output. The Technology System is the component where the tangible work gets done. In this construct, a technology is a tool used to accomplish something. A transformation is a slightly different concept, a goal reached or target achieved.

The hidden organization (or corporate culture) is a mostly invisible society of beliefs, understandings, and actions that make up the social system of your workplace. It's seen as a covert set of values, norms, guiding beliefs, and understandings that are shared by members of an organization and seen as correct. It is where unwritten knowledge is shared, and where trust, relationships, and the *quid pro quo* happen. The hidden organization is many times a complex and confusing place yet is essential for your success.

A communication network controls how information flows within an organization. The network can be both regular and irregular patterns of relationships and tools through which information moves and is consumed. That information must be managed, regulated, and structured on a continuum from an effective, formal system to an equally effective but informal, covert arrangement.

The key role of ubiquitous communication is to hold the other four components together in a spherical, 360-degree construct. When thinking about ubiquity, the whole idea of "maximum" communications vs. "optimum" communications comes into play. More is not always better. Information overload and unacceptable quality is a real danger that must be purposely designed out of the system, leaving us to communicate only correct, timely, and meaningful information.

So, we've gotten through our tour of the Inner Core, our introduction to the heart of the Essential Engineering Framework. What you've just read are how things *should* work. And you know what's coming next: a detailed look at how things might actually work once we apply those unspoken, messy factors like popular common practice, the impact of workers' varying experience, the fallibility of systems, and just the plain old stuff that surprises us each day.

Let the games begin.

NOTES

1. Bain and Company. 2018. Mission and Vision Statements. https://www.bain.com/insights/management-tools-mission-and-vision-statements/#.
2. Kenny, Graham. 2014. Your Company's Purpose Is Not Its Vision, Mission, or Values. *Harvard Business Review.* https://hbr.org/2014/09/your-companys-purpose-is-not-its-vision-mission-or-values.
3. Ibid.
4. Polman, Paul and Winston, Andrew. 2021. *Net Positive.* Brighton, MA: Harvard Business Review Press.
5. Business Roundtable. 2021. Business Roundtable Marks Second Anniversary of Statement on the Purpose of a Corporation. Washington, D.C. https://www.businessroundtable.org/business-roundtable-marks-second-anniversary-of-statement-on-the-purpose-of-a-corporation.
6. Razzetti, Gustavo. 2020. How to Define Company Values, and Why They Matter. *Fearless Culture.* https://www.fearlessculture.design/blog-posts/how-to-define-company-values-and-why-it-matters.
7. Anubhav. 2021. Organizational Objectives: Definition, Meaning, and How to Achieve Them. Keka Company. https://www.keka.com/organizational-objectives.
8. Warner, Alfred. 2011. *Strategic Analysis and Choice: A Structured Approach.* Hampton, NJ: Business Expert Press.
9. Madu, Boniface. 2013. A Vision: The Relationship between a Firm's Strategy and Business Model. *Journal of Behavioral Studies in Business.13(1): 23-32*
10. Daft, Richard. 2010. *Organizational Theory and Design.* Mason, OH: Cengage Learning.
11. Sadri and Lees. 2001. Developing Corporate Culture as a Competitive Advantage. Journal of Management Development. 20(9/10): 853-860.
12. Greenbaum, Howard. 1974. The Audit of Organizational Communication. The Academy of Management Journal. 17(4): 739-754.

Part Three

Operating within the Inner Core

6 Navigating the Formal Organization

6.1 UNDERSTANDING THE FORMAL ORGANIZATION

At this point, we've introduced the Essential Engineering Framework, dissected its various elements, toured its Inner Core, and completed a basic examination of each component within that Core. It's now time to begin the real work: uncover, correct, and prevent the common mistakes and misunderstandings you may encounter as we carefully inch our way over this barrier called the Great Divide.

Let's begin with the first component, the formal organization, shown in Figure 6.1. As covered in Chapter 5, the characteristics of formal organizations include highly defined organizational structures, detailed organization charts, prescribed methods of communication, rigid vertical and horizontal linkages, written policies, and standard operating procedures. Each of these characteristics is well defined and normally codified within the organization's operations and policy websites. Yet what is shown on these websites is a crumb compared to what is truly needed; that is, a very large part of formal organizational knowledge is missing.

This chapter will focus on the fundamental topics that will help provide those missing insights. These topics cover managing the relationships with your first- and second-level management, the paradox of organizational timing, and the perceived security of the formal organization chart. Other discussions will explore the role of contingency in setting goals, methods, and structures, operating around power, politics, and conflict, and "soft" engineering standards. Let's start with managing the relationship with your immediate supervisor and their boss.

6.2 MANAGEMENT OF MANAGEMENT

New employees like yourself generally begin their professional lives focusing on developing and delivering individual projects or assignments. Hitting the due dates, learning the technology, and figuring out how to get from here to there are all appropriate and necessary to achieve initial success. Here the mindset is "self": focusing on the tasks that you as a new employee can directly control and have clear responsibility for.

At this point, the idea of managing "up," that is, concentrating on the viewpoints, wants, and needs of your immediate supervisor and your supervisor's boss may not have occurred yet. Early on, this first- and second-level management territory tends to be a void, and no map is available to even begin to navigate through it.

This brings us to some key questions about your work. What is the goal of your efforts? Why are you doing it? Yes, a common goal is to complete your assignments

DOI: 10.1201/9781003214397-9

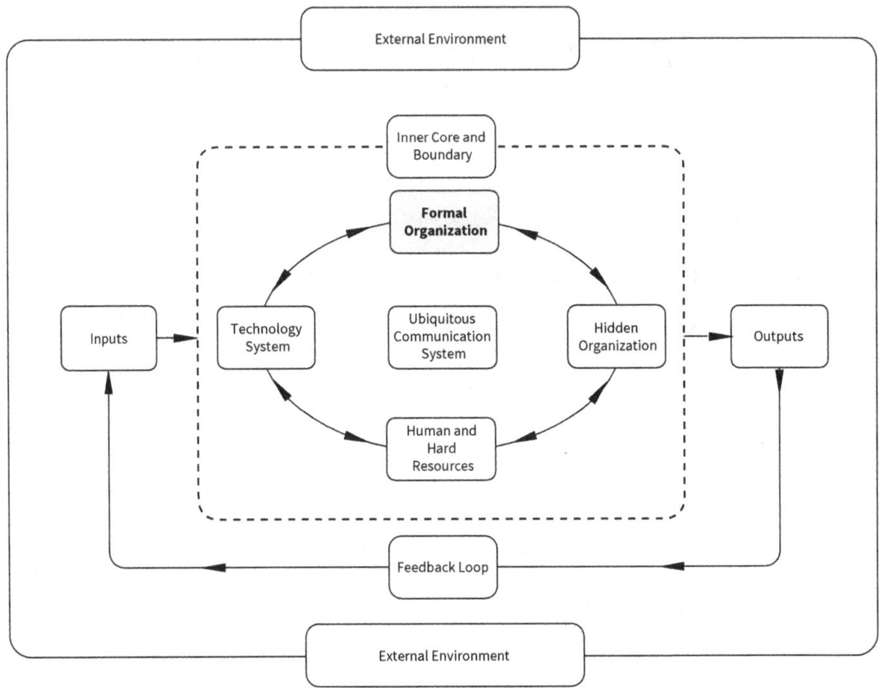

FIGURE 6.1 Location of formal organization within the Inner Core.

successfully. Certainly, you are doing it to create some awesome technologies and solve problems, to make some money, gain position and security, and perhaps earn the respect of your peers and superiors.

But some will argue those goals are not primary; they are secondary. They will contend your basic, primary goal, the reason you are there, is to solve your boss's problems. And your bosses' purpose is to solve their own bosses' problems. No more and no less. All the individual goals mentioned above are subordinate to serving the goals of your management, be it local, divisional, or even at the enterprise level. Solving their problems solves your problems.

Consider the world through your management's eyes. As supervisors, managers, and directors of their organizations, their problems are nearly infinite. They are probably receiving poor direction, unclear objectives, and continuously changing timing. They are short of time, funding, and perhaps energy. They are stressed, and it is your goal to uncover exactly what are their problems: the immediate and longer term, the important and the trivial, the new and the reoccurring. It is the responsibility of every group member to search for clues to your bosses' problems and needs, to uncover their worries and concerns, and to aim their work to fulfill those needs. In short, a new hire needs to understand the pressures on their local management and respond accordingly. This is not easy, but fortunately for new engineers, management will communicate some of their needs through the initial assignments they give you. Keep in mind, for every assignment you receive, your supervisor is attempting to solve their own related, underlying problem. As time moves on, you will become more skilled at uncovering your bosses' concerns and help alleviate them. But for now, begin to develop this skill.

One important caveat will condition your investigations into management needs. Each boss has a worldview, a certain set of beliefs, facts, and experiences that impact how they see their environment and, consequently, how they behave and react to it. For example, some bosses may have had their early formative experiences in a highly formal or bureaucratic organization, where personal conversations are limited, work is intense, and a certain interpersonal distance is always maintained. They may have learned this is the "proper" way to run a department. Or perhaps they learned in a more relaxed setting, where the atmosphere is lighter but excellence in the work product is still expected. Or some management may favor a more unusual, highly inventive, unstructured way of working, as personified by the Googleplex and Apple Park headquarters buildings, whose designs are purposely meant to foster new, informal ways of working.

As you already know, basic organizational behaviors tend to divide people into fundamental types. These types are combinations of emotional and rational characteristics that constitute an individual's worldview, and these can vary widely.

Hopefully, your supervisor or manager will know and understand their own worldview and personal characteristics and their resulting approaches to management. They may even share them with you without you having to investigate. However, this probably will not happen as the natural tendency of first-level supervision is to compartmentalize information into convenient chunks that can be easily concealed. Yet understanding your supervisor or manger's worldview is foundational in understanding your workplace. This means practicing observing and analyzing your management's actions, beliefs, and personal characteristics over time.

6.2.1 Mapping Your Management's Worldview

This brings us to the actual impact of your management's style or worldview on your own daily work. A few initial thoughts. First, employees need to attend to two levels of management: not just the immediate supervisor but also the supervisor's boss, usually a manager. Remember, employees are there to solve their supervisor's problems, and most of the supervisor's problems come from the supervisor's own boss. Second, as touched on above, employees need to understand their management's worldview and resulting attitudes. These tend to be bundled into three categories of management mindset, which we can call Normal, Proper, and Unusual:

1) Normal: A mainstream and balanced approach to their colleagues, staff, and work products. These are typically people you can relate to, and they get things done. Generally characterized by being easy to interact with.
2) Proper: A "by the book," formal and rigid approach to their work and their direct reports. "Propers" are more methodical and stick to all the rules

meticulously. For example, if there is a company logoed T-shirt they would naturally wear it as many days as possible.

3) Unusual: Unpredictable in their perceptions, work results, ways of processing information, and interpersonal behavior, tending to keep you a bit off-balance. Another description of an Unusual is "none of" the previous two definitions.

Table 6.1 shows a basic guide to understanding your bosses for whatever combination of management worldviews they exhibit. Say you have a "normal" (mainstream) worldview and are placed in a department with both levels of supervision are also judged as "normal." Good for you; this is an easily manageable situation. Expectations are reasonable, communication will probably be good, and surprises are rare; a good fit results. Not so when both bosses are "proper" while you tend to be unusual. A clear misfit, and your unusual characteristics will have to be submerged while at work, where in effect you become a façade of someone you are not. You can probably do this in the short run, but it can become an untenable and unhappy situation in the long term.

The overarching point here is that the "fit" (in this case, the worldview) held by you, your first-level supervisor, and second-level manager must be reasonable. Not perfect, but good enough to be able to work together in pursuit of a common goal. Some matches will just not work; these happen all the time and result in someone moving on (perhaps you can guess who). But this assessment tool can help figure out why your management took some action or asked for a certain assignment and guide your response to it.[1]

Another clue surrounding your management's worldview comes from Massey's Generational Sociology concept. Any significant emotional event sparks a calcification of people's attitudes, beliefs, and values at that moment in time. Today, it's easy to predict the 2020–2022 global COVID-19 pandemic is the defining emotional event for many lives. Your values may be strongly linked to this event. But chances are your supervisor or manager will have a different set of significant emotional events leading to a worldview that may be very different from what you hold. The 9/11 generation (terrorism as a significant threat) or the Great Recession generation of 2008–2011 (the risk of losing all retirement savings) may be what drives your manager's attitudes and actions. Whatever the specifics, there are most likely significant unconscious emotional differences between you and your management. The challenge is to assess their values set and respond accordingly, which means delivering solutions that meet your own definition of good or right while honoring the value set of your management. One good trend is that supervisors and managers are probably closer in age to you than senior management, so it's probably easier to connect to them than with their own seniors.

No discussion of management would be complete without touching on that most dreaded type of boss: the micromanager. Micromanagers have been famous for centuries, inflicting their way of thinking on untold millions of innocent workers around the world.

As you probably already guess, a *micromanager* is someone who is compelled to direct an employee to do something and then closely, continuously, and suffocatingly

TABLE 6.1
Guide to Management Worldviews

If Your Boss's Boss Is:	And If Your Boss Is:	And You Are: "Normal"	"Proper"	"Unusual"
"Proper"	"Proper"	Survivable Walk the straight and narrow line	Paradise Need anything be said?	Bad Depending on your unusualness, you may fail
"Proper"	"Unusual"	Tough Better mind your Ps and Qs. Still, better than the Unusual–Proper combination	OK Don't expect the Proper to help very often	Tough but Survivable This depends on the Unusual and how strong the Proper is.
"Unusual"	"Proper"	Tough The Unusual makes the Proper look like an Unusual. Inconsistent	OK Support the Proper. Expect a lot of work and changing direction	OK The Proper will be in your way. Be as careful as your unusualness allows
"Unusual"	"Unusual"	Bad Worst of all situations. Survival is in question	Tough Obey orders and keep your head down	Unknown It all depends on a long queue of characteristics
"Normal"	"Normal"	Paradise If you fail, you only have yourself to blame	OK If you fail, you only have yourself to blame	OK If you fail, you only have yourself to blame
"Proper"	"Normal"	Good Be sure that the Normal protects you if you are "improper"	Paradise Can have some awkward moments	OK Keep your eye on the Proper
"Normal"	"Proper"	Good Be proper and don't let the Normal mislead you	Good The proper will be your ally. Need to support the Proper but carefully	Survivable You're on your own. The Normal is no help, while the Proper restricts
"Unusual"	"Normal"	OK Need to support the Normal. Time consuming	Survivable The Normal will help by their nature, but you'll work hard	Good But don't "blind side" the Normal
"Normal"	"Unusual"	Survivable Need to support the Unusual. Stressful and time consuming	Tough Nearly as tough as Unusual. Keep your head down.	OK/Good This depends on the Unusual. Could be good if the Unusual's nature is yours

watches them do the work. Their corrective comments are continuous, and anxiety fills the air. I once had a micromanager who stood over me and dictated what they wanted my report to say. Finding dictation too slow, my manager literally moved me out of my chair, then sat down at my computer, and typed the assignment themselves. Apparently, hovering over my shoulder wasn't enough.

A good supervisor should give direction on what they want, but unless the employee is very new, not specify how it should be done: that is the employee's role. Micromanagers are the opposite. They tell you what they want, how they want you to do it, where they want it, and exactly what it should look like, and then check with you every half-hour on your progress.

Dealing with the micromanager is hard, and strategies abound on how to get them to stop this behavior. But like any way of managing your management, it's highly contingent on who you are dealing with. The good news is with some observation and just a little experience with that particular boss, a nicely communicated reassurance, repeatedly given, should reduce their behavior with low consequences.

Or maybe not.

What about trust? This should be obvious. An experienced engineer (or anyone, really) will state that trust is a major prerequisite to having the boss accept your work and, by extension, you. In engineering groups, as in all of society, trust is the "coin of the realm." High trust means others give more weight to your words and results, leading to more personal impact, resulting in management gaining more respect and comfort with you.

Good managers know that success in guiding an organization is all about the people on their staff, and trust is a major component defining excellent people. A manager looks for three things when establishing trust:

1) A manager must be able to believe what you tell them.
2) A manager must have confidence in the correctness and accuracy of what you are telling them.
3) You must honestly tell the manager what you don't know about the topic at hand. This is critical, and as important as the information you do know.

This third point needs to be emphasized. There are two types of "not knowing." The first is when briefing your management and a question is posed with an answer that is truly not known by anyone. As no one knows, it's easy to state so. The second type covers when you are asked a question that may have known answers, but it's just *you* who doesn't know the information. There is rarely shame in not knowing an answer to a question posed in a conversation or meeting. The landmine here is attempting to pretend you know the answer when you don't. This is very easy to see through and rarely works. Instead, immediately acknowledge you don't know and *offer to get back to them with the answer* as soon as possible. Then you must keep your commitment and immediately provide the information to the requester after the conversation. While this is apparent, it amazes bosses how many first-level engineers (and others) fail to act on these guidelines. The simple lesson on building trust can be summarized in the old adage, "Say what you do and do what you say."

Of course, there are some relationship principles between you and your management which are essentially universal. One of these is keeping your management informed and in the right way. Another is seeing what needs to be done and just doing it. That is a wonderful thing for the boss and potentially for you. However, the skill here is deciding what things you can do on your own without informing the boss and what things you need to let your management know about before doing, and, of course, then keeping them appropriately informed as you proceed.

Let's consider Steve Anderson, who, like most other new employees, wanted to take advantage of his company's ongoing education program. This program would cover the cost of a master's degree or other continuing education study for approved employees. Steve looked up the program information on his company's website, saw the link to the application, filled it out, and submitted it to the company's human resources department. Unfortunately, Steve neglected to do one thing: he did not talk to his manager prior to making the application. As the funding for the program came from the manager's home organization, Steve essentially spent his boss's money without their knowledge. His manager was totally surprised when some of his budget disappeared, which resulted in a friendly parent–child chat in the conference room.

Managers and supervisors need to be informed of all appropriate activity going on both inside and outside their local organization. The key word is *appropriate*. It is your judgment to decide what is appropriate to inform your management and what is not. A good rule of thumb says if your potential action impacts people or organizations outside your local group, involves spending money, or in some way impacts the reputation of your department, be sure to check with your supervisor or manager before going ahead. This is different than the "Seeing what needs to be done and doing it" approach, which at this point is reserved for internal group business. A good way to judge which action to take is to simply put yourself in your management's shoes. Managers and supervisors hate surprises, and if a proposed action will surprise them, it's best to let them know. When in doubt, just tell them.

6.3 THE PARADOX OF ORGANIZATIONAL TIMING

One thing all engineers can agree upon is the certainty of time. Time is a known quality, steady and constant, defined and measured with amazing accuracy by the best scientists and physicists around the globe. Beginning with the ancient astronomers, measuring time by the movement of the sun, moon, and stars thousands of years ago, time is now measured by the most sophisticated technologies available. Today, the U.S. National Institute of Standards and Technology defines the second as the amount of time radiation takes to go through 9,192,631,770 cycles at the frequency emitted by Cesium atoms when making the transition from one state to another. It is said these Cesium clocks are so accurate that they will be off by only one second over about 30 million years. Amazing.

Yet there are other, completely different definitions of time. When it comes to an engineering organization's concept of time, it is not the precise and irrefutable measurement of frequency. The concept of "organizational time" (i.e., how time is perceived and applied within organizations) has multiple forms, none having to do with the cycles of Cesium.

Organizational time is a paradox. It is highly contingent; it is variable. It disappears, reappears, and disappears again. It stretches and contracts. It is a commodity, yet it has substantial individual value. It can be hoarded like gold or wasted like a discarded sheet of paper. It can pass blindingly fast or drag on with horrible slowness. It has multiple meanings, totally dependent upon the understanding of its users. In short, time is distressingly chaotic.

New engineers need to embrace this paradox of organizational time, understand its relevancy to other engineers and non-engineers alike, and, most critically, solve problems based on this unpredictable paradox.

Context is key. Consider the case where you need to visit an innovative equipment supplier, a startup designing a new Lidar sensor for an autonomous package delivery robot. The supplier is located in a loft in mid-town Manhattan, so your plan is to fly into LaGuardia airport, take a Lyft from LaGuardia to mid-town, and conduct the meeting at the startup's loft. To plan the trip, you need to know, "How far is LaGuardia from mid-town?" You probably will consult your navigation app and discover the answer is "8.9 miles." Yet if you ask your chief engineer's administrative assistant helping to plan the trip, the answer will be "41 minutes." And ask the startup's CEO, and they will say, "$47." All three answers are correct; it's the context of miles, time, or cost that changes the answer completely.

Examples of variable organizational time abound. Say you are given a 24-hour deadline by your director to complete a power-use analysis for a new integrated circuit. The tight deadline forces you to make some simplifying assumptions that turn out to be wrong. At the 24-hour review, those incorrect assumptions come to light and the analysis gets discredited. The director asks you to redo the analysis and report back in another 24 hours. The takeaway: there may be no time available to do something right, but there is always time available to do it again.

Or take the case where you are calculating the rolling resistance of a unique butadiene rubber compound for a new automobile tire. You are given one week to report the results. You already know the calculation will only take three days. Chances are, you will complete the analysis in the three days but not report it until the week is up, using the extra time to catch up on other work. The upshot: work always expands to fill the time allotted.

The paradox of organizational time can unexpectedly create unusual and irrational behaviors. Take the idea of "face time." Here, an employee spends perhaps 10 or even 12 hours per day at work in a bid to demonstrate to the boss they are a dedicated worker. Unfortunately, these workers tend not to accomplish very much during those extra hours at work; the goal here is to spend time being seen as present. Investing in face time is a common office behavior; in fact, in some cultures this behavior is particularly widespread and customary.

The question of deadlines is particularly vexing. Organization members, especially management, have a firm belief that deadlines are solid and inviolate and will judge you negatively for a miss. In general they sincerely believe in this, as a deadline is something that can and should be "managed to." These deadlines come from multiple sources:

1) Your management's management, based on any number of reasons, rational or not.
2) A belief that good supervision, in service of the firm, will always push people to their maximum, normally expressed as deadlines. The thinking is that efficiency means fast, and quick results must mean the group is efficient and performing well. The deadline is established as a guess by the boss but then rationalized as correct after the deadline is set.
3) As a new employee, the supervisor may wish to test you by purposely giving you a tough deadline.
4) The deadline is absolutely real and correct, and meaningful consequences will happen for a miss.

The lesson here is that not all deadlines are set in stone, but not all deadlines are arbitrary. The secret is knowing which is which, and judging when and if you can negotiate the deadline if needed. This means understanding the balance between work quality, completeness, and correctness vs. time allotted.

Another variable time concept practiced by management is the "shrinking time syndrome." Here, the supervisor unwittingly shrinks the time a previous routine assignment took, even tasks that they themselves did as a younger engineer. Say you are asked to conduct a standardized test for lead contamination in a municipal water supply. It takes 72 hours to prepare the sample, run the test, verify the answer, and report. The supervisor is well versed in conducting the test, having done it themselves for many years, but that was 8–10 years ago. Unfortunately, over that decade, their memory of the test changes until they are convinced it only took, say, 48 hours, and that is the due date you receive. The test still takes 72 hours; the only thing that shrunk is the bosses' memory. In these types of situations, it is important to bring it to the supervisor's attention and correct the due date immediately. Ignoring or "letting it go" will only cause the problem to reoccur and become the new standard. Best to correct the issue as soon as possible, as a little short-term discomfort prevents longer-term frustration.

An interesting situation regards changing or adding goals in the middle of a task without time relief. *Mission Creep* is a term used to describe a request to increase the size or number of tasks within the same amount of time originally given. Say you are given an assignment and agree to the timing. Then as the task proceeds, your boss adds more related assignments to the original task, with timing held to the original date (and no additional staff to help). In other words, many added tasks and no resource relief.

The rub comes when you have these multiple tasks and a set deadline, all stacked up where the chances of doing a good job on all of them become slim. The result is some tasks are not completed or some tasks are finished but with poor or unacceptable quality.

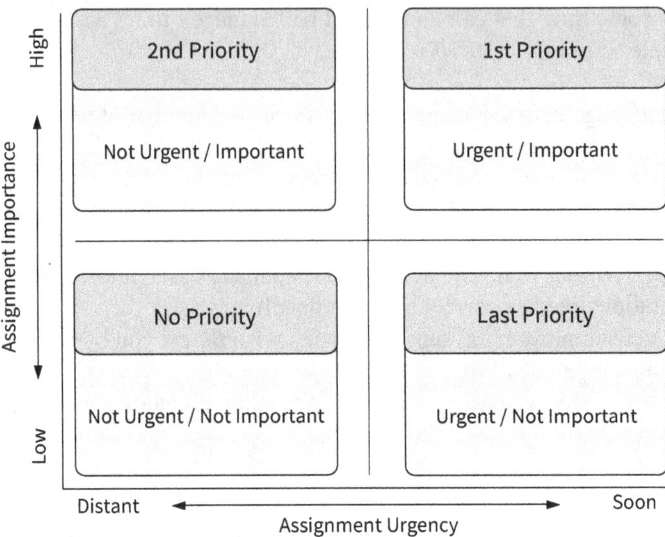

FIGURE 6.2 Eisenhower assignment prioritization matrix.

Here, one way to resolve this dilemma is through the Eisenhower matrix. First popularized by General (later U.S. president) Dwight D. Eisenhower, the matrix is a simple and effective way to shift out competing assignments, to help guide the choice of where and when an assignment should be prioritized, thus helping regain control of your time. This is shown in Figure 6.2.

The idea is simple. Tasks can be categorized in a two-by-two matrix based on urgency (i.e., timing) and importance. These tasks are assessed as one of four combinations: urgent/important, not urgent/important, urgent/not important, and not urgent/not important. As the figure shows, the first to fourth priorities fall out quite easily, with the only caveat being knowing that discarding any not urgent/not important assignment will require a conversation with your supervision.

You may have noticed this entire discussion regarding time has a single underlying assumption when applied in an engineering (or any other) large firm or organization: speed is paramount. Speed to act, speed to decide, speed to market, speed to build: time is always short. It becomes obvious that speed of action is valued far beyond over attributes. This assumption drives your time and the time of everyone around you, and that won't be changing anytime soon.

6.4 THE LURE OF THE FORMAL ORGANIZATION CHART

On arrival, new engineers like yourself may search for something, anything really, to serve as a touchstone or signpost to navigate this strange new world you've entered. The very first document you will receive (probably within 15 minutes of arrival) is a formal organization chart. Relieved to have something in hand that tells you anything, you may clutch this paper like a modern-day Rosetta Stone.

Gazing upon it are boxes, ovals, circles, solid lines, dotted lines, arrows, and text within boxes with strange initials and codes. In fact, some of these charts may look more like the labyrinth in Greek mythology than a chart of your new organization. So, the first order of business during this initial meeting: have the supervisor decode the organization chart.

After this initial confusion, you may see the formal organization chart as something safe and certain. Somehow these lines and boxes on a sheet of paper give a sense of safety, security, and certainty, especially since it came from the supervisor. And at this point, that's good enough.

But note that a formal organization chart is meant to be a map of reality, a holistic guide to department relationships, communication nodes, and power levels defining how the enterprise or department should work. As detailed in Chapter 5, it should be the result of a careful process incorporating the company's vision, goals, and objectives. Instead, it usually is the first and only step created in an urgent process; its creators seemingly always having a due date of two or three days hence and no time to consider any key factors.

But here's a secret: the organization chart is merely a map of what management *hopes* the organization is, but not what it really is. A formal organization chart contains very little real information. In the grossest sense it represents only two nuggets of information, and poorly at that. First, it attempts to definitively explain the formal authority lines between members of the organization, and second, provide a hint of the relative responsibility between those individuals. For a new employee like yourself, it may give an initially warm feeling, but really it acts more as a palliative for management than a trusted map. Unfortunately, as time moves on, this first introduction to the chart becomes calcified in the employee's mind, creating a permanent misunderstanding of its usefulness.

The reason for this disappointment is simple. Management needs an uncomplicated and clear way to communicate the organization's makeup or characteristics. Yet organizations are much, much more complex and interesting than that. The traditional organization chart does not indicate true individual power. It does not show influence, or funding approval, or talent. It does not show true responsibility, or true authority, and may not even show the most current organization. And it certainly doesn't show the informal (i.e., hidden) organization present within every group. Frankly, it doesn't show much. An organization chart only designates the simplest of formal reporting relationships, including the number of levels in the hierarchy and the managers' and supervisors' span of control, and identifies the grouping together of individuals into departments, and departments into the total organization. At best, the formal organization chart shows only a gross approximation of the group's reality.

Then why have it? After all, it does not include the systems that ensure effective communication, coordination, and integration of effort across departments. However, these organization charts are required by four groups: middle and upper management (who need a simple map of the larger enterprise), legal (to fend off potential lawsuits), finance (to assess lending risk), and external regulatory and quality control organizations. For example, the International Organization for Standardization (ISO) 9001 certification process for engineering organizations requires organization charts. So do Wall Street and Main Street. Essentially any outside stakeholder needs the firm's organization chart as part of their normal due diligence.

So, if these charts are extremely simple and of limited usefulness, how can taking the organization chart at face value cause trouble? As these charts are only approximations of the organization, using them *exclusively* as a tool to operate from will at best cause misunderstandings, or, worse, failure in reaching your goals. This can happen because:

1) Job titles are mostly incorrect, inaccurate, purposely misleading, or just plain wrong, causing misunderstanding as to who is the decision-maker and who does what.
2) Responsibility or real power sources do not necessarily correlate to chart position.
3) Guidance into essential informal or underground relationships is missing.
4) The chart may be (or probably is) out of date.

Consider organizational titles. Several decades ago, job titles, printed on embossed business cards, were something that were coveted, to be distributed as a token of influence and power. They certified that the bearer was indeed a member of the organization in good standing, someone to be taken seriously and entitled to respect. Considered valuable in their own right, these little cardboard totems were sometimes even collected by recipients and kept as a trophy, stored in their own specially designed faux-leather case.

Over time, this business etiquette became obsolete as title inflation became the norm. Today, a person may email or call a virtual bank to discuss a personal loan and be put in contact with an "Executive Director of Consumer Accounts." Are they truly an executive director? No, they are simply the customer service representative at the call center.

The upshot here is to never confuse organization rank with an organizational or position title. To find out the true level of an individual, compare indicators of relative rank. These can include annual salary (if available), span of control, organizational budget authorization, or financial signatory authority and operational, tactical, or strategic-level control of the organization.

Overall, relying strictly on an organization chart for intelligence can be problematic. For example, say you have been assigned to assess the feasibility of adding a new feature to an existing consumer product, in this case a device to predict the impending failure of a surge protector within a household emergency power generator. The assessment must answer:

1) Will the new device work reliably once placed into the existing design?
2) Will any new customers now buy the product once this new feature is included?
3) Will the incremental margin increase if the feature is now included?

The first question is yours: the engineer is responsible for determining functionality, reliability, resultant quality, and related engineering questions. However, the second and third questions are not yours but reside within the Marketing and Business Office organizations. You must reach out to these unknown entities, explain the assignment, get their agreement to accomplish their portion of the assignment, and then receive the results on time and check to see if their assessment is reasonable. In effect, you are now being asked not to answer these questions but to manage them.

Your first step is naturally to consult the Marketing and Business Office organization charts, pick off the name of the top manager or director of that activity, contact them, arrange an in-person or virtual meeting, explain your problem, ask for their help, tell them when you need it, and obtain their legitimate commitment to deliver. This is fine, if you've selected the correct manager or director from the chart.

Unfortunately, this is often not the case. Their title will probably not match their functional assignment. Instead, the person selected may not cover that particular product, or, if they do, may be overwhelmed with their own core activities, may think your request as unworthy, may have a bad perception of your engineering activity based on prior experience, or think of the request as a "black hole" that needs to be actively avoided.

A better path to success is to use the chart as a list of names to get you *somewhere in the neighborhood* of the relevant person you need. To get help with the assignment, you need to identify and partner with the correct decision-maker, the person with the power, authority, and responsibility to help in your quest. This is rarely the manager or director, but often a regular-level employee within that activity. Reach out to the director's administrative assistant (if there is one) or a lower-level employee, someone who resides in the general vicinity, and start an investigation into who is the correct decision-maker, based on reality, not chart position. There may be more than one.

A formal organization implies control, stability, a defined and stable external environment, and day-to-day certainty that is static, safe, and secure. It is not. The formal organization chart is a paper tiger, promising insight that it cannot deliver. Examine it, know that it exists, but keep in mind that it is a very weak tool for understanding your new organization.

6.5 DIFFERENT GOALS, DIFFERENT METHODS, DIFFERENT STRUCTURES

As stated earlier, when asked to design a new organization structure, engineering managers don't go to their policy deployment training websites. They immediately grab a paper and pencil and begin to draw boxes and lines. What results will undoubtedly be a hieratical or pyramid-shaped drawing that takes about seven minutes to complete. You and your co-workers will all be placed in that pyramid. But your placement may have little or nothing to do with what you are actually responsible for. The engineering manager will probably be more concerned with getting the drawing to come out in a nice symmetrical shape, with a predetermined span of control that is not what it should be.

Like organization charts, organizational structures suffer from a disconnect between what is perceived and what is reality. Structures should be developed as the last step of a careful process. Instead, the structure is usually the first and only step in a rushed process.

The key understanding is that organization structures should be explicitly designed for the type of work the engineering group is expected to produce. An oil refinery should have a continuous flow structure; a product development firm should be organized in a matrix arrangement. Without a structure design consistent with the output or product, effectiveness and efficiency both suffer. Most low-level (i.e., operational) supervision does not do this. Instead, they default to the traditional hierarchy, or perhaps at most a matrix, or may simply copy whatever the previous supervisor did. Why? Frankly, designing an organization correctly is labor- and time-intensive, yet again results are normally expected immediately.

So, here's the fallout. At the operational and tactical levels, the assumptions used in developing an organizational structure include:

1) The existing group already has the talent capable of performing the work assigned. The new structure must employ the same people as before ("rearranging the deck chairs on the Titanic"). No or little additional talent is available, or, if so, it is expected to be short-term agency help.
2) The structure must be completed with minimal effort.
3) The manager or director responsible will have no idea if the new structure is successful or not until substantial time has passed.

As you examine the typical organization chart upward from the operational to tactical to strategic levels and horizontally across functional or product areas, contingency theory dictates that each local department's goals are different, their work methods are different, and their external environment is different. Yet their organization structures are probably the same and thus disconnected from their goals.

Meanwhile, there is an additional twist: the external environment changes at an ever-accelerating pace. What in the past was a well-behaved, stable set of environmental factors are now accelerating into a churning caldron over time, and the organization is usually ill-equipped to change rapidly in response.

So, what does this mean for a new engineer? The formal organization is neither static nor stable. It is dynamic over time, though the organization chart is represented as stable and authoritative. Current working relationships and networks will have limited shelf-life. Assignments, work methods, overall goals, and even bosses will change more rapidly than anticipated. Those who expect stability may have a difficult transition into their new work environment. As long as a new hire expects this churn and flexes with the new conditions, these changes can have a less significant impact on their work.

6.6 MEET CERBERUS, THE THREE-HEADED BEAST

For those familiar with Greek mythology (or maybe Dante's *Inferno*), you will know Cerberus, the three-headed dog that guards the gates of Hades. In the myth, one head of the dog represents the past, one the present, and the third the future. Organizations have their own Cerberus, luckily not charged as the gatekeeper of Hell, but an interrelated trio of heads representing the organizational concepts of power, politics, and conflict. Each concept has a relationship with the others, and we need to understand what they are, how they interact, and, most importantly, how they can affect any new employee.

6.6.1 Operating Around Power

We all know what power is. We've seen it operate toward both good and bad ends, applied by individuals or groups, governments, or commercial companies. It is present in any situation where two or more people congregate to achieve a common purpose. But power is not a single idea or "thing." Power is a multidimensional entity that can be described in several ways and conditioned by several caveats. There is personal power and organizational power. There is power derived from internal or external power and combinations thereof.

The classic understanding of power comes from the sociologist Max Weber. Weber developed three basic modalities (or types) of power. The first is *Bureaucratic Power*, where obedience is given to a set of "rational" rules (such as a legal system in a city or state) rather than to a particular person. An example might be obeying a judge's or police officer's direction. The next is *Traditional Power*, which is obedience given to a particular person based upon societal customs and habits. It's given because it has always been given in the past, such as to royalty or religion. The king of England and the Roman Catholic pope are good examples. The third (and one of the most interesting) source of power is *Charismatic Power*, which forms when special or unique needs occur that cannot be satisfied by ordinary rules, but only by special people with special gifts "of body and mind."[2]

For our purposes, we'll define power somewhat more narrowly than Weber and focus strictly on organizational power, a subset of bureaucratic power. *Organizational power* is the ability of one person or department in an organization to influence other people or departments to bring about desired outcomes, even when others disagree. There are certain characteristics of power in organizations. First, it exists only in a relationship with two or more people and is the result of conflict (a direct interference with goal achievement) and competition (the rivalry in pursuit of a common achievement).[3]

Power within an organization can exist as two general types: *individual* power, which is created due to certain characteristics of a particular individual, and *organizational* power, which is derived from a group's position within a hierarchy.

One of my favorite stories regarding individual power has to do with a manager named Ray Jessop. Jessop came to my company's advanced engineering activity by way of an air force career, retiring as a full bird colonel. Ray carried himself with a military bearing: a commanding presence, a steady gaze, and always impeccably dressed.

Ray was a strategist. Each action he took during his workday supported one strategy or other. There was always a goal in mind with every word and deed. In fact, our joke about Ray was "the brand of cornflakes Ray had for breakfast this morning was a strategy."

Visiting Ray's office was a textbook example of exhibiting power symbols. On entering you could not miss the 11″ × 14″ full-color photo of astronaut Neil

Armstrong walking on the moon. Inscribed on the photo was Armstrong's signature and the inscription "To Ray – Best wishes from your friend Neil Armstrong."

Sitting in his visitor chair, you couldn't help but notice an unusual totem. Mounted on a small wooden plaque was an odd-looking key. Beneath the key were the words "Operation Thunder Sky, 23 June 1974." Of course, the plaque always draws a comment from every visitor. On being asked, Ray would nonchalantly say,

> Oh, that. That's the launch key from a practice Minuteman II missile launch back in '74. You turn that key at the same time with another key to launch a nuclear strike against the bad guys. That's the key I turned during our launch.

Sitting there, you come to realize that Ray Jessop had the power to launch a nuclear ICBM across the Pacific Ocean. You came away with the feeling Ray was telling you he was some minor American Prometheus.

Ray's display of his personal power extended far beyond mere objects. Ray could surprise you in public meetings with demonstrations of his power, to your detriment. In one case, Ray was leading an effort to build an advanced aerodynamic concept vehicle. He was using an unconventional and, in my view, risky method to develop the basic shape, and I mentioned this to one of Ray's engineers in the hallway. Several days later, I was urgently summoned to a meeting, with no indication of what it was about, who would be there, or why I was needed. Duly walking into the meeting, I found about 25 of his direct reports sitting around a large conference table. As I entered, Ray said to me directly: "I understand you have some problem with what we're doing here. Tell us what your problem is." I was floored. Ray was using this meeting to publicly show he would slap down any criticism of his engineering method. I stammered out a few words and left the meeting, bright red. In short, this was a trap. The purpose of my summons was to elevate and enhance Ray's power in front of his direct reports.

I learned three nuggets of wisdom that day. First, for any unusual meeting invitation or management summons, find out in advance the "what" but also the "who and why" of your invitation. Second, realize that some people will use you as a tool for their own gain at your expense, especially when you might not expect it.

The third nugget is my responsibility. Rather than sharing my complaint with a random engineer in Ray's organization, I should have set up a one-on-one meeting with Ray in his office and voice my concern directly rather than in a hallway. Criticism is a private matter, to be expressed privately. Ray was teaching me a lesson about operating around power.

Let's spend a bit more time on power. Power can be exercised both vertically and horizontally, meaning power is essentially everywhere in your company and, once sensitized, you'll see it in use continuously.

As the name implies, vertical power sources are dominance factors that individuals at the top of an organization use to wield power. Common techniques include formal position, resources, control of decision processes, and information control. In contrast, horizontal sources of power are how departments obtain power over competing departments. This occurs through methods such as a department possessing organizational centrality, the inability to replace a unique, critical department, or the ability of cope with uncertainty.[4]

What about authority? If power is the ability to achieve desired outcomes, then authority is a force for achieving these outcomes that are determined by an organization's formal hierarchy and reporting relationship. Thus, authority is a subset of vertical power more limited than generalized power.

There are several rules regarding authority. Authority is vested in organizational positions that are preestablished when you enter the group. Because of that, authority has legitimacy and is traditionally accepted by subordinates. Importantly, authority only flows down the vertical hierarchy. Thus, power can be exercised upward, downward, and horizontally while authority can only be exercised downward.

It may be surprising, but lower levels of an organization also have power. Obviously, upper levels have power, but how would lower levels gather their own power?

Actually, lower levels have many possibilities available which tend to fall into two main categories: gaining power for themselves or limiting power available to others. A person or department may place themselves at the center of an information flow, positioning themselves in a "gatekeeping" role, such as the chief of staff to a CEO or an administrative assistant to a chief engineer. Perhaps you may be part of a committee reviewing corporate policies or rules. After all, knowing the rules better than anyone else is a powerful tool, using them as technicalities against opposition. Expertise that is in short supply, such as the ability to code in IO, Racket, or Erlang, could potentially command a higher salary or snag more interesting work. Perhaps being a technician on an expensive, highly technical, and rare piece of equipment would gain you more status and cache.

The second category, power limiting actions, is how employees without power can limit those with power. These can include retaliation (a revolt against higher powers), regulation (whistleblowers and other appeals to existing rules), apathy (disengagement from the work), or creating an opposing coalition (such as a labor union).

Keep in mind that management has its own sources of power. How managers achieve power within their departments is plentiful. They include the ability to bestow rewards (i.e., "I can give you a good recommendation"), the ability to exact or recommend punishment (i.e., "I can fire you"), legitimate, which is another word for authority and derived from formal position (i.e., "I am the boss"), or expert, derived from skill or knowledge (i.e., "I am the expert and you are not.")

Figure 6.3 represents the locations and forms of these various power sources.

6.6.2 Individual Managerial Power

We all have observed powerful people. We may have even met one or two. And you may wonder why they are powerful: after all, in a group of people it's relatively easy to pick out the people with personal power. So what is going on?

There are two types of personal power in organizations. The first is power derived solely from the person's position in the organization. This is easy to see. This power can be shown by the size of an office (if they exist) in a workplace. It's the amount of budget the person controls or the amount that can be approved by that person. Non-organizational personal power is a different animal. This

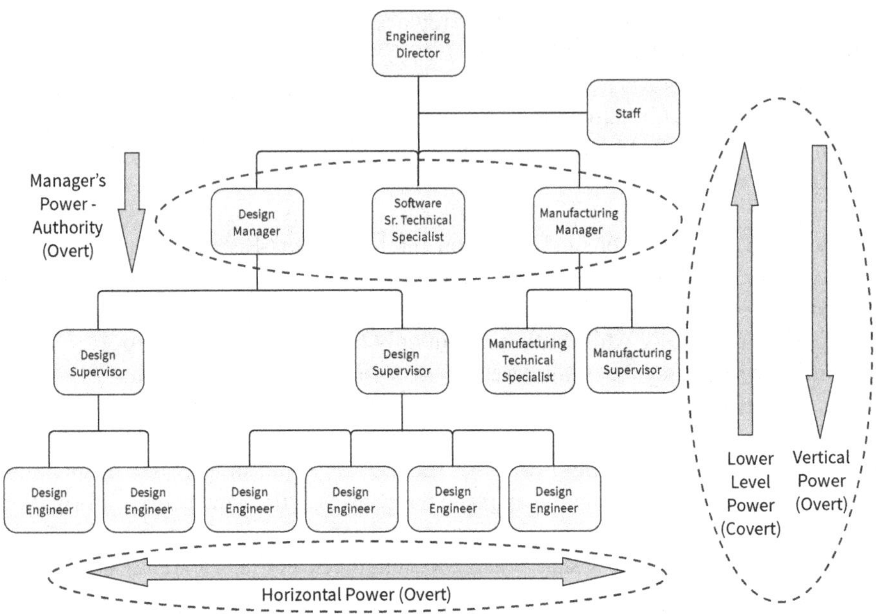

FIGURE 6.3 Organizational power structure.

personal power is not bestowed on the individual because of their organizational position; their power comes from attributes outside the organization. An interesting example is what we might call in the United States the "national genius." For many decades now, the U.S. media has tended to assign one of the country's top scientists the unofficial title of national genius. In the 1930s and 1940s it was Albert Einstein. In the 1950s it was Robert Oppenheimer and Enrico Fermi. In the 1960s it was Margaret Mead, and in the 1970s and 1980s it was Richard Feynman. The 1990s brought us Stephen Hawking and today Elon Musk fills the role. The power of their opinions could move national policies, establish new scientific fields, and generally push scientific and social positions and global finance in their preferred direction. The interesting thing is each of these individuals initially emerged from an organization, mainly universities and startups. However, over time their influence extended far beyond their original organizations: they became a power center unto themselves.

Speaking of power and its expression, another expression of managerial power is *totems*. Totems are small objects that are symbols of personal or organizational power. They are meant to be displayed and must be seen to be effective in conveying the power of the individual wearing it.

"Belonging" totems are the most common. A company lapel pin is a common totem, as is a laptop case sporting a company logo. A certain color of the company badge denoting access to confidential information is highly desired. A windbreaker or hoodie with a project team's name over the left breast (embroidered, of course) is another example. Your organization will have specific ones that are meaningful only in your group. You may see people with certain items on their desks or hanging from their cubicle walls. Certificates denoting a training course completed, quality awards, thank-you notes signed by top company officials, and other tchotchkes all indicate relative power.

It can be enjoyable looking for power totems in your group or department. Some are obvious, and no training is needed to understand their meaning.

But other power totems are harder to see. Well-known or prominent individuals may require a personal security staff during public events or meetings. One U.S. Cabinet Secretary security detail uses a small metal lapel pin with a blue letter "R" placed within a white box. Barely visible, this pin is a subtle security badge. All members of the staff wear this pin, an indicator that they have been cleared to be inside the inner security perimeter. Without that pin, the security detail will remove anyone inside the perimeter by whatever means necessary. While the actual pin is small, anyone who possesses it certainly feels like they are important indeed.

But here is a key point. Many people, including top executives, will mistake organizational power for personal power, to their detriment. William Broadbank was the chief engineer of a major engineering corporation, heading up a global firm employing some 13,000 engineers and technical staff. Bill was a superstar and he knew it, with most of a 35-year career at the top echelons of the company. He was abrupt, demanding, and more than a little dictatorial. And the first thought anyone scheduled to meet with him was a nervous, "Oh, boy. What will Bill think?"

Unfortunately, Bill had forgotten that his power, his ability to command people at any hour of the day or night, was based solely on corporate power, not personal: he had conflated the two. And after many years of presumptuous and authoritative behavior Bill was asked to leave the company. Within two hours, the joke in the halls of the company was, "Bill who?" Where thousands of engineers used to think, "What will Bill think?" became, "Who cares what Bill thinks?" The transition from powerful to powerless was stunning, instantaneous and complete.

CASE EXAMPLE 6.1 THE RECOVERY OF THE *APOLLO 8* SPACECRAFT

One of the most interesting examples of demonstrating organizational vs. personal power came during the preparations for the first trip to the moon by the *Apollo 8* astronauts in December 1968. As flight director Christopher C. Kraft tells in his book *Flight: My Life in Mission Control*, the pressure was on to fulfill President Kennedy's goal of landing a man on the moon and returning him safely to earth by the end of the decade. A critical step was to first fly to and then orbit the moon. The only time this flight could be made was over the 1968 Christmas holiday, and the U.S. Navy needed to provide a large network of ships to recover the spacecraft after its flight. Unfortunately, many thousands of sailors had already been given leave to be home for Christmas, and that was needed to change if the flight was to be successful. It was Kraft's job to convince the admiral in charge of the Pacific Fleet, Admiral John McCain, to cancel Christmas leave and deploy the fleet. This required an in-person meeting between Kraft and McCain to request the change.

McCain had a deserved reputation for being tough, hard questioning and no-nonsense. As Kraft tells the story:

> "They took us to amphitheater conference room, nicely paneled and with about 100 seats. We sat down in front of the damnedest bunch of brass I've ever seen, all captains and admirals and even a couple of four-star Army and Air Force generals filtered in and filled every seat."
>
> "At 10:30 AM sharp someone yelled 'Attention!' and in walks four-star Admiral John McCain, commander-in-chief of the Pacific Fleet, who was a couple of inches shorter than me and full to the brim with military bearing. Someone said 'Be seated' as McCain took his chair right down front and looked me in the eye. 'Okay young man,' he barked in a loud voice, 'What do you got to say?'"
>
> "McCain's eyes only left my face to look at my charts and graphs. I don't remember many of my exact words. I simply ran through the mission and told him what we wanted to do. I know that I stressed the importance of the flight and its risk, and that the greatness of the United States of America was about to be tested in space."
>
> "Then I got to the real point. We had to land the crew in the Pacific. I memorized exactly how I wanted to put it."
>
> "Admiral, I realize that the Navy has made its Christmas plans and I'm asking you to change them. I'm here to request the Navy support us and have ships out there before we launch and through Christmas. We need you."
>
> "There was complete silence in the room for maybe five seconds. McCain was smoking this big long cigar, and all of a sudden he stood up and threw down it on the table."
>
> "'Best damn briefing I've ever had,' he said, loud enough for everyone to hear. 'Give this young man anything he wants'. And he walked out."
>
> "Apollo 8 was go."[5]

A really great story, but there is more here than just getting the go-ahead to support the mission. Notice what McCain did. Before the meeting was ever held, there was no question McCain knew he was going to have to supply the ships and men to support the Christmas mission. No way was the president of the United States going to postpone this historic flight with so much national prestige at stake just so some sailors could go on vacation. Knowing he had no choice, McCain turned the "decision meeting" into active power theater. Instead of a private meeting in his office, McCain filled an entire auditorium with 100 or more staff so they could hear that the mighty NASA space program needed *him*. In short, McCain used this meeting to reinforce his personal power over his people. A brilliant demonstration of individual power.

6.6.3 Navigating Conflict

It's time to look at the second head of our friend Cerberus, the head dealing with conflict. In a word, wherever there is power, there is conflict. *Conflict* is the natural and inevitable outcome of the close interaction of people who may have different beliefs, values, objectives, or varying access to information and resources within an organization. Just as with power, we are taught that conflict can exist in two forms: organizational conflict (conflict between two or more organizations or groups of varying size) and individual conflict (conflict between two or more individuals interacting with another).

Conflict can occur due to several reasons. Perhaps different goals, ill-fitting skill sets, personalities, or cultures, departments depending on each other, or the struggle for limited resources, which can be significant.

Our idea of conflict creates two counterintuitive points. First, no matter how we parse the terms "organizational" or "individual" conflict, at its root all conflict stems from individuals. Organizational conflict is just conflict between groups of individuals driven by a person in charge. Individuals have conflicts in the name of organizations. That means the tools used to resolve individual conflict are the same tools that can be applied to the person in charge of a group. The takeaway here is that all conflict is, at its core, individual.

Our second counterintuitive point is conflict in organizations is not necessarily bad. Conflict challenges the status quo, encourages new ideas and approaches, and can lead to positive change. It occurs in all relationships: individual to individual, group to group, and group to individual. Of course, we are talking about reasonable conflict: too much conflict can be detrimental, requiring a change in approach if unresolved over time.

This idea might be a bit uncomfortable. After all, haven't we been raised to "play nicely with others" and avoid conflict at all costs? Depending on the organization you work in, there are various degrees of conflict that are deemed acceptable and even encouraged. Manufacturing plants may have a reputation for being high-conflict organizations, where "telling it like it is" is not only acceptable but admired. Software houses may practice an inclusive, high-respect culture (sometimes disparagingly called "kumbaya") where polite interaction is valued and conflict is to be avoided. No matter how it is expressed, many corporate cultures believe tension and conflict are necessary ingredients for any successful organization.

It is inevitable that you will be involved in conflict. Your conflict may be with members of a different organization you partner with, or with colleagues in your workgroup or perhaps your management. You may have a conventional perspective where conflict is bad and should be eliminated, that conflict should not occur in the first place. If so, you may wish to consider the pluralistic perspective, where conflict is good and should be encouraged and managed with respect. Instead of being passive about conflict, it's important to understand it and develop *in advance* a process or strategy to work through it where no party is injured. Let's see how this might be done.

Each of us has a personal conflict style, a predetermined approach to perceiving, processing, and acting in conflict situations. A good personal goal is to evaluate and modify as required a current personal conflict style to drive toward more mutually beneficial resolutions.

Knowing this, what does this mean for you personally? Since organizational conflict still comes down to individual people, in a way conflict is indeed "all about you," and that means you must again understand yourself when it comes to dealing with it.

Seeing a need, Kenneth Thomas and Ralph Kilmann created the Thomas–Kilmann Conflict Mode Assessment (TKI), a personal evaluation tool that identifies conflict management styles and provides solution techniques. It identifies five common conflict management modes, based on two overall conflict modes: *assertiveness*, the extent to which an individual attempts to satisfy their own concerns, and *cooperativeness*, how much a person attempts to satisfy the opposing person's issues. This produces five conflict management modes: Competing, Collaborating, Compromising, Avoiding, and Accommodating, as mapped in Figure 6.4. These are situational and time-dependent, and are tendencies, not absolutes.[6]

At its root, conflict is a form of negotiation. Unfortunately, many new engineers have a fundamental discomfort with negotiation and little experience as well. Numerous new graduates (not just engineers) struggle to become more assertive. A key benefit of the TKI is to help raise awareness of the reflexive responses to conflict that can impede resolution. The insights gained from taking the TKI can be quickly used, as it helps to realize your dominant response to conflict situations.

Remember, no single approach to conflict negotiation is best; contingency theory applies here. But there are four general lessons:

1) In a conflict, a negotiator is not just stuck with their reflexes. They have a choice in resolving the conflict between value claiming ("mine") and value creating ("ours").

FIGURE 6.4 Thomas–Kilmann Conflict Mode Assessment.

2) A negotiator needs to know their reflexive response to conflict so they are more mindful of the choices they make.
3) Departing from intuitive reflexes requires energy: preparation, planning, mindfulness, and conscious effort.
4) Adaptability is needed and desirable. A negotiator might move from one TKI "type" to another as a negotiation progresses.[7]

Since collaboration is widely seen as the best method in resolving conflict, many subject matter experts have offered suggestions on improving the chance of success, as in Table 6.2.

It's simple to say that collaboration is the preferred technique in resolving conflict as it can produce "win–win" outcomes if applied correctly. How to achieve this happy ending involves the thoughtful application of various methods, too many to address adequately in this book. There are many good guides available; an online search and review is an obvious place to start.

Unfortunately, collaboration will probably not be achieved 100% time, so a negotiator will have to decide which of the remaining paths to pursue. Compromise is always a good choice, but as far as the next level, a discussion with your supervision is required.[8]

Now here's the most important point in conflict resolution a new engineer needs to absorb. Never, never, never express anger inappropriately. We're not talking about never showing anger but showing anger inappropriately by shouting, moving, or otherwise acting out in a way that causes fear, insult, or other denigrating actions. One instance of becoming unsuitably angry will do more to damage a reputation than any mistake in daily work. It is very easy to become known as a hothead, to be seen as unreliable or as immature.

We all know this already, yet the temptation to vent or act inappropriately can sometimes overcome better judgment. Remember, no one appears powerful when angry; they just look uncontrolled and immature.

6.6.4 Politics Lurking around the Corner

Some portion of operating in a technical organization will have to do with politics and political tactics. For those starting out, this is a high-risk area. The expression of politics and resulting political gain in any organization is so highly contingent that hard and fast rules are dangerous and ill-advised. Yet seeing politics as practiced in your firm is an important skill, so let's look at some tactics in action and become aware of their use and power. But please note, political activity is like playing with live ammunition: no one would recommend to actually use political tactics at this point in your professional development.

Politics is the exercise of power and influence when conflict exists. It is expressed as self-serving behavior within an organization or achieving a natural organizational decision process when conflict is high. Political tactics are used when uncertainty or disagreement exists within an organization.[9]

There are three common areas or domains of political activity. It can be through an organization's structural change, where managers may negotiate to maintain their

TABLE 6.2
The TKI Conflict Instrument Summary

Conflict Resolution Strategy	Resulting Behavior	When to Apply
Competition	• Creating win–lose situations • Using rivalry • Using power plays to reach one's ends • Forcing submission	• When quick/decisive action is vital, such as emergencies • On important issues where unpopular actions need implementing • When mission is vital to company welfare • When you know you're right • Against people who take advantage of noncompetitive behavior
Avoiding	• Ignoring conflicts • Putting problems under consideration or on hold invoking slow procedures to stifle conflict • Using secrecy to avoid confrontation • Appealing to bureaucratic rules as a source of conflict resolution	• When an issue is trivial or more important issues are pressing • When there is no chance of satisfying your concerns • A potential disruption outweighs the benefits of resolution • To let people cool down and regain perspective • When gathering information supersedes immediate decision • When others can resolve the conflict more effectively • When issues seem tangential or systematic of other issues
Compromise	• Negotiating; looking for deals and trade-offs • Finding satisfactory workarounds or acceptable solutions	• When goals are important but not worth the effort or disruption of more assertive models • When opponents are committed to mutually exclusive goals • To achieve temporary settlements to complex issues • To arrive at expedient solutions under time pressure • As a backup when collaboration or competition is unsuccessful
Accommodation	• Giving way • Submitting and complying	• Allowing a better position to be heard • To learn and to show your reasonableness • When issues are more important to others than to yourself, to satisfy others and maintain cooperation • To build social credits for later issues • To minimize loss when you are outmatched and losing • When harmony and stability are especially important • To allow subordinates to develop by learning from mistakes

(Continued)

TABLE 6.2 (CONTINUED)
The TKI Conflict Instrument Summary

Conflict Resolution Strategy	Resulting Behavior	When to Apply
Collaboration	• Problem-solving attitude • Confronting differences and sharing ideas and information • Searching for integrated solutions • Finding solutions where all can win	• To find an integrated solution when both sets of concerns are too important be compromised • To merge insights from people with different perspectives • To gain commitment by incorporating concerns into a consensus

personal power bases. It can be the management of group secession, which is bargaining over who gets what positions in the organization or who may be promoted. The third is the allocation of resources: money, equipment, and talent, as well as the methods of awarding good or bad performance.

There are also political tactics for increasing an individual's power by making themselves more valuable to the organization. They can take high-risk assignments, create dependent behaviors, or provide scarce resources. And they can negotiate for what they want through building power bases.

One condition of politics at the group level deserves a mention. Say a group's working environment is stable, where goals are consistent, power and control are centralized, decisions are orderly and rational, and norms are based on rational measures. Then conflict is low, and so political activity is low. But when the environment is unstable, goals become inconsistent and changing, power and control become decentralized with shifting coalitions and interest groups, and disorderly decision processes emerge as the result of bargaining among interests rather than rational means. Then conflict is high, and political activity becomes common. The upshot is to be aware of any increased group conflict over time and prepare for incipient political behavior as a result. Knowing in advance how to deal with emerging political behavior can provide protection and more certainty in your daily work.

To sum up, let's return once more to Cerberus. Power, conflict, and politics really are a three-headed beast, with each head of the creature separate yet interrelated. And as in Dante's *Inferno*, this beast can't be ignored, but it can be managed and successfully dealt with a little forethought and savvy. But it's your job to prepare for Cerberus' arrival.

6.7 LEADERSHIP IS NOT MANAGEMENT

Be forewarned: the argument between what is management versus leadership is a decades-old, unresolved, and emotional question, certain to cause red faces and acid indigestion among anyone trying to definitively compare and contrast the two.

Yet it's important we attempt to clarify these two terms and the difference between them. It's safe to say there are as many definitions of these two concepts as there are management consultants for hire on the web. We certainly can't settle the dispute here, nor do we want to. Instead, our goal is to introduce the idea of management as opposed to leadership and provide a basis for you to personally decide what it is and how to benefit from it.

In practice these two terms have become synonymous, a shorthand version for "the boss." Both are concerned with people, yet leadership and management are two distinctly different things, two different concepts. And your interaction with these two ideas will vary accordingly.

Like so many topics in engineering organizations, the idea of management vs. leadership is highly contingent and rather fuzzy and soft. Rather than spending excessive time struggling with this, let's just start with a single core idea from Michael Maccoby: *Management is a function while leadership is a relationship.*[10] What this means is that the centrality of leadership is ambiguity while the center of management is certainty. Leadership is hazy; variation is its central tenant. Management is control and measurement; discipline is at its core and variation is the enemy.

Warren Bennis further defines these contrasting terms in Table 6.3.[11]

Similarly, management thinker Peter Drucker famously defined organizational management as a function that must be exercised in any organization, while leadership is a relationship between "the leader and the led" that can energize an organization to action. Drucker defined management as five specific activities: planning, organizing, staffing, directing, and controlling. Management is about control of people and processes, certainty of the results and "keeping things on track."[12]

Defining leadership gets a little more slippery. We've already said that leadership is a relationship. It involves selecting talent, motivating employees, coaching those employees, and building trust between the leader and the organization they lead.

TABLE 6.3
Management vs. Leadership Characteristics

Characteristics	
Management	**Leadership**
Administrator	Innovate
Ask how and when	Ask what and why
Focus on systems	Focus on people
Do things right	Do the right things
Maintain	Develop
Short-term perspective	Longer-term perspective
Imitate	Originate
Are a copy	Are original

Most employees assume that leadership means the top of the organization, and for many reasons this is true. However, recently there is a movement toward "leaders at every level." These proponents contend that leadership is a relationship that can exist at whatever level the leader is placed within the organization. Yet a key difference is that while it can be a relationship, contingency theory must be satisfied, meaning the nature and content of that relationship will vary for each level.

A simple way of slicing this multilevel leader idea is saying there are two types of leaders: Strategic (high-level) leaders and operational (low-level) leadership. But still, at the core of each leadership example resides the relationship; just the specifics of the relationship vary by level.

6.8 STAYING SAFE VS. STEPPING OUT

Formal organizations tend to be safe. In fact, safety is one of the main characteristics of the formal organization and provides one of its critical benefits. Here, *safe* means that if you follow the written (and unwritten) rules of the organization, you won't get into major trouble. Ambiguity is low, and policies (i.e., the rules) are clear and generally well documented. Standard operating procedures are common, and resulting employee anxiety is low. And what if you don't comply with the rules? Well, there's an old proverb: the nail that sticks up gets hammered down.

A policy is a pre-made decision, covering situations that repeatedly occur within the organization that do not warrant revisiting. If situation A occurs, then action B is the immediate response. The key here is that proper policies are always documented in writing and retained in some easy-to-find location. A policy not written down is not a policy; it's an opinion.

Remember that very first meeting with your new supervisor on day one? He or she may have covered perhaps two or three basic policies you needed to remember. Perhaps it's the policy regarding not accepting gifts above a certain nominal value from a supplier. Or maybe it covers confidentiality restrictions in meetings, or care of sensitive information, or allowed expenses during business trips. Alarmingly, there are dozens and dozens of written policies you will not know about until you break one of them. And unfortunately, there will probably be no central website or database showing all these policies in one place, and if there is, no one will know where it is.

Meanwhile, in large, long-established engineering organizations, many rules may seem arbitrary and irrelevant to you as a new engineer, reflecting a world from many years ago but unsuited to the 21st century. Remember that written rules are policy, and policies reflect the organization's reality at a point in its history, not necessarily today. The pace of change in the organization always outruns the rules, replacing their relevance with insignificance over time.

This is not a new problem. The author Martin Lindstrom in his book *The Ministry of Common Sense* examines company battles to fight bureaucratic nonsense. Silly rules are often caused by new policies being introduced without a "sunset clause" or considering unanticipated short- and long-term implications of the rule. Why does this happen? One reason is that bureaucracy tends to endlessly multiply without a mechanism to retire old, irrelevant rules.[13]

While you may (correctly) believe certain rules are wasted energy or nonsensical, it's important to think of the policies through the bosses' eyes. In addition to many other responsibilities, your supervisor or manager is accountable for policy compliance within their group or department. Outside organizations (such as Finance, Human Resources, Quality, or Legal) require compliance for their interest area and will attempt to ensure it through periodic audits. And it is your management who must certify that your group complies by providing hard-to-find evidence. And if the boss must visit a new employee and remind them about the missing information, it's just one more onerous and irritating task they must do. Best not to be an irritant and stay safe: perform compliance.

Obviously, formal policies carry a cost in money, energy expended, and time lost. The time impact is particularly common. What if you have a core assignment under tight timing, and the time required to fulfill some policy requirement will cause the assignment to be late? What should you do? Fulfill the policy rule and take the timing hit, or ignore the policy rule for now and complete the assignment, hoping the penalty is not too severe? Here, advance notice is your friend. Simply speak to your supervisor immediately after you discover the assignment conflict and await their decision. Here, the key is giving the supervisor or manager enough time to investigate the relative importance of each course and render a low-stress decision.

One of the most famous and far-reaching "staying safe vs. stepping out" decisions in recent history is the story of Mr. Ben Sliney, a former U.S. Federal Aviation Administration (FAA) National Operations Manager. On the morning of September 11, 2001, Sliney was starting his first day of a new job overseeing the national U.S. airspace, managing some 4,200 aircraft simultaneously flying over the entire country.

About two hours into his first day, two aircraft crashed into the twin towers of the World Trade Center in New York and a third into the Pentagon in Washington, D.C. Chaos grew as other planes were now suspected of being used as terrorist weapons. This fear was warranted: one additional hijacked plane crashed into a Pennsylvania field later that morning.

According to the *National Commission on Terrorist Attacks Upon the United States*, faced with intense uncertainty, urgency, lack of data, and unknown casualties, Sliney alerted his superiors as to his understanding of the situation, who then went into a conference. Sliney then decided he could not wait for his superiors and, without approval, ordered over 4,000 aircraft to immediately land, reasoning that any aircraft still flying must be under terrorist control.

Sliney did the unprecedented: on his own he gave the order to immediately land every aircraft in the air over the United States, shutting down the entire U.S. airspace. The 9/11 Commission later cited this action as an important and decisive moment in that morning's chaos, and completely justified.[14]

How could he do this? The background information is important. While Sliney made the decision on his own initiative, he had the advice of an experienced staff of air traffic controllers and traffic managers within reach. And while it was his first day in charge, Sliney had over 25 years' experience in air traffic management within the FAA. He had held positions as an Operations Manager and Traffic Management Officer in New York, a Traffic Management Specialist, National Operations Manager, Tactical Operations Manager, and Regional Office Manager for the Airspace and

Procedures Branch for the Eastern Region. He was a hugely knowledgeable individual who not only had the experience but also the confidence to assume command and decide on the best action to protect the United States. For Sliney, stepping out was a relatively low-risk decision.

Let's look at something much less dramatic. Take the example of Noah, a new electrical engineer charged with testing a batch of 15 next-generation touch screens for uniform brightness. His reservation in his building's test facility is at 3:00 pm, meaning his screens must arrive there by 2:30 pm for preparation. Being a large building, certain job classifications are performed by union-represented workers, who work to the contract negotiated by management and their union. One contract rule specified that moving large objects (large being defined as requiring two hands to carry safely) must only be performed by union workers, called millwrights. Any non-union worker caught violating this rule would have a "grievance" filed against them, resulting in the millwright who discovered the violation receiving a cash payment equaling four hours of overtime pay.

At 2:15 pm, Noah and his test articles are in the hallway, waiting for the millwright to arrive and move the articles. At 2:20 pm, Tom (a very experienced engineer) walked by and asks what Noah is doing:

Noah: "I'm waiting for the millwright to move my stuff down to the test lab."
Tom: "What? Pick it up and carry it down there."
Noah: "I can't. I'll get written up by the millwrights for a two-handed carry."
Tom: "Jeez, pick it up and carry it. Any engineer who doesn't have at least four outstanding grievances isn't doing their job."

Now Noah doesn't want to be late to test, and Tom, an "old hand," says having a grievance is expected and no big deal, but Noah doesn't what to do something explicitly wrong, especially so early in his tenure.

So, what should Noah do? Once again, contingency theory rules. If Noah was in good standing, had substantial experience, had a known and positive track record over time and perhaps some gray hair, ignoring minor rules may not cause him much grief. In this case, the negative of ignoring the policy (or knowingly breaking a rule) is often forgiven by prior respect and past successful results.

Unfortunately, a new hire like Noah has none of these positives. As a new engineer he possessed no *social capital*. The concept of social capital is important to all employees, no matter the organization's size, longevity, or marketplace success. Social capital is the potential of individuals to secure benefits and invent solutions to problems through membership in social networks. It revolves around three dimensions: interconnected networks of relationships between individuals and groups, levels of trust that characterize these ties, and resources or benefits that are both gained and transferred by virtue of these social ties and networks.

So, based on the amount of social capital available, Noah should not take the older engineer's advice. Far better to wait for the millwright and be late (which can normally be rectified at the engineer's level) vs. getting the grievance which will bring the unwanted attention of supervision. And, of course, plan extra time for the next test.

But here's another important variation of contingency theory applied to social capital: even older engineers with a large cache of social capital can be mistaken. For example, say an engineering company (a large, traditional manufacturing firm) has a long-standing rule requiring all engineers to be at their workstations each morning at 8:00 am. However, as a new employee, you see many colleagues arriving as late as 9:30 am or as early as 7:00 am, adjusting their quitting time accordingly. Obviously, the start time is not an enforced rule. Then you notice an experienced colleague (who is a constant 9:30 am arriver) having long-running, work-related conflicts with their supervisor. After several months, the colleague is let go. The stated reason for the release: the colleague was breaking work policy due to chronic tardiness. The unstated reason: the ongoing, unresolved conflict between the colleague and supervisor. The lesson: if you are in good standing, certain minor work rules can be safely ignored or modified. However, if your status with management changes for the worse due to any reason, those same ignored rules can and will be retroactively applied as cause for reprimand or dismissal. Here, breaking rules is the ammunition to create delayed, self-inflicted wounds.

What does all this mean? Stockpiling social capital is an investment to attain future flexibility in dealing with some rules and situations, a buffer to absorb the mistakes of omission and commission we all make.

Stockpiling social capital can include, but is not limited to:

1) Avoiding creating extra work for your supervisor.
2) As much as possible, perform favors or actively help colleagues who need it. Be seen as a helpful individual.
3) Building trust that your judgment is sound.
4) Doing the low value-added work with the same enthusiasm as more significant assignments.
5) Avoiding squandering your capital on minor situations, reserving your social capital for important uses only.

Sometimes, you just can't stay safe. Take, for example, a former manager of mine, a crusty British engineering director named Ken Cookson. Ken was a firm believer in the old school of honoring and obeying the hierarchy of engineering organizations. He was aggressive, extremely self-confident and his natural inclination was to "shoot first and ask questions later." One day, my manager Jack, an engineer of the more modern school, asked me what I would prefer my next job assignment to be. He asked me to discuss it with several veteran engineers within our local group, which I did. Thinking about it, on my own I also approached another director outside my department with equal rank to Ken; in fact, he and Ken shared an office. That director was happy to advise me.

The next day, Jack asked for a word, took me into our conference room, and shut the door:

> "Bob, let me give you a bit of parental advice. When I suggested you ask the engineers in our group for their thoughts, that's all I wanted you to do. That's my style. Ken is

> old school and will severely criticize both you and especially me for even asking for your opinion. I want your thoughts, but in Ken's eyes your feelings or preferences have absolutely nothing to do with his decision. He does not want your views. For him, they do not matter. And by the way, he doesn't want you sharing his internal business with outsiders. Please remember this: old school directors can be dictators, and they protect their turf. Remember that."

I learned that when stepping out, I should first anticipate any unintended consequences of my actions. So, think ahead and attempt to look for any surprises when crossing a boundary.

Let's examine one more example of staying safe vs. stepping out. As described in Case Example 3.3, during the historic economic recession of 2008–2010, the finances of the United States (and the rest of the world) teetered on the edge of chaos. No one alive had seen this situation before, at least since the Great Depression of 1930–1933. The rules no longer applied. Barack Obama had just been elected the U.S. president and had to address this problem swiftly and effectively. Decisive action was called for.

Instead of staying safe at this historic, high-risk moment, the administration stepped out in an unprecedented manner, fundamentally changing American (and global) economic policy while simultaneously establishing a new national healthcare system. In describing these actions, Obama's Chief of Staff, Rahm Emanuel famously said: "Never let a good crisis go to waste." What he meant was large crises called for (in fact demanded) large responses unconstrained by the status quo. It means when in uncharted waters: the standard rules are gone, normal performance standards become irrelevant and management may not know what to do. These major events are an outstanding opportunity to step out as opposed to remaining safe.

6.8.1 An Imperative

Of course, there are many examples where stepping out is not only acceptable but imperative.

Surprisingly, certain cultures allow and even condone mental and other abuse to occur from supervisor to employee. According to recent reporting, this workplace bullying shows itself through verbal, emotional, and psychological abuse cases in different businesses and social strata. This abuse is common enough that it has a special term in some countries: *pawahara*, or power harassment. In Japan, worker objections to this behavior reached some 82,000 complaints in 2018, leading the government to pass two anti-harassment legislation bills. Japanese firms are now required to have clear policies and internal systems for reporting and verifying abuse.[15]

This is also happening in South Korea. South Korea's business culture can be notoriously punishing. Workers in South Korea work some of the longest hours in first-world economies, where they have little control over their time and a small chance to escape from harassment. There is also a special word in South Korea for this authoritarian attitude: *gapjil*. Gapjil includes shouting at underlings, requiring unpaid all-nighters, and requiring some employees to handle their managers' personal errands. Even at new South Korean startups based on America's Silicon Valley meritocracy model, some bosses are still tending toward this behavior.[16]

Obviously, through the Western lens, this type of behavior is unacceptable, and in many of the locations mentioned safeguards are now in place to provide a pathway to protection. In these companies, reporting abuse is imperative and relatively low risk. Firms in countries without these protections are a different story.

6.9 STANDARDS ARE NOT NECESSARILY STANDARD

As part of their training in engineering design, new engineers will be told explicitly that their work must follow "the requirements." And for good reason. Engineering requirements are the result of many years of sometimes bitter experience in practicing engineering correctly.

In this book, *engineering requirements* are defined as a collection of standards, codes, and specifications that new and existing designs must meet. All three are extremely important, often essential, in engineering and related technical fields.

A *technical standard* is an established norm or requirement. There is usually a formal document that establishes uniform engineering or technical criteria, methods, processes, and practices for a given job. These documents, prepared by a professional group or committee, specify proper engineering practices and contain mandatory requirements.

A *code* is a set of rules, specifications, or systematic procedures for design, fabrication, installation, and inspection, prepared in such a manner that can be adopted by legal jurisdictions. Codes are generally approved by local, state, or federal governments and can carry the force of law. The main purpose of codes is to protect the public by setting up the minimum acceptable levels of safety for buildings, products, and processes.[17]

Specifications are quantitative, measurable criteria that a product is designed to satisfy. To be measurable and unambiguous, specifications must contain a metric, a target value, and defined engineering units for the target value. A *metric* is the characteristic of the product that will be measured. A *target value* is the acceptable range of the metric, such as "less than 6 seconds" or "between 20 and 50 pounds." An *engineering unit* is an amount of a physical quantity defined and adopted by convention or law that is used as a standard for measurement, such as minutes, degrees Celsius, and kilograms.[18]

All these requirements define a level of compulsory engineering performance. It's simple: your design or project must meet its requirements or the work doesn't proceed.

It's easy to love requirements. Clean, clear, definitive, and inarguable, they are the bedrock that any engineering design or project rests on. But occasionally that bedrock is not as solid as you might think.

Requirements are the result of hard-won experience derived from engineering projects in the past. They are the equivalent of saying, "we shouldn't do that again; somebody write that down." Requirements capture the judgments of engineering experts that set the boundary of a discipline's state of the art at that moment. Some requirements (like codes), once established, may remain in effect for decades: their strong tie to the force of law makes them very externally focused and stable. For example, civil engineering projects are highly invested in codes.

Consumer products are a different matter. External codes for these products can be minimal. A child's toy may have only a few governmental code requirements: no small parts a child can swallow, use nonflammable materials, and prevent electrical shock. Yet the toy's specifications may actually require a higher level of performance than the applicable external codes. This is normally driven by both consumer competitive pressures and conservative legal positions: the toy's part size may be double the governmental safety requirement, the fireproof materials may be much better quality than the competition, and higher-cost waterproof materials be installed on all electrical connectors. The goal is to avoid legal challenges while providing a competitive advantage.

But here's the flip side of the coin. Internal company requirements are established in response to lessons learned on past engineering projects. Many times, these requirements must be satisfied on all subsequent projects via formally signed written documentation. Yet these lessons tend to be contingent on the past project's time and place: the specific conditions causing the issue. And due to the quickly changing external environment, these once-viable requirements quickly become irrelevant and obsolete. Yet the requirement remains.

Take the design of a new two-door, all-electric sports car. For a traditional gas engine coupe, the desired high acceleration and top speed require maximum engine cooling performance, and the company's decades-old cooling requirement is for the car's front-end grill opening to be at least 35% of the vehicle's total frontal area. But all-electric vehicles don't need traditional cooling: there is no grill on the car at all. What to do?

The answer is the deviation. A *deviation* is a formal document, signed by all top engineering management, that allows the current project to deviate from the standard in a specified manner. It documents under what conditions a project will not meet company standards, and why.

Yet the deviation is not actually a solution. The problem is twofold. First, *obsolete standards are rarely retired*: they stay in full force well into the future. The second is that upper-level engineering management is loath to sign deviations. No one wants to sign a document certifying their new product is "not up to standards," no matter how irrelevant that standard is. What results is a new project needing dozens of obsolete deviations signed off before the first prototype is built, signed by management that doesn't want to do it.

Why not retire the obsolete rules? It makes sense, except rarely is there a single responsible group charged with performing this task, plus the unproven fear of a current condition somewhere that the obsolete deviation still covers. The result is to shy away from the situation and let the next director address the problem. Best to just sign the deviation and get on with the job.

What this means is this: You will undoubtedly be working with standards, codes, and specifications quite early on. You need to know what requirements are legitimate and which are not; what are governmental and which are company-based. Which are obsolete and which are fully in force. Once you fully understand these requirements early on, you can proceed with minimum rework, which results in higher efficiency, certainty, and confidence.

6.10 REALIZATIONS

So, what does all this tell us? On the surface, the formal organization appears stable and certain, guided by written policies, bounded by management rules that are accurate, fair, and provide clear-cut expectations so you will know where you stand at all times. In certain stable environments this can be true. However, the rest of us operate under different conditions. We've learned that management's worldview may not match yours; that deadlines and timing are variable. We know that organization structures vary across the enterprise; that their charts tell us surprisingly little, and insignificant guidance exists on when you should step forward or hold back. We know that management is not leadership, that power, conflict, and politics are interrelated, and that standards are less certain that expected.

We've discussed some causes of these realities and hopefully offered some insights and suggestions on how to respond. The best response is to adopt an "observe and listen" approach. Be that engineering archeologist we've discussed earlier. Then assemble the pieces you've observed to create a rough mental map of the formal organization and adjust your work product to suit.

Just remember that the formal organization is a frame of mind; it is the hope of management. Be aware that management's hope is not necessarily what the formal organization really is.

That said, our next stop is in the area of resources, in the form of both human talent and hard objects. A different set of operating principles apply here. With this in mind, let's examine how to operate with the resources available to you.

NOTES

1. Bolling, G. Fredric. Professor, University of Michigan, Dearborn. Interview by Robert M. Santer, March 3, 1999. Transcript.
2. Weber, Max. 1968. *Economy and Society: An Outline of Interpretive Sociology*. New York: Bedminster Press.
3. Daft, Richard. 2010. *Organizational Theory and Design*. Mason, OH: Cengage Learning.
4. Ibid.
5. Kraft, Christopher. 2001. *Flight: My Life in Mission Control*. New York: Dunton
6. Shell, G. Richard. 2001. Bargaining Styles and Negotiation: The Thomas-Kilmann Conflict Mode Instrument in Negotiation Training. *Negotiation Journal* 17: 155–174
7. Brown, Jennifer. 2012. Empowering Students to Create and Claim Value through the Thomas-Kilmann Conflict Mode Instrument. *Negotiation Journal* 28(1): 79–91.
8. Ibid., Shell, 169.
9. Ibid., Daft, 510–511.
10. Maccoby, Michael. 2000. Understanding the Difference between Management and Leadership. *Research Technology Management* 43(1): 57–59.
11. Bennis, Warren. 1977. *On Becoming a Leader.* Reading, MA: The MIT Press.

12. Drucker, Peter. 1954. *The Practice of Management*. New York: Harper & Brothers.
13. Lindstrom, Martin. 2021. *The Ministry of Common Sense: How to Eliminate Bureaucratic Red Tape, Bad Excuses, and Corporate BS*. Boston, MA: Mariner Books.
14. National Commission on Terrorist Attacks Upon the United States. 2004. *9–11 Commission Report*. Washington D.C.: U.S. Government Printing Office.
15. Economist. 2020. Anger Management: Japan's Bullying Bosses. *The Economist*. June 11, 2020. https://www.economist.com/business/2020/06/11/japans-bullying-bosses.
16. Economist. 2021. South Korean Tech Workers Are Having a Lousy Time at Work. *The Economist*. July 10, 2021. https://www.economist.com/asia/2021/07/08/south-korean-tech-workers-are-having-a-lousy-time-at-work.
17. Sölken, Werner. 2022. What are Engineering Standards? http://www.wermac.org/documents/ engineering_standards_what_are.html.
18. Farris, John. 2022. New Product Development Engineering Specifications. https://npd-book.com/engineering-specifications/.

7 Understanding Human and Hard Resources

7.1 COMPREHENDING RESOURCES

Getting a handle on both human and hard resources is a bigger challenge than you might think. As outlined in Chapter 5, human and hard resources are two separate, but related, commodities, each of high value (I'm sorry, but in large commercial organizations humans are sometimes seen as commodities). As both resources are firmly entrenched within the Inner Core, as shown in Figure 7.1, it's worthwhile to spend some time dealing with the concepts behind these types of resources. Let's talk about human resources first, as this is definitely the more complex portion of the discussion.

7.2 HUMAN RESOURCES

Human resources is defined as the personnel of a business or organization. They are the talent and skill that an engineer brings to the organization to assist in fulfilling the firm's goals.

Just as with the subject of management, Google can provide you the names of thousands of volumes dealing with organizational human resources. In fact, one of the largest professional associations in the United States is the Society for Human Resource Management (SHRM). With over 300,000 members, SHRM exists to improve the professional practice of human resource management. There is no way I can (or will) address the breadth and depth of the various human resources systems in existence, but I do want to touch on a number of topics that directly affect new engineers. (Note: in this book the phrase "human resources" in lowercase denotes an actual person as a resource. Human Resources capitalized or written "HR" refers to the organizational department that administers the company's talent.)

7.2.1 The Mystery of the HR Office

For those new employees in larger engineering organizations, a trip to the HR office is a mysterious journey. What goes on in there? Who lives in those offices? What do they do, and does it mean bad things for me? Probably not, but I don't know any new engineer who doesn't view a trip down the hallway to HR without a bit of low-grade anxiety.

The purpose of the Human Resources department is twofold. The first is to administer the written HR policies of the organizations fairly, quickly and without further problem. The second is much more creative. These HR people are there to

DOI: 10.1201/9781003214397-10

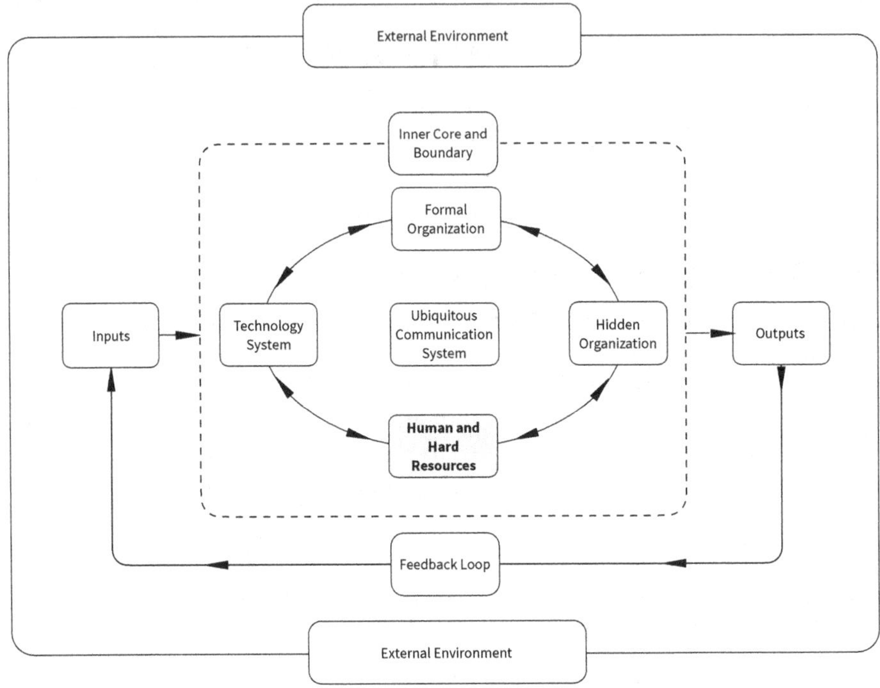

FIGURE 7.1 Human and hard resources within inner core.

solve unique, one-of-a-kind problems related to personnel issues. As a result, larger organizations have two types of HR employees.

One is the individual whose job is to strictly enforce those written policies and impose compliance. Tracking performance against required organizational metrics, processing paperwork, and handling 95% of personnel issues that are routine is their job. These employees can identify unique personnel issues, but their role is to refer these one-off problems to HR management, not necessarily solve them. These issues go to the HR managers as their role is to apply experience, creativity, or modify existing policies to solve the problem.

An example of this occurred to Mary McAllister, a relatively new engineer eager to continue her learning through her company's continuing education program. This program, in place for many years, pays for advanced degrees for qualified engineers while they are employed: most firms have programs of this type and are based on part-time, evening classes.

Mary applied to this program and was approved. As the average time it takes to complete the degree is about three years, Mary knew she was committed to that timeframe and planned accordingly. Unfortunately, halfway through the program her firm suffered a revenue downturn and quick action was needed to reverse the losses. One of their first actions was immediately stopping the continuing education program, with no reinstatement date. Mary, now halfway through, was told that her education would no longer be paid by the company. Mary felt this was reneging on the explicit and implicit social contract between her and the company. She made an appointment with the Human Relations department and spoke with a low-level employee. This employee told Mary this was the new policy and she was financially liable for the rest of her degree: the company would no longer pay. Luckily, an experienced co-worker told Mary to make an appointment with the HR director and explain the problem to them. Mary did so, and the director agreed that the company had made a commitment to her: she should be "grandfathered" in and have her entire degree paid for as promised. As this was a policy exception, the director took the issue to the vice president, who agreed and personally approved Mary's program for the remainder of her studies. This is a perfect example of how a lower-level HR employee will only enforce written policy, while an upper-level HR manager will solve more judgment-based problems. Dealing with the ranking HR employee directly solved the problem: Mary could keep attending class, the company's promise to pay was kept, and Mary strengthened her positive attitude toward her firm. Without the HR director's decision to keep the commitment, Mary's attitude would certainly have deteriorated.

Dealing with HR is an extremely contingent activity. Outcomes can be quite good, such as Mary's experience, or perhaps frustrating if no true resolution of a concern occurs. Because of the wide variation in HR department policies, rules, and procedures, dealing with HR requires a modicum of savvy. Luckily, there is a very simple, common-sense approach to help solve problems related to personnel issues:

1) Become educated. First research the HR-related policy documents and other information available related to your concern. While you won't know everything, you should be on reasonably solid ground relative to whatever the written policy and process specifies.
2) If practical, solve interpersonal issues directly with the person or persons involved. If the issue elevates, the first question you will be asked is: "Have you spoken to the other person yet?"
3) If unresolved after an initial outreach, approach your local supervision for assistance. Part of their responsibility is to solve concerns such as these.
4) If still unsatisfactory, be sure to follow all written process steps the firm specifies before making your next move. Whether to approach your manager or other person or group, be sure to follow each process step completely and to the best of your ability. Compliance to the policy and its process must be demonstrated throughout.
5) When in compliance, go to the HR office. Notice that visiting HR is the fifth step in your checklist, not the first. Once HR is involved, your issue becomes "official"; documented and highly visible. HR is required to address it through their own system, meaning you now lose control of the outcome. While necessary, HR will view resolution in the company's best interest, *whatever that is.*

In short: be deliberate, be educated, and know the territory in advance.

7.2.2 What Human Resources Does (and Doesn't) Do for You

There is general agreement that there are six basic functions of a Human Resources department:

1) Recruitment and Selection: using different selection instruments in a reduction process to identify and decide upon the best candidate for a given company position.
2) Performance Management: a set of processes and systems aimed at evaluating an employee so they perform their role to the best of their ability.
3) Learning and Development: a systematic process to enhance an employee's skills, knowledge, and competency, resulting in better work over time.
4) Succession Planning: the process of planning for various contingencies if and when critical employees leave the company.
5) Compensation and Benefits: the monetary and nonmonetary payments a firm provides to its employees in exchange for their labor.
6) Information Systems and Data Analytics: a digital system used to collect, store, manage, and mine data on an organization's employees.

For an engineer now employed, the three relevant HR areas impacting them are performance management, learning and skills development, and compensation and benefits. Succession planning is strictly a role for management, while human resources information systems and data analytics are back-office functions of the HR department supporting management reporting and analysis.[1]

Underlying these three HR areas is an unstated goal. The HR department operates with the relentless goal of moving the routine functions of administration to "expert" digital systems, making these three functions a self-service operation. Remember that HR is viewed strictly as a cost center, and these costs must be continuously reduced.

This means that certain functions traditionally performed by HR are being moved to the employee's desk. Routine, periodic data collection and reporting is now in the employee's hands vs. HR. A major movement has been to push career advising and planning completely onto the worker's shoulders: don't look for career advice in the HR office. As career planning is 100% the employee's responsibility, this means new hires must find their own support system if advancement is the goal.

7.2.3 Your Objective Set

In Chapter 5, we defined Policy Deployment as the cascading down of goals and objectives to ever-lower organizational units. The natural extension of this system down to an individual engineer's level is the *personal objective set.* This set is the collection of tasks, normally negotiated and agreed on with your immediate supervision, that you will accomplish by a specified time and at the correct quality level.

There is no doubt you want to do the best job possible. Believe me, your management wants the same thing. But what does this really mean, and how is it accomplished? While there is not one best way to tackle this (remember our friend contingency theory), a good bet is through your objective set.

We now come to the point of setting your own personal objectives. Most companies use objective setting methods based on the *SMART* technique, a little kitschy memory device that allows you and your supervisor to create, track, and complete both your short-term and longer-term goals.

Objective setting can be complicated and onerous; the SMART method simplifies it. SMART is an acronym for *Specific, Measurable, Achievable, Relevant*, and *Timely*. Each letter of the acronym reinforces the goal to create an objective set that is clear, intentional, and can be tracked.

You may have set goals in your past that were difficult to achieve because they were too vague, aggressive, or poorly framed. Working toward a poorly crafted goal can feel daunting and unachievable, and creating SMART goals can help solve this problem. Let's consider each portion of the method:

S: Specific

Specific is an explicit statement of what is desired. It means being as clear and exacting as possible with what you want to achieve. The more narrow the goal and more clear the boundary surrounding it, the more you'll understand the steps necessary to achieve it.

M: Measurable

Measurable is metric-based evidence to prove you're making progress toward your goal. This usually means setting up measured milestones to track progress over time toward the end goal.

A: Achievable

Is the goal you wish to attain achievable? Reasonable goals provide structure and appropriate feasibility going forward. Before setting a particular goal, decide if it's achievable now or if additional preliminary steps are needed.

R: Relevant

When setting goals, determine if they are relevant. As with policy deployment, your goals should align with the organization's longer-term objectives and strategy. Avoid goals that don't contribute toward these broader objectives.

T: Time-based

The date you intend to complete your objective. A hard end-date helps provide motivation and prioritization. When setting a certain goal, you also set an estimated completion date. If the goal hasn't been achieved in that timeframe, it might have been unrealistic or just have been unachievable.[2]

Table 7.1 summarizes each of the SMART categories.

TABLE 7.1
SMART Objective Setting Definitions and Examples

SMART Objective Name	Definition	Desired Quality	Example
Specific	A clear and explicit statement of what you want to achieve.	Should reasonably stretch your professional capability throughout the timeframe desired.	"I want to earn a position managing a development team for the company's new sensor group."
Measurable	Means a valid metric assigned to the objective. What evidence will prove you're making progress toward your goal?	Should be actually measurable using agreed-to measurement techniques.	"In one year, reduce overall departmental manufacturing waste by 17%, and polycarbonate waste by 22%."
Achievable	Objective should be realistic. While your general objective will be set by your supervisor, you will be in negotiation to set the specifics of that objective.	Agreement is important; it means both supervisor and you truly agree the objectives can be attained.	"I believe, to the best of my judgement, that these objectives can be reached with reasonable effort and knowledge."
Relevant	Goal should support or relate to the overall strategic, tactical, or operational direction of the organization.	The objectives you are setting should cleanly fit policy deployment direction.	"This new chemical technique will replace our existing CO2 measurement protocol."
Timely	The time frame specified to complete your goal.	Complete objective as defined by its metrics within the range of delivery time specified.	"I will complete the voltage regulation test between 3 to 4 weeks after testing initially begins."

7.2.4 So What Does Your Supervisor Think?

A fundamental assumption in this book is that you wish for a successful first professional work experience. That you desire to demonstrate talent, intelligence, and technical savvy within your new engineering organization; to establish a good (or even great) reputation. All this is admirable, but also begs an age-old question: How do you know if you are doing a good job? That you are well regarded within your organization? Most importantly, how does your supervisor judge you so early in your tenure?

Difficult questions, and there are not simple, clean answers. But there does exist a basic approach to employee performance assessment that (to some extent) is universal in engineering organizations, and you need to understand how it operates.

Fundamentally, management judges you as an employee not with a single tool or method, but using several approaches, designed to "triangulate in" on a performance rating and thus your value to the group. These approaches can be divided into two types: objective methods and subjective methods.

Objective methods are formal performance measurement systems or tools featuring numerical inputs, tightly defined terms and reassuring algorithms to try to ensure a fair comparative rating. These are the performance ratings systems your company will describe in their employee manuals and orientation sessions. Objective methods are generally well designed, widely used, and low risk, having been court-tested over time.

Subjective methods tend to be more covert in nature, used by managers and supervisors as their personally preferred evaluation methods they have found to be useful over time. The reason for the widespread use of these covert methods is simple. A core role of management is making meaningful decisions and judgments, a role that bosses are not keen to give up. While the HR department provides a safe evaluation method, management still wants to make their own assessments and preserve their own decision-making authority. Let's touch on an example of both objective and subjective systems and see how they might work.

Let's talk first about a typical overt, objective performance measurement system. Obviously, performance is the "getting it done" phase of the job and is the reason for your existence within the firm. Your performance is what your management is interested in. Now, contingency theory dictates *how* your performance is measured and assessed. It depends upon the type of assignment you have, the characteristics of that assignment, its criticality, the company's culture, and a host of other factors.

These objective performance systems normally consist of two parts. The first is an assessment of actual accomplishment; how well were your assigned tasks performed. The second addresses what the corporation or company believes is acceptable behavior and attitude, and how your behavior measures up against these standards. These two factors can be represented by a two-by-two graphic as shown in Figure 7.2. Both horizontal and vertical axes have precise definitions attached to them, attempting to minimize confusion on what constitutes good or bad performance and acceptable or unacceptable behavior. This system means an engineer can have excellent performance but be perceived as having poor behavior. Alternatively, you can have very poor performance but be perceived as having excellent intra-group skills and overall "fit."

Normally, performance is captured by a simple process of clearly stating the individual objective, a response statement on what was accomplished relative to that objective, and finally a judgment whether the accomplishment met, did not meet, or exceeded that objective. Many times this judgment is then converted into a numerical score.

The behavioral axis means you want to ensure you fit into your local organization. Many organizations specifically define what those behaviors look like. Trying to codify something as Jell-O-like as behavior is not trivial; it requires a fair amount of verbiage to pin it down. Figure 7.2 shares a sample assessment for a Fortune 500 firm, both performance and behavior within four major categories.

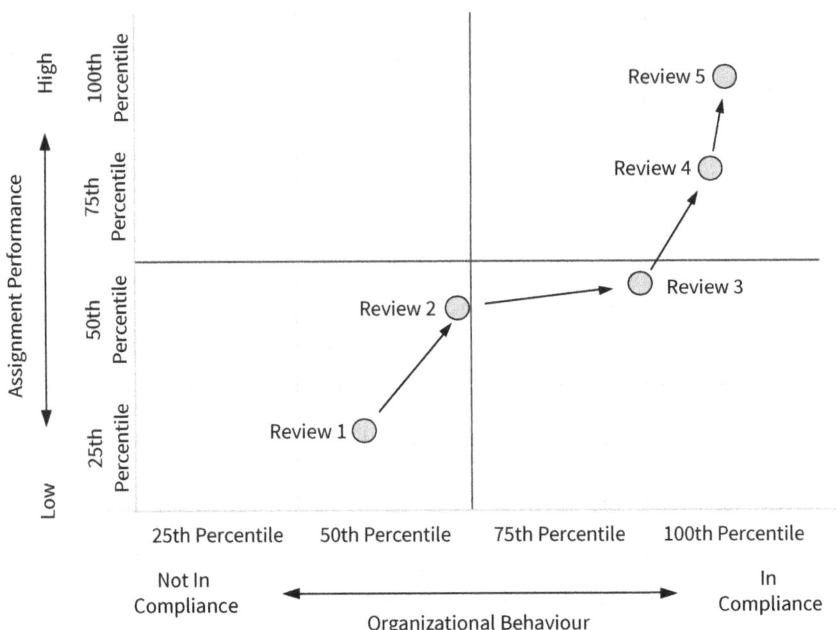

FIGURE 7.2 Two-factor employee performance review structure.

Note the goal is for you to move your assessment continuously up and to the right. This indicates improvement over time, which is what your supervision is looking for.

A person's location on the graphic is the result of a preloaded algorithm with numerical data input by the manager or supervisor, and the goal is to move the assessment up and to the right. Very neat and tidy.

The result of this objective analysis is then compared to other engineers in a cohort of similarly grouped employees, perhaps based on job role, level in the company, their home department, experience, or any other of the dozens of ways management can group individuals. From there, a relative (or "forced") ranking can be constructed, with engineers falling into "clusters" with other similarly rated employees, with pay and other benefits distributed on a sliding scale. Clean, but a somewhat mechanical way to judge someone's worth.

This system requires good communication and clear ground rules between the employee and management to avoid needless conflict. The first ground rule is to ensure you understand how accomplishment is measured and that understanding is consistent between yourself and your management. This means knowing the company's official definitions of performance and behavior. The second ground rule specifies that performance discussions should not be a once-a-year event, but a continuous and ongoing informal, verbal occurrence so you always know where you stand. This ensures performance reviews are not surprises but are consistent with what you've been hearing all year long. A third ground rule is to be sure to ask for ongoing feedback, as most supervisors and managers will not give it unless asked. Requesting continuous feedback is a small investment in effort that will pay big dividends at your yearly review. No matter the method, the fundamental point here is to communicate repeatedly and well with your management.

An important insight about objective performance systems is their strong link and dependency on the organization's corporate culture. Several years ago, a Fortune 25 multinational attempted to move its existing performance system to the then-popular (and somewhat infamous) General Electric "A/B/C" ratings system. This system, relentlessly performance-based, forced supervisors to rank employees into three categories of achievement: "A" for the top 10% of workers, "B" for the next 80%, and "C" for the remaining bottom 10% of employees. This 10–80–10 split was compulsory; all employees were required to be ranked. The rub came with the "C" employees. They were given one year to improve to at least a "B" level; if not, they were released. Of course, if one employee is elevated to a "B," then another employee is demoted to a "C." This Darwinian rating system apparently worked within General Electric's hard-nosed corporate culture. After moving to the General Electric system, that same Fortune 25 multinational, being a family-owned company, found this approach unacceptable to their culture and scrapped the system in less than two years, a remarkable move. The General Electric system was a substantial failure for that company: it did not fit the culture.

To do these objective methods well requires the supervisor to have a very good knowledge of the individual, complete notes of positive and negative work examples, and careful wording of the official document, least the employee dispute the ranking. This takes time and effort, meaning it is commonly done only once a year. But management needs a useful yet informal system of rating their people, developing them, and ultimately deploying them to best advantage.

Fortunately, there is a model that does this, and it's useful to understand it. Developed by Ken Blanchard and D. Johnson et al., and called the *Situational Management Model*, this concept allows supervisory-level management to informally assess your worth to the group by evaluating *confidence* in the technical quality and correctness of your work and *comfort* with you as a growing professional and with your fit within the group.[3]

Here's how it works. Supervision tends to bin their employees into one of four categories, based on the combination of an engineer's willingness to perform required technical tasks and their ability to actually accomplish the work. As shown in Table 7.2 and Figure 7.3, this combination results in four evaluation classifications that are easy to evaluate:

The Situational Management Model is simple, widely applicable, and easy to remember. In short, it works.

We've mentioned the rule of seeing what needs to be done and just doing it. It's time to expand that rule a little bit. Early on, you will be assigned jobs and assignments that cleanly fit into your published role. But it's important to demonstrate a willingness to comfortably volunteer to help with assignments outside your assigned role. And this means meaningful tasks, not planning the company picnic.

Department-level operations are not all the time well planned out. Many assignments come in like meteors: they are identified only at the last moment before their impact causes a lot of damage. If your colleagues are not ready for them, they will be scrambling. And in crisis situations, there will always be some assignments that

TABLE 7.2
Situational Management Model Characteristics

Supervisor/Manager Action	New Engineer Characteristics: Unable and Unwilling	Unable and Willing	Able and Unwilling	Able and Willing
Supervisor assesses current capability of engineer on their Willingness vs. Ability to do the job (Figure 7.3a)	This is a nightmare group for the Supervisor. Not only does the new engineer lack ability to perform the job, but attitude toward it is also poor. Whatever the reason, the Supervisor perceives a poor fit between the new engineer and their department. This situation would be equivalent to a military draftee: they don't want to be there and are not willing to invest in the job.	This is where almost all new employees will start. Enthusiasm toward the job and its tasks is evident, but they do not yet have the technical skills to meaningfully contribute. This is the classic new employee starting point.	This is an interesting case. Here, the engineer is perfectly able to do the task assigned; they just don't want to do it. This is the archetype of the burned-out employee. Normally longer-term, experienced employees, they are no longer interested in performing the job, even though they may still be very good at it. Boredom, lack of other opportunities or unpleasant past experiences can drive an engineer into this zone.	The golden engineer. This is a high-performing employee who demonstrates both excellent technical skills in their job performance and high level of enthusiasm and social "fit" inside the local group. These engineers can almost supervise themselves and are in high demand both inside and outside the organization.
Results in Supervisor performing a Comfort vs. Confidence assessment of the engineer. (Figure 7.3b)	Normally results in Supervisor being personally uncomfortable with the employee and not confident in their work.	Results in Supervisor being personally comfortable with the employee but not yet confident in their work.	Results in Supervisor being personally uncomfortable with the employee but confident in their work product.	Supervisor is both personally comfortable with the employee and confident in the quality of their work.

(Continued)

TABLE 7.2 (CONTINUED)
Situational Management Model Characteristics

Supervisor/ Manager Action	New Engineer Characteristics			
	Unable and Unwilling	Unable and Willing	Able and Unwilling	Able and Willing
Results in Supervisor coaching and counseling engineer through creating direction and support plan. (Figure 7.3c)	The supervisor gives very specific direction on how to complete assignments, but will not invest in explaining why it's important, nor explaining why it should be done as specified. Just do it and be quiet.	Supervisor gives high direction though detailed instructions and training on the assignment, and also provides high support through sharing their time and detailed reasoning regarding how the assignment fits into the organization's operations.	For this burned-out engineer, Supervisor would provide low direction (since they already know how to perform the task) and high support in an attempt to re-motivate this experienced employee. Means understanding their mindset while respecting them as a professional. Hopefully, this support moves them into a different quadrant.	These ideal engineers need very little direction, as they already know their technical tasks, and very little support, as they are already motivated to do the task with quality and timeliness. The supervisor just stands back and admires their performance.
Ultimately results in Supervisor taking final action of: (Figure 7.3d)	Supervisor not investing in employee. Encourage employee to self-select out. Take decisive action if employee declines.	Support engineer's move to Able and Willing if deserved. Guard against negative influence from Able and Unwilling.	If performance plan succeeds, move to Able and Willing. If not, terminate or transfer to more appropriate department.	Provide opportunities for promotion and enhancement. Help identify career path, support employee plans. Advocate for promotion if deserved.

don't require specialized technical skill. What they need is somebody to step up, take it on, and do it fast. That's where you come in. Be sure to get to the person in charge and volunteer to help with anything. And I mean *anything*. If it means creating some simple slides, do it. If it means collecting telephone numbers or email addresses for a Zoom call, do it. If it means setting up a conference room and ensuring the communication links are working, just do it.

I'm continuously amazed how certain new engineers hesitate when asked to do something that's not "engineering." Your value is not only in your formal role but in your willingness to do anything at any time, provided it is not illegal or immoral.

So jump in.

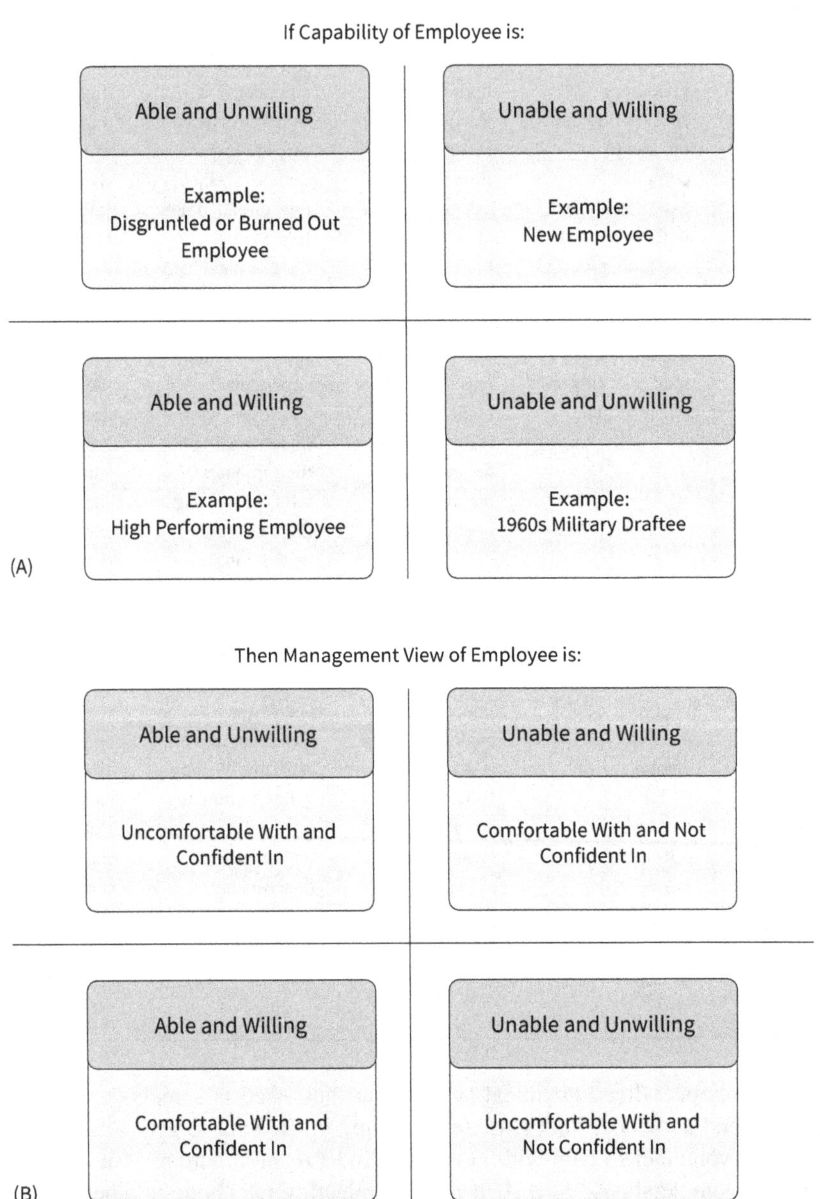

FIGURE 7.3 General Situational Management Model. (A) Situational Model starting point. (B) Management assessment of employee.

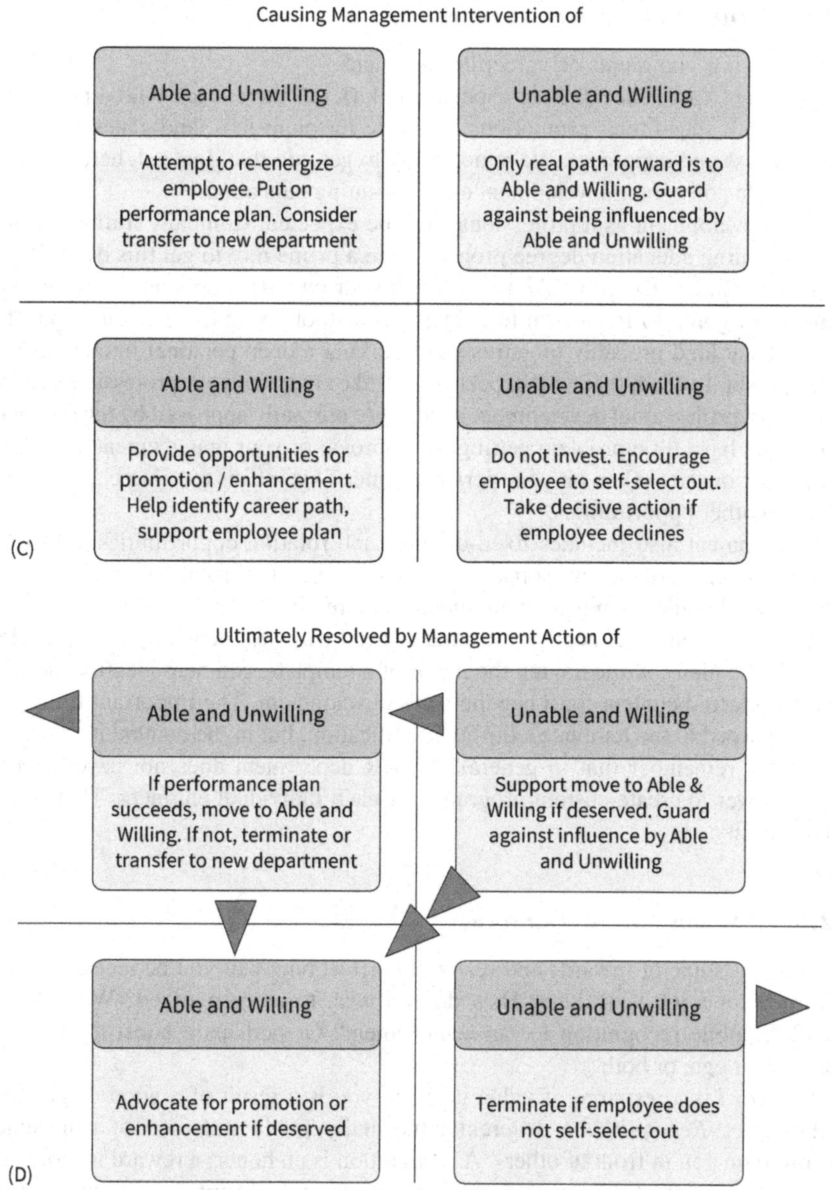

Causing Management Intervention of

Able and Unwilling

Attempt to re-energize employee. Put on performance plan. Consider transfer to new department

Unable and Willing

Only real path forward is to Able and Willing. Guard against being influenced by Able and Unwilling

Able and Willing

Provide opportunities for promotion / enhancement. Help identify career path, support employee plan

Unable and Unwilling

Do not invest. Encourage employee to self-select out. Take decisive action if employee declines

(C)

Ultimately Resolved by Management Action of

Able and Unwilling

If performance plan succeeds, move to Able and Willing. If not, terminate or transfer to new department

Unable and Willing

Support move to Able & Willing if deserved. Guard against influence by Able and Unwilling

Able and Willing

Advocate for promotion or enhancement if deserved

Unable and Unwilling

Terminate if employee does not self-select out

(D)

FIGURE 7.3 General Situational Management Model. (C) Management intervention choices. (D) Final management resolution.

7.2.5 Notes on Career Development

Just a very few comments on career development.

Armies of career development specialists lurk out there, each with a promise to the reader to speed their promotions, compete for plum jobs, and generally ensure their success and happiness. We're not going to get into that. Instead, here are just a few thoughts on career development and continuing education.

Your development as a professional will be expected. Company training classes and continuing education degree programs are a prime tool to get this done. But be aware you must personally take the lead for your own development. Your management is not going to invite you to enhance your toolbox or formal education: they are too busy (and probably too stressed) to taking a deep personal interest in your development. Instead, they will expect you to take charge of your professional future. Since most professional development actions are normally approved by the company, you should have little problem getting the approval of your management, especially if these actions are taken on your personal time. This means evening classes, seminars, and other opportunities.

Development also includes fixed-duration job rotation opportunities, taken at a relatively decent frequency. Some companies have formal rotation programs. This may be worthwhile but normally are not strictly required.

HR departments frequently have preapproved paths for new engineer's development. These plans, while having the flavor of a template, can help identify the company's preferred content for a new person's development. The important thing is to not be chained to the template's timing specification, but in the content it provides.

Finally, remember that in general, the HR department does not have the time or manpower to create custom programs for each individual engineer. That is your responsibility.

7.2.6 Rewards and Recognition

Ah, yes, the topic of rewards and recognition. Just how will you be recognized and rewarded for a job well done? How do you wish to be recognized? What do you want? A public recognition for an achievement? Or perhaps a boost to your paycheck, your ego, or both?

A *reward* is a resource of value given to you is a result of a good or great job performance. *Recognition* is different, a (normally) public statement of achievement bestowed on you in front of others. A recognition is an honor; a reward is more concrete. A reward can be given as part of a recognition. Unfortunately, many times (or perhaps even in a majority of instances) rewards and recognitions are misapplied, as they are generally not well understood by management.

All engineering companies will say that recognition and reward is a necessary and important part of their human resources toolbox and is a legitimate and sincere expression of achievement. Yet how this expression is articulated is contingent. For large firms, reward programs are normally outsourced to companies that specialize in administering these programs on behalf of the organization. Because your management may not necessarily see reward and recognition as a core part of a firm's engineering activity, recognition is offloaded to someone else and made as fast and efficient (i.e., cheap) as possible. A typical example is the popular "catalog rewards program" used by hundreds of large companies in North America and elsewhere.

Say you did an outstanding job on an urgent, high-profile assignment and your supervisor wishes to reward you for your performance. With these catalog programs, your supervisor accesses a website, types in your name and employee number, selects an amount, and presses return, an investment of approximately 53 seconds. Within a week, a pro forma email arrives informing you of your honor, with the facsimile signature of the vice president of Human Resources attached. In the email is a code number with an amount. You are instructed to visit a certain website, select a gift of some type, and then use the code to order the gift which comes in the mail. Enclosed with the gift is an auto pen signature from the CEO, assuring you that the company's success is a direct result of your contribution. Obviously, this system is not true reward and recognition. This is a low-cost way of keeping management from spending time on true admiration and celebration.

Meaningful recognition is (like most things) contingent on both the person receiving the recognition and the achievement being highlighted. Types of rewards are many and very diverse. They can range from money (increased wages, one-time bonuses, or profit-sharing), enhanced career attainment, promotion, better amenities (perhaps a more capable computer or other digital device), additional vacation, recognition through praise or awards valued by the employee or intrinsically meaningful work assignments. Untold other forms of recognition exist. The key is the recognition should be both motivating and enriching to the recipient, not the company.

Tom Siligato was a supervisor of engineering during the design and development of a new luxury vehicle program. Tom's most valuable engineer was Rob Mierle, and together were performing feasibility on the interior and exterior of this vehicle (feasibility being the analysis of all appearance parts so they can be manufactured and assembled with quality). Their overall boss was Terry Carlson, director of project management who ran the entire program. Carlson was demanding and looked to Tom and Rob for fast analysis and high-quality results over an extended period of time. Carlson placed tremendous pressure on them to accelerate their feasibility timing even faster than the schedule, as he wanted to polish his own reputation as an overachieving, "get it done" director. So even though they were on schedule, they were (in Carlson's eyes) late.

Tom had booked his annual vacation in the middle of this intense period. Tom assigned Rob to act in his behalf for the two weeks of Tom's absence. Unfortunately, a critical review and approval event was scheduled to take place while Tom was away. In this review of about 60 people, Rob was representing Tom and their group. When it came time to review and approve their feasibility work, Rob reported the truth: while they were on time against the baseline schedule, they were not on time or ready to be approved for the accelerated schedule.

What happened next became a "war story" told for years afterward. Carlson publicly berated Rob for being late, incompetent, and not capable of being part of this program. Some people present say Carlson even called Rob "stupid." All told, it was a brutal dressing down.

When Tom returned from vacation, several people immediately told Tom what happened. Rob had performed admirably and endured this harsh public reprimand well. On hearing the news, Tom sat down, and with different colored construction paper fashioned a paper "Medal of Honor." About 3 inches wide and 6 inches long, this paper medal was decorated with words like "Great Job" and "Courage under Fire." Tom then called a meeting of their group, asked Rob to come forward, and in front of everyone pinned this paper medal to his shirt.

Rob loved it. He had it framed and for the next 18 months proudly displayed this paper medal on his desk. The lesson here was by taking the time to create a unique, one-of-a-kind reward and publicly honoring him for the bullet he took on Tom's behalf, Rob received a truly meaningful recognition. And he valued that medal more than any gift from some nameless catalog. Tom really appreciated Rob, and Rob really appreciated the gesture.

While meaningful recognition is a good management practice your bosses should use, there is no such thing as a "perfect" reward or recognition. And it is certainly impossible to please everybody at one time. What is important is the perceived equity between the accomplishment and the recognition, not absolute sums. Reward systems should be tailored to the individual, and clear, simple expressions of appreciation are better than "make it fit," rote gestures.

7.2.7 Flattening the Organization: Job Promotion vs. Enhancement

I'm willing to bet that, early in the recruiting process, you had a discussion about promotions: the time it takes to get one, how much would any pay increase be, what do you have to do to earn one; all the questions every employee asks going back to antiquity. And those are very good questions. Yet over the past several decades a purposeful and premeditated trend among large organizations, including engineering firms, is the idea of "flattening" the organization. You may have heard the term in describing "driving decisions down" into the organization or increasing employee responsibility. Whatever name is used and whatever rationale is given for this flattening trend, one outcome is obvious: there will be less opportunities for promotion. If the span of control in a firm was 6 to 1, and flattening now makes the span 10 to 1, it just means fewer open positions at the next level and greater competition for the spots that remain.

This is not a secret to anyone. Supervisors, managers, and directors all have had these conversations over the past decade with their employees. One of the more creative solutions to this reduced promotion opportunity is called "job enhancement." Here, the employee stays in the same location on the organization chart, but the nature of the work changes. Now, this is not just another phrase for "doing more work for the same pay." Job enhancement is different. Insightful managers and supervisors know their people. They know what each individual engineer gets excited about, the assignments that give particular satisfaction to their employees. In short, good management knows what their direct reports like and what motivates them.

Knowing that there may be a longer time between promotions, supervisors will adjust assignments for each engineer, sending them desirable, unique assignments as available. In other words, these engineers receive more meaningful, important, and higher-status assignments, chosen by what is truly important to them.

Let's take our friend Greg Gullette, a highly regarded engineer hired about 18 months ago. Greg's specialty is data visualization, and his professional interest extends to designing new visualization techniques that simply and accurately convey sophisticated data quickly. Greg's supervisor judges Greg to be a high-potential employee and would like to reinforce Greg's success so far. Knowing it's too soon for Greg to qualify for promotion, his supervisor instead selects an assignment that might've gone to a more experienced engineer: designing a visualization tool using "4D" analytics, a new field that promises a great leap forward in engineering analysis. As no one in the company currently works in this area, Greg has the opportunity to become the company's internal subject matter expert on 4D. The status and importance of the assignment reduces the emphasis on waiting for a promotion. Greg's job is enhanced.

7.2.8 The Dual Ladder

It is almost certain that you, at some point, will be part of an organization employing many technical researchers, engineers, and scientists at a number of levels. Normally located in separate Research and Development or Product Development groups, these technical specialists thrive on inventing new materials, innovating new digital methodologies and investigating whatever their curiosity demands. These technical specialists are the company's engineering elite.

Yet this means you may face a dilemma in your long-term career direction, a future choice that can be both stimulating to you personally and productive for the organization.

A Technical Specialist means just that: a specialist in inventing, innovating, developing, and applying new technologies. Their skill set is deep and very specific: fulfilling a niche in a wide technological territory. In theory, technical professionals are motivated by desire to contribute to their fields of knowledge and to establish distinguished reputations within their disciplines. They are not the same as management, those supervisors and managers who direct the organization. Their skill sets are different, their goals are different, and their worldview is very different. Management desires upward mobility in the organizational hierarchy by focusing more on achievement of company goals and acquisition of organizational approval than technical excellence. Technical Specialists covet the opposite.

For decades, traditional companies with embedded technical subgroups had but one career path open to engineers: a traditional management path that moves high-potential engineers ever upward, performing less and less technical work and more and more purely management functions. Technical work falls away at each level until the engineer becomes a full-time administrator.

But what if your professional aspirations are technical in nature? What if your goals are in conflict with the organization's available career paths, especially if you are working in a large organization with many non-technical employees? In these cases, going into management creates a built-in conflict of values, where engineers with a technical value set are placed into nontechnical occupations, with values and definitions of success that differ significantly from those in technical organizations.

The *Dual Ladder* system is an organizational arrangement designed to solve this problem by providing meaningful rewards and alternative career paths for technical professionals.

The dual ladder is a specific corporate policy of providing promotions and rewards along two parallel hierarchies: advancement in a traditional managerial career path or advancement as a technical professional. The purpose here is to provide more equal status and rewards at equivalent levels in both the technical and traditional managerial hierarchies. In short, the dual ladder rewards technical professionals (especially scientists and engineers) for outstanding scientific and technical performance without having to remove them from their professional discipline. The major difference between the two is that traditional managers always have engineers reporting to them: pure technical specialists do not normally have staff working directly for them. This organization arrangement is shown in Figure 7.4.

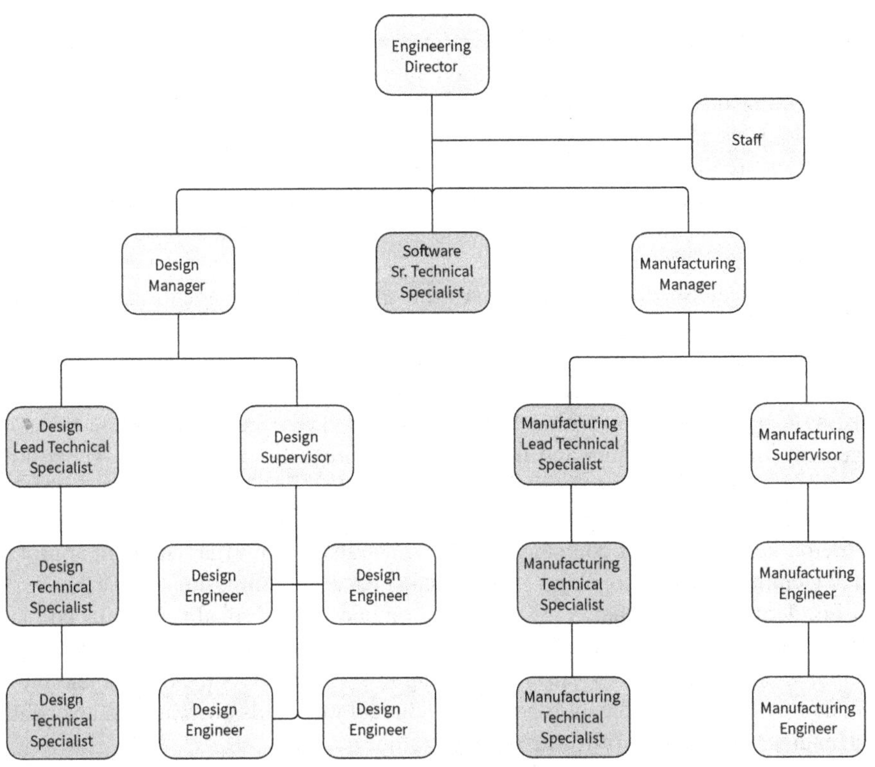

FIGURE 7.4 Dual ladder organizational structure.

While this sounds pretty good, there are some cautions to consider when considering following the dual ladder. A new engineer will probably not have a well-defined notion of success in their early days, so their first-level supervisor will play a strong role in shaping the new hire's view of the dual ladder system. Technical specialists will sometimes bias their advice toward the technical route. The opposite is also true: supervisors who are on the managerial ladder may unintendedly influence an employee to follow that path. This is natural: people advocate for what they know best. Just be aware that some subtle bias may occur during your career discussions.

Another caution is the tendency for inertia to occur in your career development when you are placed on one side of the ladder or the other. Technical new hires tend to be placed in technical specialties populated by other technical specialists, providing a bit of a cocoon that keeps the technologist within the technical community. Mirroring that, engineers placed on the managerial side of the ladder may tend to be kept on the managerial track. If this is your desire, all is well. If not, if you intend to move between managerial and technical tracks, you must establish linkages across the two tracks so that you are known to both sides. Being known is the key to moving back and forth. Just like the actor who takes only villainous roles or just comedic scripts, you will become "typecast" in your role in the organization. And you want to be sure you are typecast in the role that is right for you.

7.2.9 Thee versus We: Your Place in the Team

The word *team* is loaded with meaning, and endless books, blogs, and articles discuss what is their purpose, how teams should be organized, how they work together and generally get along. Your entire university career was undoubtedly filled from morning to night with teams: project teams, study teams, presentation teams, design contest teams, and probably teams washing the dishes in your dorm cafeteria. That's fine and I recommend you check out any of the many, many books on "how to do" teams. At this point, it doesn't matter which one.

Here, though, we will take a little bit different view regarding teams. Not all teams are created equal. In many organizations any group of employees are called a team, even when they are not. For a team to be legitimate, we must be very careful in defining what we mean, what we want from them, and how we will operate within them. This is hard, and very, very few groups of engineers can truly be called teams.

Let's approach teams as a tier concept, where each level in the tier refers to a type of team. Dependent on the level, the appropriate approach, way of operating, and expected result will vary. We can think of teams as populating four basic tiers:

The Pseudo Team

Also known as the "One Team," this is a group of people (sometimes a very large group) that are a team in name only. None of the typical characteristics of a real team are present. An example of this is calling an entire corporation or company "Team 3M," "Team Cisco," or "Team Tata." The pseudo-team may be very large, but it doesn't matter except for the number of T-shirts ordered: these are simply corporate advertising slogans and are not teams by any definition.

The Project Team

A project team normally takes the name of the project or product they are working on. Typical examples might be "Team Cyborg" or "Team Deep Magic." These can also be very large, but at least they have some sort of goal or identity to muster around. A project team tries to include everyone involved and is acceptable in some situations such as when building a strong subculture.

The Near-Ideal Team

These tend to be small and have expansive characteristics that actually promote a valid team attitude and action. These teams should be flexible, create mutual trust, and create a subculture where everyone truly pulls their weight. Their link to corporate management is loose and more freedom of action is their advantage. A good clue to its existence is if the team leader knows everyone on the team.

The Ideal Team

The rarest type by far, it is small, confident, and creates high anxiety for their management. Known for self-direction, autonomy, and creativity, these are highly sought-after positions where everyone knows everyone, and trust is strong and exists everywhere. The ideal team is like a Los Angeles street gang. It sets its own goals. Its members decide how to achieve those goals. It decides who is on the team and elects its own leader. And, most importantly, it is self-funded. If you believe you are part of an ideal team, just apply the street gang test: you may be surprised.

These four descriptions do not begin to describe the characteristics, uses, and appropriateness of each team type. Teams are complex creatures, and Table 7.3 more completely describes each team's characteristics.[4]

Everyone knows you will be assigned to at least one, if not several teams immediately upon arrival. While not initially assigned as the leader of a team, you will need to judge what kind of team you are placed in, how does the team operate, what do they need, and, most importantly, is the team set up for success or not. Making these judgments helps you adjust the way to contribute to the team in a legitimate and meaningful way.

Note that ideal or near-ideal teams can have another name, even though that name is very rarely used today. In engineering it's known as a *task force* or a *tiger team.* These are defined as special groups brought together for an urgent, distinct, singular purpose (read this as "crisis") and are expected to disband immediately after completing their task. The subculture of a task force tends to be hard-edged. The goal is paramount: little slack is cut for extraneous discussion, rambling speeches, or other wasted time or effort. The common goal of tiger teams or task forces is getting the assignment done extremely fast: time is of the essence. It also implies that tiger teams only select the most talented or knowledgeable people to be members, creating their own subculture.To be selected for a task force is normally seen as an honor.

Teams can also experience *teamicide*, the destruction of a work team due to any number of reasons. These reasons can be secrecy, defensive management, strong bureaucracy, or fragmentation of people's time or effort. It can be caused by physical separation, silly

TABLE 7.3
Characteristics of Engineering Team Types

Characteristic	Pseudo Team	Project Team	Near Ideal Team	Ideal Team
Size	Can be very large	May be very large due to effort to "include everyone"	Will be small. Depends on leader at least knowing everyone on team	Small, could just be two or up to seven. Only as a stretch have up to twelve. Everyone knows everyone
How Started or Determined	Depends on organizational philosophy / structure	Depends on task to be completed	Company culture is oriented to teams. A chicken or egg question: teams lead to culture change or culture change leads to teams?	Teams are the only mode of operation. Teams appear by themselves
Source of Members	People selected without much choice	People slotted according to project needs	Ranging from some to much self-selection	Self-selection
Leaders	Chosen on an organizational basis	Chosen for suitability to project, on technical basis	A visible leader may be selected by some of the team	Selected by team
Mix of Men and Women	Set to obey EOC guidelines	To satisfy project	To satisfy team. Best person for the task	To satisfy team
Location	Historical. Altered to bring in people if budgets allow	If people not located together, then arrange for key periodic meetings and digital linkages	Together	Together but immaterial
Goals	Set by company. May be unrealistic	Set by company, generally realistic	Realistic	Set by team. Enthusiastic and realistic adoption
Budgets	Goals sometimes subordinated to preordained budgets	For project	Difficult but project related	Difficult for team and management to obtain. Needs special attention
Lifetime	To suit organization's needs	To suit project's needs	Generally for a project or task need, but not iron-clad	Unlimited until team self-dissolves or an individual team member is "unelected"

(Continued)

TABLE 7.3 (CONTINUED)
Characteristics of Engineering Team Types

Characteristic	Pseudo Team	Project Team	Near Ideal Team	Ideal Team
Performance Reviews	Uniform company system	Uniform company system and project objectives	Varies, Project objectives with some individual recognition	Only of team, unless team selects an individual for recognition
Rewards	Individual; formalistic and organizational. Coupled to performance review	Individual, influenced by project success. Coupled to performance review	Individual, directly connected to success, but over longer time period. Coupled to performance review	Collective except when team selects an individual for special attention
Individual's Promotion	Promotions as part of system regulations	Promotions from home base per system regulations	Complex. Promotions only between team assignments	Only as suggested by the team. Somewhat unimportant
Individual's Career	Individual's performance	Mixed, but mainly individual's performance on project	With team, may be promoted away independent of PR	With team until individual elects to leave or management needs individual to impact elsewhere
Individual's Style	Closed. Fits characteristics to those expected by organizational profile for success	Mixed because of split allegiance to team and outside home base	Open to other team members, may discuss with "the outside"	Completely open within team with no local consequences. Will not reveal any of the inside workings to "the outside"
Individual's Attitude	Resigned, perhaps secure in professional / functional role	Less secure and suffers from split between needing a home base and enjoying teamwork	Enjoyment is important and success must be individually fulfilling	Fun and rewarding, or team member would not be there

deadlines, or powerful cliques. In particular, secrecy is one good way to kill a team. If team members are secretive with each other, teamicide will soon follow.

This discussion is not intended to encompass the entire territory of teams but to spark some thoughts as you are assigned to and operate within your firm's teams. Begin to learn the individual characteristics of each to make your time there productive and advantageous.

7.2.10 When Is a Mentor Not a Mentor?

Personally, the topic of mentorship drives me a bit crazy. Most HR departments provide what they call "mentorship" programs for their new employees. Colorful, stylish websites containing beautiful process charts accompany these programs, hoping to give new employees the sense that company management is truly invested in the arc of your career.

Unfortunately, these programs (which I'm sure you've already read about on that website) are not mentoring programs at all. For true mentorship has a precise and important meaning.

Mentorship is a very specific kind of employee guidance. It is special, unique, and frankly, somewhat rare in actual practice. In contrast, your company's definition of mentorship is something completely different.

A company mentorship can be boiled down to what that website talks about; a monthly meeting with a mid-level engineering manager designed to answer questions and receive company-approved information that is safe and true. This is a HR program designed to answer the decades-long complaint by new hires for career support from experienced employees. And for routine matters and surface-level advice and guidance, this is perfectly fine. After all, most of this advice consists of telling you to look at a given webpage. And it's very cheap: no real investment is made by the organization into this kind of development. In short, this type of program is convenient but lacking in the quality of its information. This is not real mentorship.

The guidance gained by true mentorship, however, is golden. *True mentorship* is a relationship between two individuals: a person asking for advice, guidance, and assistance, and an experienced employee with the capability, power, and willingness to invest in that person. In short, a mentor has one job: *to tell you the secrets of the organization.* That's it; it's letting you in on the mysteries of how things really work and provide you rich information on what in the world is going on.

Obviously, mentorship is not limited to engineering firms only. All organizations, whatever their goal or purpose, have mentors and mentees. An interesting story about mentorship comes from the book *Man of the House: The Life and Political Memoirs of Speaker Tip O'Neill.* O'Neill was the 47th Speaker of the U.S. House of Representatives. As Speaker, O'Neil was responsible for passing (or blocking) all legislation before the U.S. House, controlling which legislation was brought forward and passed and those bills that didn't. No law was passed in the U.S. Congress unless O'Neil agreed and supported it.

In the early 1980s, the American automobile manufacturer Chrysler was in deep financial trouble. Chrysler was on the edge of bankruptcy and out of financial options.

As a last resort, Chrysler was about to ask the U.S. government for $1.5 billion in loan guarantees to avoid liquidation. To that time, it would be the largest rescue package ever granted by the U.S. government to an American corporation. This was a highly contentious issue, as many House lawmakers were ideologically opposed to providing any government guarantees to private companies whatsoever. Yet other lawmakers felt to let Chrysler fail would deeply hurt the U.S. economy. At the height of this crisis Lee Iacocca, the new chairman of Chrysler, asked to see Speaker O'Neil in his Capitol Hill office to discuss a potential rescue package. O'Neil picks up the story:

> "When Iacocca came into my office to tell me about Chrysler's problems, he brought an entire entourage, including members of his board, lawyers, and lobbyists. As soon as everyone was assembled, Iacocca held forth on the various pressures affecting Chrysler and outlined his plan to bring the company back to life. When he finished, I thanked him for coming in and explaining his situation."
> "The following day a mutual friend called O'Neil to ask:
>
> *Friend:* "So what went wrong?"
> *O'Neil:* "What do you mean?"
> *Friend:* "He [Iacocca] called me and said you were the coldest bastard he ever met."
> *O'Neil:* "What did he expect? He came in with a whole damn army. Do you think I'm going to tell them how to get the job done in front of all those lawyers and lobbyists? They'll just take credit for my ideas. Tell Iacocca to come back and see me, just the two of us, head-on head, and I'll tell him what to do."

When Iacocca returned, O'Neil told him quite specifically how to pass bailout legislation that would get bipartisan support. O'Neil would only tell Iacocca the secrets of how to get legislation passed if it was done in private. Iacocca took O'Neill's advice point by point and Chrysler got their loan guarantees. A simple lesson in mentorship saved tens of thousands of U.S. jobs and helped keep Chrysler viable throughout the next three decades.[5]

So how do you know if you are in a true mentorship relationship or a HR program? It's actually quite simple to judge, as shown in Table 7.4.

There's an easy way to think about mentors and mentorships. Unlike a company program, a true mentorship is a relationship, built on trust and mutual respect. The mentor invests personally in the recipient, supports them in their goals, and takes pleasure in their success. In many cases the mentor wishes to "pay back" for help they may have received on their own personal journey. Whatever the reason, the mentorship relationship is extremely valuable and special. If you come to have one, treat it with great care and attention.

7.2.11 Hidden Testing, When You Least Expect It

Perhaps you thought that once you left university, you would never be tested again. This is certainly not true. You will be tested, and not at a time, place, or subject of your choosing. True, you will not be traditionally assessed using a web-based, multiple-choice,

TABLE 7.4
Comparison of HR Mentorship Program vs. True Mentorship

	HR Program	True Mentorship
Who assigns the mentor	Company	You find them; they may find you
Meeting location	Mentor's office or conference room	Off company property or open area
Meeting frequency and duration	Exactly 30 or 60 minutes, about once a month on a set schedule	Varies greatly in both frequency and length
Appointment reliability	Low. Frequent postponements and reschedules	Good. Fewer postponements or reschedules. Mentor has genuine desire to meet
How questions are answered	Most questions referred to company website, or "just ask so and so" person	Honestly and sincerely. Mentor will sometimes offer to find the answer themselves
Visibility and availability	Extremely low. Won't see advisor except at appointments	Frequent
Personal information exchange	Nearly zero	Appropriate and realistic amount; enough to build trust

video-enhanced site or asked to write an essay on the history of C++ coding, but you will be quizzed at least once (if not several times) a week in an ongoing oral exam. This normally happens during the well-known and beloved "one-on-one" meeting, where your boss will actively check on your work, with your answers verbally provided on the spot. Actually, this sounds much worse than it really is, as these questions and answers will be embedded in a normally pleasant conversation about your current and future results. But this meeting has exam-like characteristics, and your grade will be built up day after day, week after week in these minor, but cumulative, examinations.

This "one-on-one" or individual meeting with your boss is 95% universal. It's designed for your supervisor to handle daily business with you: to assign tasks, review your progress, and generally try to understand you as an employee and as an engineer. These normally take place each week, and, because we're using digital scheduling devices, will last precisely one hour. Your supervisor may set the agenda for this meeting (an agenda here just means a list of topics he or she would like to discuss with you, and vice versa). As time goes on and the supervisor becomes more comfortable with you, they may have you design the entire agenda for the meeting. Again, these do not need to be formal agendas, but a general list of important items that you need to touch base on.

We'll talk more about the one-on one meeting in Chapter 10, but for now please practice for these individual meetings featuring unscheduled questions and other exciting conversations. Because there is a reason: you will be assessed when you least expect it.

Sometimes you may be tested by your immediate supervisor in a way that you do not know you're being evaluated. Some of these tests will be a spur of the moment, while others will be designed in advance.

One master of the hidden test was an engineer by the name of Norm Zeiger, a very experienced, no-nonsense, and respected supervisor. Norm was a straight shooter and didn't suffer fools gladly. Because of his experience and approach to managing engineers, many new college graduates were assigned to Norm for training. And Norm did not disappoint.

Norm had a trick he used on all his new graduate employees. He would select a technical topic, say, the design of the new automotive headlamp. He would tell the new engineer to coordinate with a technical lighting specialist in an adjoining building. What Norm did not tell the engineer was the lighting specialist was a close friend. Norm would call up his buddy and tell him, "I'm sending over a new hire. Be sure to give them a hard time and let me know how they did." Inevitably, the new hire would meet with Norm's friend and be subjected to an aggressive, in-depth technical question and answer conversation. As promised, Norm's friend would report back on how the trainee performed. Norm would then hear the new employee's account of the meeting, evaluate the two different stories, and then counsel the trainee on what they may or may not have done correctly. Norm would never let the new engineer in on the secret, but each trainee certainly remembered the experience. As a result, Norm was very effective in developing his people through these tests, and the rest of us loved him for it.

7.2.12 Characters You Will Meet

We can't finish this section until we talk about some interesting characters you will undoubtedly meet in your new engineering firm. Depending on the size and age of the firm, you may not meet these people all at once, but the odds are overwhelming you will meet at least a few of these interesting personalities over time. These are my own classifications, and you will certainly create your own catalogue after bumping into these people in hallways, meeting rooms, elevators, the cafeteria, and on business trips. Here are a few brief sketches:

The Intrapreneur

Some of your most interesting engineers. We all have a good idea of the entrepreneur; a self-starting rule-breaker who looks for opportunities and takes the personal risk in developing new products or services. The intrapreneur is the corporate version of the entrepreneur. You know these people because you will never see them. They will never be in their office; their office will be an absolute mess and their voicemail will constantly be full. Why? Because they are busy looking for the next new thing, which they know is not in their office. And when you do bump into them, they are moving at the speed of light and quickly disappear. In a word, they are the ghosts of the engineering firm.

The Technician

The technician is just what you might think. Relentlessly rational and highly dedicated to whenever technology they are working on, technicians are essential "get it done" individuals who are worth more than their weight in gold when it's time to deliver a new technology. These are excellent people to get to know.

The Bean Counters, aka Pencilnecks

Oh yes, our finance friends. Finance people are overwhelmingly well intentioned, friendly, and genuinely earnest about their role in the company. Many feel a personal responsibility for minimizing company costs, be they big or little, which is certainly deserving of respect. Yet some engineers develop a bit of an attitude toward finance, and while there may be a natural conflict between the finance people and any other department within the firm, respect for their professionalism and a sense of humor really goes a long way with these folks (despite their nicknames).

The Malcontents

These people are corrosive. These are the individuals who are continuously complaining, with an overall negative outlook about the company's future, their colleagues, your company's latest products, and just about anything else having to do with your firm. Spending too much time with malcontents rubs off and can impact your own attitude, and not in a good way. Be polite and keep your distance.

The Marketing Whiz

You can easily spot the marketing whiz. These are outgoing, high-energy, engaging, and likable individuals who tend to attract attention. Get them talking about the marketing game and you may quickly discover that deep down, they might view themselves as just a little bit smarter than anyone else in the room. I suspect this view is based on a belief in their personal insight into the human condition, which is of course their job. If they could speak to engineers in more quantitative terms, they might be a bit more persuasive.

The Walking Egos, Big and Small

The name says it all. Walking Egos, no matter what their level or position, are common: in every department, in every specialty and every part of an engineering organization. They are easy to spot in any meeting, as their comments are skewed to be ever so slightly pompous. Sometimes they are very entertaining, especially when they don't mean to be.

The Court Jesters

Always good for a joke or funny story, Jesters provide a welcome break from the daily stress and seriousness caused by the precision work of many engineering organizations. But unfortunately for them, unless Jesters have a way of instantly switching to a serious, focused, and fully professional demeanor, they will suffer from low credibility, even when their engineering insights are spot on. They trade gravitas for likability, to their detriment.

The Company Boys and Girls

Enthusiastically pro-company, many new hires initially end up in this category. A positive and energizing asset to their departments, more experienced employees will give company boys and girls a pass on their behavior, as their enthusiasm is a normal phase of their integration into the firm. These are enjoyable, useful people.

The Nervous Nellies

Some engineers, including supervisors, managers, and directors, live in abject fear of making a mistake or not being in total control at all times. To them, committing any error opens them to criticism, embarrassment, or, in their mind, dismissal. In the extreme, this belief can freeze their ability to function, even in low-risk situations. See also *nervous wreck, worrywart, hand ringer, scaredy-cat, nebbish, and bag of nerves.*

7.3 HARD RESOURCES

Fortunately, understanding and using hard resources is a much easier task than dealing with its human counterpart. For one thing, hard resources are inanimate objects. They don't have personality issues or talk back or take rash actions on their own. They just sit there and wait for somebody to do something with them.

As mentioned earlier, *hard resources* are the nonhuman tools, facilities, equipment, cash, and other physical commodities that, when coupled with human resources, assist transforming inputs into desired outputs.

While we can talk about computer systems, ergonomic chairs, fiber optic networks, test equipment, prototypes, and so many other things, the common factor of each is, of course, money. All hard resources can be transformed into money: the only questions are at what value and how fast. Money is your friend. With money, all things are possible. So, let's talk about operating within the money system your firm might use.

First, let's get our terms defined. We defined hard resources as any physical object such as tools, facilities, equipment, and other physical commodities that has worth. This includes money in any of its forms; cash or other liquid securities that has value to you today. You will hear the words "funding," "money," and "budget" used interchangeably; this is incorrect. Money is not funding. Money is immediate; funding is a promise, and budget is a schedule to convert funding into money. Your finance department, immediate management, or other authority promises you funding, scheduled and specified by a budget, to receive the money needed to complete your job. This promise is conditional. It is reexamined and renegotiated periodically (perhaps annually, semiannually, or quarterly), where progress and performance are reviewed, and if your work is deemed worthy, funding will continue.

All hard resources are fungible. Fungibility refers to a hard resource's (or asset's) ability to be exchanged for something else of equal value, in this case the ability of a hard resource to change forms: from cash and its infinite variants into equipment, facilities, data, or other objects. All resources are fungible in that all can be ultimately turned into any other form of value through sale or barter. The only difference is that each change in form changes the worth of the resource: normally a reduction in value, sometime substantial. Obviously, I can sell a piece of test equipment for cash; I just won't get anywhere near the original value of the equipment, and it may take an eon of time.

Changing a hard resource from one form to another is hard work; so hard that doing it within an organization is normally a last option that smacks of desperation. It's seen as a low percentage play and thus not practiced widely in most engineering groups. Instead, the equipment or other hard resource is usually forgotten in a corner somewhere and eventually scraped as its original use fades from memory.

7.3.1 Projects and Their Funding

A word about your work and this whole world of funding. As we've mentioned before, some engineers have a natural disposition to stay away from anything that's not the development of core technology. After all, it can be hard enough to deal with advanced engineering innovation; knowledge of funding is an unneeded burden. The money questions should be left to the money people.

Unfortunately, this approach runs counter to the aim of becoming a "professional." A professional has not only mastered their projects and core work, but they also know the peripherals: finance, timing, purchasing, warranty, service, and all the other disciplines involved in a successful project. And funding is a prime peripheral that needs to be clearly understood, if nothing more than to protect the ongoing viability of the work. That said, let's look at how a typical assignment would normally be funded by this specific hard resource called money.

When it comes to funding your work, the seven different functions of engineering identified in Chapter 1 (Research, Development, Design, Test, Production, Construction, and Operations) fall into roughly two categories: project work and operational work. Project work is the creation and development of something new, an object, service, or other effort that results in something not previously in existence. Research, Development, Design, and Construction are of this type. Operational work supports an existing process, product, or activity and includes Test, Production, and Operations.

Budgets and funding for operational work tend to be stable: required funding is essentially the same year-to-year, with perhaps slight adjustments for inflation or operational improvements. For operational work, finance management uses a *relativistic* funding technique, which merely means last year's budget level is carried over to this year, with a small adjustment made as necessary. Project funding is a different animal. Projects are defined as work which focuses on completing a task. As most projects create something new, there is no history of how much money is needed to drive forward a concept's creation, development, and delivery. Funding (along with its partner, timing) is an open question where estimates are unreliable, overly optimistic, and generally wrong. For these projects, a *zero-based budgeting* method is used, where the overall project is broken down into its component parts and each part has an estimated budget assigned. Error upon error occurs, resulting in a total budget that could charitably be called "creative."

This matters, as many new engineers will initially be assigned project work. In larger companies, you may be assigned one, two, three (or even more) projects to deliver; the number being determined by each project's size, complexity, urgency, and your capabilities, keeping you more than busy throughout the day. Since you are part of a group of other engineers, your projects will become a portion of your supervisor's project portfolio, a portfolio being nothing more than a grouping of projects under the management of a single person. Each project has a cost estimate attached which projects how much money is needed to complete the project. As just mentioned, these projections are wildly inaccurate. Nonetheless, these projections, once approved, become an individual project's final budget. And combined, they become the supervisor's portfolio budget.

This "projection, approval and combination" task cycles through ever-higher levels of the organization, creating bigger and bigger bundles of projects. Each supervisor's portfolio is combined with other peer portfolios to create the manager's portfolio; managers' portfolios are combined to create the director's portfolio and onward up the organization chart. That is, until the whole list of projects (which can easily number in the hundreds) arrive at the finance department. Theoretically, here finance reviews the entire portfolio, asks (hopefully) insightful questions, attempts to test the feasibility of the project proposals, and then sets and approves the overall budget for the entire collection.

Now here's the key insight. The decision-making authority regarding the funding level of individual projects does not permenantly reside in the finance department. Once approved, portfolio management authority is delegated back down to local engineering management, and local management has the power to shift, mix, and match funding between projects and portfolios at will. If Project A is $40,000 over budget and Project C is $60,000 under, the supervisor can shift $40,000 from Project C to cover the overage of Project A. This balancing act occurs repeatedly up and down the engineering hierarchy throughout the year and is strictly internal to the local organization. Local engineering management controls the budget. Finance is not involved: they are interested only in the overall portfolio number, not individual projects.

One more important point regarding project funding. Everyone has great ideas about what technology projects to pursue. It is a "law" of technology development that there are at least 5 times more projects ready to develop than money available to start them. This creates an ongoing, eternal struggle to choose the best, most promising technologies to develop through some kind of Darwinian contest for survival. This results in a group of worthy, but unfunded, projects called something like the "next up list." If a previously approved project is canceled or stopped, there is always a new project ready to be elevated and begun.

Why this is important is its linkage to the financial balancing act above. If a manager's entire portfolio is spending at a rate slower than initially projected (and it normally is), or certain projects are canceled during development for whatever reason, the local organization would then not spend their entire approved budget. Knowing the amount of the shortfall as early as possible is very useful, as management can now pull ahead some "next up" projects and maximize their portfolio's effectiveness.

Of course, this begs the reoccurring question of "spending to your budget," the desire to match actual spending to that wildly inaccurate projection made a year or so ago.

Surprisingly, bosses may not necessarily try to spend to their budgets exactly. This is because spending to a budget target level becomes a message to their management that falls into any of three categories:

1) Spending beyond the budget by some small amount. Many managers want to come in at some nominal value over budget (say, 3% or 5%) every year. This message says they are highly effective in using their resources (no

money is left on the table), and they need every penny of their existing budget (plus a bit more) for next year. It sets them up to argue for an increase the next budget year.

2) Hitting the budget exactly. This boss wants to be seen as absolutely in control of their work and thus be perceived as a highly effective manager. Yet hitting a budget exactly never, never happens. After all, the entire system is unstable, so reporting an exact hit is highly suspect. It leads some to believe some book-cooking might be involved. It strains credibility.
3) Spending substantially below the budget in a non-crisis year. Some managers think they are a hero by underspending their budget by a significant amount (say 10%) in a given year, hoping they will be seen as a "star" team player. In a noncrisis year, this may be actually counterproductive. It shows one or more of the following:
   a. Their initial budget projection was substantially padded, and the excess money could have been used for other projects.
   b. There is a major risk of having next year's budget permanently cut by the amount of the underspend.
   c. Their project objective was too conservative, indicating a reluctance to stretch their goals to an appropriate level.

There is one caveat here, having to do with the "noncrisis" condition. If the organization is not in a financial or other operational crisis, this message holds. If an organizational crisis exists, financial or otherwise, all bets are off.

7.3.2 Budgets and Budgets

So far, we've been talking a lot about management's tactics regarding funding. But what about your own approach to this whole world of money?

Say you've submitted a budget proposal for next year's projects. It's been approved and you're ready to go, looking forward to delivering that technology breakthrough you've long dreamed about. Perhaps that funding approval was difficult to land, but now that's in the past.

Unfortunately, you're nowhere near home yet. Approved budgets are unstable. Remember, an approved budget is not money; it's a promise of money. And like your unreliable friend who promised to pay you back that $50 three months ago but hasn't, an approved budget isn't any good until the funds transfer actually occurs. Finance (or management) have an endless assortment of reasons why they can, at any moment, delay your funds transfer, reduce the amount (the most common form of instability), or outright rescind their approval (rarer, but always there) for your funding.

Remember, project funding normally occurs in groups or tranches, and your project or task is bundled with other projects and managed as a group. And since approved engineering budgets are unstable, traditional technical firms have "annual" budgets that can be revised something like three to six times a year. If your project spending isn't proceeding swiftly after approval, your project can get caught in one of these revised budget exercises. The threat to your funding is real, as if you haven't spent it quickly, management or finance may think it must not be that important. Their response is to commandeer your unspent funding for something more "important" (or perhaps to a project with an engineer who is a better advocate than you). The lesson is clear: the instant your project is approved for funding, spend your budget.

What about the situation where your project is short of funding and you need an infusion of cash? This commonly happens, as we already know the entire budget estimating process is inaccurate. If your boss is supportive and backstops your shortfall, no problem. If not (for whatever reason), you may experience various excuses explaining their lack of support.

One technique a supervisor or manager may practice is the blanket "we don't have the budget" statement. Say you require $25,000 to purchase an updated software load for your project you didn't anticipate initially. Your budget cycle begins on January 1st and ends on December 31st. You discover you need the new software by May 28th. You go to your management and request the $25,000 supplemental funding. They immediately turn you down, cold. The stated reason? "We don't have the budget," i.e., "It's not anyone's fault, we just don't have the money."

This is as old as Methuselah. The truthful statement is: "We choose not to give you the money." Management may decide it's too much trouble to process, your project may not be deemed that important, or it may be that they don't trust you yet or you're too new to be taken seriously. Whatever the reason, "we don't have the budget" avoids uncomfortable conversations. And for the amounts you're dealing with, it is a nearly universally false statement as we already know funds can be transferred from other projects if desired.

There is a simple (but a very provocative, high-risk) way to test this reason: ask to see the budget and the year-to-date (YTD) spending against that budget. Management probably won't have an accurate YTD summary on hand, and the budget may be several months out of date. Of course, even if they do have accurate data, they probably won't show it to you anyway. Budget matters are very sensitive, and asking to see the budget and associated spending will be a major sign that you don't trust what they are saying, as well as indicating you want to rummage through (what they feel is) their own private business. And overall, pressing for funding too hard will run the risk of burning bridges. Best to withdraw and come back another day. This gives your management time to regroup and eventually find the money.

7.3.3 Economy and False Economy

Obviously, every technical (and other) organization, no matter how large or small, is interested in minimizing their cost of doing business. And it's fair to say all employees would sign up to this ground rule. After all, low expenses mean higher profits for a given revenue level, leading to (hopefully) all employees sharing in that profit. But there are areas of company financial policy where the pursuit of true savings leads to the fantasy land of false economy.

Formal financial policies are a way of imposing economic discipline on a system, and the corporate travel department has a huge number of these policies in place.

Proclamations require employees to do this, don't do that, only book travel from certain websites, only stay at these hotels, don't exceed this daily allowance for meals; these rules are pretty much universal. If you travel for your job, you will see this immediately.

Why does this happen? Because even though a company's total travel costs are generally minuscule compared to other corporate expenses, they are closely watched for two reasons. First, management may tend to think business trips are fun and subject to abuse, and second, travel costs are extremely easy to track. Thus, travel costs are the number one item to audit during normal times and cut during any business downturn. Its influence is far beyond the actual impact it has on the corporate bottom line.

One example of pursuing false economy had to do with a major factory automation firm named ARGON Systems, whose specialty was providing automated assembly systems for appliance makers. ARGON had a strict policy: travel expenses (especially for foreign trips) would be rigorously controlled.

Michelle McQuaid, an engineer for ARGON based in Atlanta, Georgia, was tasked with negotiating a new $21 million contract with their tier-one supplier located in Cologne, Germany. To finish the contract, Michelle was required to travel from the United States to Germany and negotiate a 9.7% reduction in the cost of the agreement. Michelle, though well versed in the technical details of the contract, was not a trained negotiator, and she knew she was going to have to be "on her game" for this face-to-face meeting.

The meeting was set for 9:00 on a Thursday morning, Cologne local time. To prepare, Michelle submitted a travel plan for her to leave Tuesday evening, fly east through the night, and arrive at 7:00 Wednesday morning. This would give her 24 hours of recovery time from jet lag and obtain last-minute intelligence from her local Cologne staff prior to the negotiation.

You can imagine Michelle's surprise when management turned down her travel plan. Instead of leaving Tuesday night, she was required to leave Wednesday evening, fly all night, arrive at Cologne at 7:00 AM, and immediately take a Lyft to the meeting at 9:00 AM. The reason? By policy she was not allowed to spend an extra night on "non-core company business." The fact that this plan would put Michelle into an exhausted state, unprepared, jetlagged, and otherwise poorly functioning was trumped by the need to save $207.52 for one night's lodging and $87.43 for meals.

The meeting was a mess. In her groggy state, Michelle missed a newly inserted clause that negated any hope of a reduction and, under certain circumstances, would actually increase the cost of the contract by 1.3%. Also, she misunderstood a clause stating the timing of a certain critical component would be extended by six weeks. All told, instead of achieving a 9.7% reduction, she increased the contract cost by 11% and put her product timing at risk.

This could have easily been avoided if the company offered a realistic, flexible policy that allowed exceptions due to the importance of the activity. The only thing needed was to empower her supervisor or manager to make these kinds of decisions on their own. Taking away responsibility and decision-making power from local management and making it automatic and nonnegotiable negates the purpose of the control policy. A simple solution, but not implemented.

You will see many, many examples of this kind of false economy. Your responsibility is to point this out and, by asking questions, urge supervision to address it. You may or not be successful, but as they say, "the key to change is repetition." Keep bringing these examples up when appropriate. A single win in the future begins with the effort today.

7.4 REALIZATIONS

Dealing with resources can be complicated. While human and hard resources are two separate, but related, commodities, both are of high value. Hard resources are relatively easy to deal with, while human resources require skill, patience, empathy, and an arsenal of understanding to navigate this mysterious world.

The purpose of the Human Resources department is to administer the policies of the organization and creatively solve unique, one-of-a-kind problems related to personnel issues.

These policies can include the rules controlling performance, behavior, and numerous other topics stemming from the messy interaction between an organization and its members. Since this messiness can easily spark conflict and contention, then an excellent strategy toward dealing with HR, your colleagues, and management means always leaving room for subjective judgment and understanding.

While hard resources are easier to deal with, certain hard resources are closely monitored by finance and associated purchasing departments, meaning the difficulty in obtaining and using these is more than you might think. And ensuring enough hard resources for what you need to do may take a surprising amount of your energy and time.

Yet while hard resources are inanimate objects, these objects are still controlled at some level by humans, which can sometimes create behaviors that are defensive or well intentioned but difficult. Funding is the most common sticking point, as it is fungible and viewed as the most important category of hard resource. Various individuals and departments have over time developed strategies to control and manipulate funding. Some are common, and some are specific to the organization.

When dealing with funding and budgets, just remember our discussion from Chapter 3: *nobody knows anything*, and that includes budget estimates. Management knows this. The key action in these situations is to apply creative and innovative thinking to overcome these common barriers.

At this point, let's change our perspective and address an area a little closer to home and perhaps more enjoyable: the Technical System.

NOTES

1. von Vulpen, Erik. 2022. Seven Human Resource Management Basics Every HR Professional Should Know. *Academy to Innovate HR*. https://www.aihr.com/blog/human-resource-basics/#HRMBasics.
2. Indeed Editorial Team. 2022. SMART Goals: Definition and Examples. https://www.indeed.com/career-advice/career-development/smart-goals.

3. Hersey, P., Blanchard, K. H. and Johnson, D. E. 2007. *Management of Organizational Behavior: Leading Human Resources*. Hoboken, NJ: Prentice Hall.
4. Bolling, G. Fredric. Professor, University of Michigan, Dearborn. Interview by Robert M. Santer, April 17, 2000. Transcript.
5. O'Neill, Thomas and Novak, William. *Man of the House: The Life and Political Memoirs of Speaker Tip O'Neil*. New York: Random House.

8 Exercising the Technical System

8.1 A UNIVERSE OF SYSTEMS

It's a good assumption that the Technical System is where you will begin your career. This is where your technical training from university is finally applied. For you, operating within this space is the most straightforward and comforting of all the components we've outlined, and where your performance is best demonstrated and most easily measured.

As located in Figure 8.1, the *Technical System* refers to the work processes, techniques, machines, actions, analysis, tools, and all other objects used to transform organizational inputs into outputs.[1]

The Technical System is probably the easiest way to begin to understand how your organization works and your work within it. As discussed in Chapter 5, most technical jobs are well-defined, repeatable, relatively easy to do with appropriate training, and straightforward to manage. And you will not be alone there: working directly within the technical core is where the overwhelming number of new graduates will be placed. Technical system jobs cover the waterfront: big and small, narrow or broad, so it is relatively easy for management to match the appropriate job to new engineers just entering the company.

What the technical core does is simple. Just as the team manager in the movie *Bull Durham* lectured his ballclub after a terrible start to the season:

> This ... is a simple game.
> You throw the ball.
> You hit the ball.
> You catch the ball.
> You got it?

The creation, development, and sale of a technical product or service at the enterprise level is the same thing:

> You design the product.
> You make the product.
> You sell the product.
> You got it?

A new engineer's placement into the technical core is something we could call "wide or deep." Simply put, technical jobs can be classified as jobs where the knowledge

DOI: 10.1201/9781003214397-11

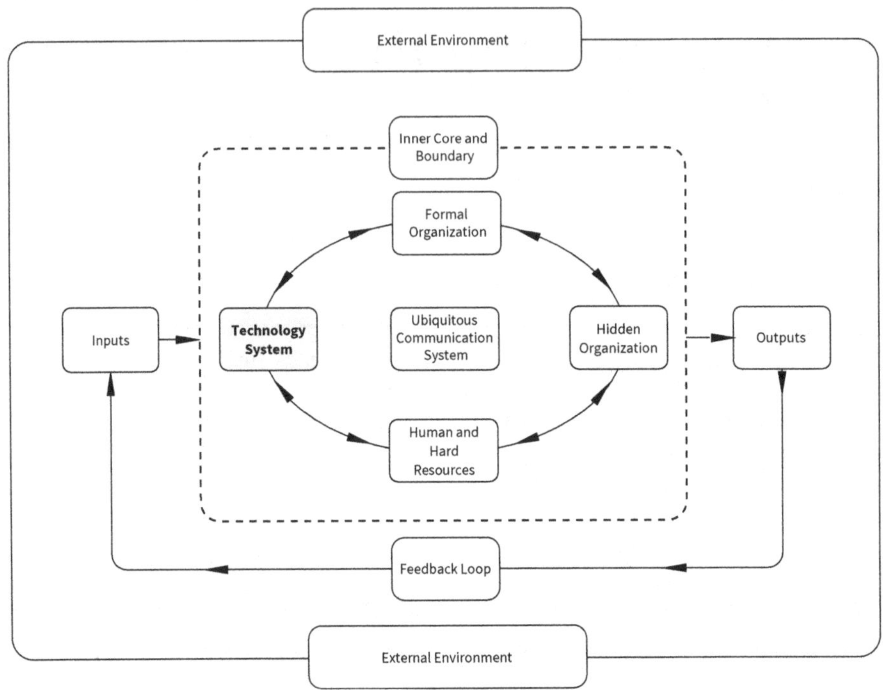

FIGURE 8.1 Technology system position within Inner Core.

required is either "wide" (i.e., basic knowledge across a wide variety of tasks) or "deep" (i.e., substantial expert knowledge of a particular task but little elsewhere). There's an old joke about experts vs. generalists. The expert learns more and more about less and less until they know everything about nothing, while the generalist knows less and less about more and more until they know nothing about everything. A clever saying, but containing a nugget of truth.

Let's consider an example of a new engineer's placement in the technical core.

Meet Julia Müller, a graduate engineer from the École Polytechnique Fédérale de Lausanne in Switzerland. Julia achieved a master's degree in acoustic design and analysis, a notoriously difficult specialty within engineering. She was soon hired to work on the aural design of future products for the dominant European headphone manufacturer. This is a plum job, and Julia was delighted to apply her technical knowledge on behalf of this well-known and respected firm.

During her first week, Julia was assigned to the production quality assurance organization. The purpose of this group was to test sound quality performance for new headphone systems to be sold to the general public. As acoustics is a highly advanced field that encompasses many technical concepts as well as subjective customer preferences, Julia was first placed in a subsection of this acoustic testing area: frequency response measurement. From here, Julia experienced an unexpected path through the technical core.

In her first position, she applied the basic measurement techniques in acoustics, such as DB(A), DB(B), DB(C), and sones. She learned to handle the delicate instrumentation needing to be calibrated and installed into the test headphones to gain repeatable and accurate measurements. This took her months of training to achieve reliable results.

At this point she had traveled deeply into a very narrow subset of acoustics. Once she demonstrated her capability in testing, her job expanded to handle more subjective parts of the evaluation discipline. This entailed analysis of frequency response curves, design of new testing techniques, and correlation of microphone measurements versus customer preference data. At one point she published a professional paper for the Acoustical Society of America on a new acoustic test technique recently adopted by her firm. This slow expansion of her technical skills (from first replicating standard operating procedures to interpreting those results to mastering test procedures to finally creating new knowledge) is a classic example of learning a technical system. Eventually, Julie became known within her own organization as a reliable professional in "all things testing." Soon after, Julie was tapped as a key member of the advanced design and implementation group. The general job description communicated during her hiring had finally come true, but only after an extensive journey through the technical core.

The point is that technical core activities are the normal and expected route into the company, and making yourself known as a respected engineer or technical expert is a classic way to establish yourself within the firm. Here, the depth of your technical knowledge is your competitive advantage while you increase the breadth of your skills by becoming educated in adjacencies, i.e., fields related to your normal technical specialty. Career advancement at this early stage depends upon establishing your baseline skills, and then expanding your technical prowess into adjacent areas, all with the purpose of demonstrating your bona fides. This general path of narrow knowledge transitioning to wider skill in the technical core is probably the most common path forward for new engineers.

In Chapter 7 we discussed the idea of hidden testing, the "pop quiz" of professional engineering. In the technical core, management's desire to conduct ever deeper and more detailed questioning sometimes can't be resisted.

At his first job at Thyssenkrupp Steel Europe AG, Keith Penski, a recent engineering hire, had a director by the name of Gerhard Schmidt. Gerhard was not a "touchy-feely" kind of guy. He was one tough Swiss, taught in the hard knocks school of engineering. Aggressive and with a sharp tongue, he was intimidating.

One day Gerhard commanded Keith into his office and, out of the blue, asked him if either aluminum or stainless steel should ever be anodized. Keith was caught completely flat-footed, and rather than saying he don't know (or offering to look up the answer), he took a guess and said yes to both. "Are you sure?" Keith said yes, even though he wasn't. Ken stared at him, said nothing, and abruptly dismissed him. To this day, Keith is certain Ken knew he was guessing. He is also certain that he was tested that day, and failed his test. Keith learned he needed to have his technical knowledge down cold, and never try to fake an answer (by the way, the answer is yes for aluminum and no for stainless).

8.2 DECISION TECHNIQUES AS TOOLS

One of the core functions of an engineer is making choices. During your career you will be asked to make hundreds (if not thousands) of decisions, be they individual or organizational, small or large, short-term or long-lasting in their impact. And your home organization will make even more choices directly impacting you, many of which will be hidden from view or sometimes nonsensical.

As an engineer, decisions are what you will make almost daily, and hopefully you have already had a taste of some decision-making techniques at university. But as decision theory is a huge topic, your decision training may have been limited to a single portion of the decision-making universe: the purely rational, quantified optimization model. In this chapter, we're going to cast our net a bit wider and touch on several other models, not meant as an in-depth discussion to make you an expert, but allow you to recognize what decision technique may be used in what application.

The first key in understanding decisions is that the choice of a decision tool is a major example of contingency theory in action. Decision theory is massive, with scores of methods, techniques and other tools that can drive a conclusion in any number of directions. Each decision depends on dozens of factors, making outcomes highly individualistic. In short, contingency rules decision-making.

Second, addressing engineering or technical decision theory in a book of this type is high-risk. Making decisions, especially in large technical organizations, is like entering a large and frightening swamp: a territory full of poisonous snakes, alligators, and a universe of other creatures just waiting to pull you under. In-depth arguments about which method to use, what assumptions are in place, and the validity of the data can consume hours, if not days. In no way does this book attempt to address that detailed level of decision theory. Instead, we'll limit ourselves to a basic understanding of the field. With that in mind, let's define a simple topology of decision theory and introduce some fundamentals.

First, let's agree that *decision-making* is defined as the process of identifying and solving problems.[2] There are several classifications of decisions. One classification is by the complexity of the decision tool used: *simple applications*, requiring only basically trained individuals or groups to successfully apply the tool, and *complex applications*, needing an individual expert or several well-trained specialists to correctly use the technique.

A second classification is by the decision method employed. Again, there are two general groups of methods or techniques: *engineering* and *organizational*. Engineering decision methods are those techniques dependent on high-quality quantitative data, using calculation-based, maximization and optimization approaches. Conversely, organizational decision models must be able to handle much more subjective, qualitative, and complex information. This requirement means organizational methods will vary significantly from the engineering approaches. It also means organizational decision-making processes are more complicated, as conflicts often arise from disagreement among influential members or factions within the organization.

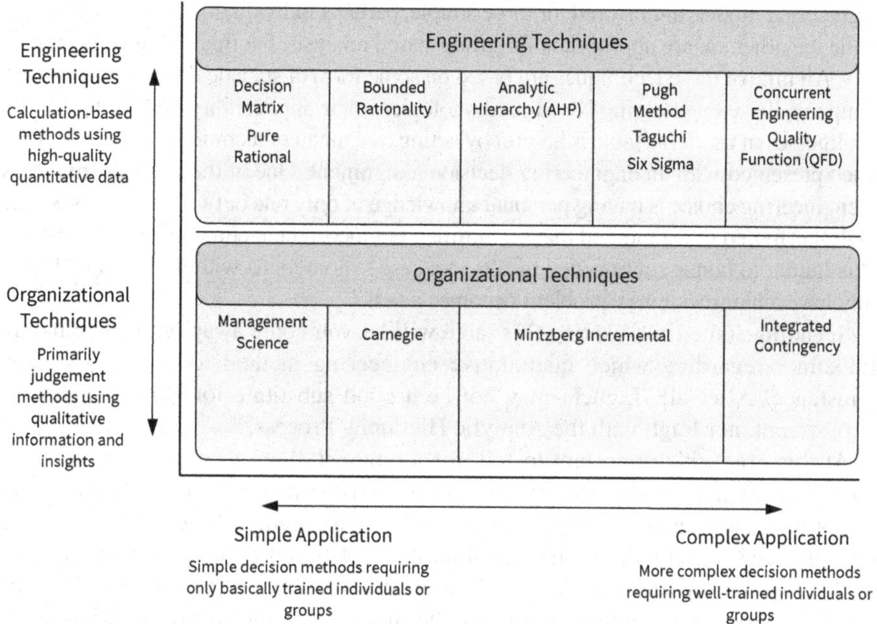

FIGURE 8.2 Decision use based on technique and application.

Luckily, we can categorize decision understanding into a matrix represented by both the basic engineering and organizational technique used and by the complexity of using that method. This provides four categories of decision tools on a continuum: Simple Engineering, Complex Engineering, Simple Organizational, and Complex Organizational, as mapped in Figure 8.2.

Let's discuss the engineering techniques first, as these may be more familiar.

8.2.1 Engineering-Based Decision Methods

As mentioned above, engineering decision methods are those techniques dependent on applying optimization approaches to numerical data. A landmark study by the National Academies of Sciences, Engineering, and Medicine identified the major decision techniques specifically developed for engineering, science, and medical use. This study aimed to:

> [First] … identify the strengths and limitations of tools currently used in engineering design as they relate to decision making and issues of risk and values. Secondly, to identify approaches to decision making in other fields, such as operations research, economics, and management sciences, that address issues of risk and value.[3]

The resulting nine engineering techniques are all intended to analyze detailed quantitative situations. Generally speaking, while these models are placed on a continuum, three engineering models are favored for more simple, perhaps individual-performed, analysis, while the other six are appropriate for group-based analysis, but this split is not compulsory. All nine of these techniques are based on some form of statistical analysis. Table 8.1 compares these engineering-based methodologies, their applicability and strengths and weaknesses in use. This table is helpful by acting as a menu of techniques to choose from when presented with an engineering decision assignment. One of the common missteps in engineering choice is having personal knowledge of only one or two techniques, so the analyst is forced into a method that's a familiar, but incorrect, technique for the situation. This failure to honor contingency results in a case synonymous with the saying, "If you only have a hammer, every problem becomes a nail."

Spending some time visiting this table will let you come away with some handy guidelines regarding which quantitative engineering method to use in what circumstance. After all, Taguchi may not be a good substitute for Quality Function Deployment, nor Pugh with the Analytic Hierarchy Process.

At this point, it's important to talk for a moment about what may be the typical engineer's perception of a certain kind of decision model. Many new engineers do not have substantial decision training and, when asked, will default to only one kind of decision method: rational, quantitative models. There are two types of these models: the purely rational view and the bounded rationality perspective. Both of these center on the systematic analysis of the choices available for that solution. They are numerically based, data-driven, and relentlessly logical, employing calculation as the primary means of analysis. The *purely rational view* does not consider any exceptions to the logic of the solution, even when there is substantial evidence that local conditions make rational-only analysis inappropriate.

The substantial limitations of the purely rational point of view drive the importance and realism of the bounded rationality perspective. *Bounded rationality* means just that: a more credible judgment on the limits and assumptions of the problem impacting the result. For instance, say you work in the product engineering division of a manufacturing firm. Your engineering division makes up approximately 10% of the company's total workforce. Your director is pondering whether to offer free tuition for training in Taguchi methods. This executive decides to ask the workforce for their views. Here's the boundary question for the survey: should management ask the entire workforce or only the 10% of technical employees? Obviously, if only the technical employees are asked, free tuition would win easily. Yet if everyone is asked, the probable answer will be the engineers voting yes and the remaining 90% of the total workforce saying no, as they probably believe the engineers are getting special treatment: the training is denied. The validity of the entire decision rests on where the decision boundary is, so setting an appropriate boundary is critically important. This brings us to an essential insight. Bounded rationality explains the importance of both contingency theory and the pervasiveness of intuitive decision making; those decisions are based on experience and gut feelings rather than a strictly logical sequence of steps.

Early on, these two rational decision techniques will make up a large majority of your decision-making needs, but only for a while. But what about the more complex decision situations you will face?

TABLE 8.1
Engineering-Based Decision Methodologies

Title	Definition	Primary Use	Complexity in Applying	Environment Where Used	Advantages	Disadvantages	Ratings (a)	
							Current Utilization	Ease of Use
Pure Rational	The systematic analysis of the choices available for a given decision situation. Is numerically based, data-driven, and relentlessly logical, calculation as the singular means of analysis	Quantitative decision-making only	Simplistic situations with clear cause and effect relationships	Highly controlled and stable environments	Simple to use in most situations	Normally incorrect results due to poor cause and effect understanding and incorrect input data	4	4
Bounded Rationality	A method attempting to apply quantitative decision methodologies to complex and ill-defined situations. Forces substantial limits on assumptions, reducing number of choice scenarios	Primarily quantitative decision-making	Moderately simplistic situations with generalized cause and effect relationships	Stable to moderately stable environments	Assumptions used may simplify decision-making	Assumptions may be incorrect, making decisions invalid or useless	4	3
Concurrent Engineering	A systematic approach to the integrated, simultaneous design of products and their related processes	Multiple product design decisions in parallel workstreams	Moderate complexity as individual decisions are highly interdependent	Stable to moderately stable environments	Can shorten product development times, improve customer satisfaction	Many decisions are made without complete information, complexity in use	4	1
Decision Matrix	Used to define multiple design attributes and provide relative ranking of these alternatives	Decision tool to obtain stakeholder input	Moderate, depending on number of alternatives to be evaluated	Stable to moderately stable environments	Promotes team interactions and speed of decision-making	Subjective weighting technique used may not capture customer views, can falsely rationalize decisions	4	5

(*Continued*)

TABLE 8.1 (CONTINUED)
Engineering-Based Decision Methodologies

Title	Definition	Primary Use	Complexity in Applying	Environment Where Used	Advantages	Disadvantages	Ratings (a) Current Utilization	Ease of Use
Pugh Method	A matrix containing both design concepts and criteria, used in iteration to identify appropriate design options	Decision tool to obtain stakeholder input	Highly qualitative due to its reliance on experience and judgment	Stable environments	Simple technique	Not recommended for relative ranking of design alternatives	3	2
Quality Function Deployment (QFD)	A decision tool meant to translate customer wants and needs into actions that can be measured, assessed and improved	Decision tool to obtain stakeholder input	Highly complex due to detailed matrix creation in multiple steps	Stable to moderately stable environments	Highly disciplined and detailed approach builds confidence in its results	Slow results due to deep dive approach and large number of experts required	2	1
AHP Analytic Hierarchy Process (AHP)	Methodology using multi-criteria analysis to evaluate and prioritize solutions to complex problems	Decision tool prioritizing solutions to complex questions	Highly complex	Moderately stable to moderately unstable environments	Independent judgments of experts expands solution set	Difficult to use under higher uncertainty situations	3	NA
Taguchi Method	Application of statistical decision methods throughout the entire engineering design process, from product concept to customer usage	Decision tool to address variability, quality and uncertainty	Moderate to high complexity dependent on number of statistical factors employed	Moderately stable environments	Applies sensitivity analysis to design decisions, improving decision robustness	Training in Taguchi techniques is required, with solid prior education in statistical methods	4	2
Six Sigma	Application of statistical decision methods using problem statement development, brainstorming, fishbone diagrams, process mapping hypothesis testing and many other design support tools	Decision tool to address variability, quality and uncertainty	Substantial due to wide range of potential applications for single decision	Stable environments	Emphasizes quality as a fundamental driver of decisions	Requires substantial training in its use. Specialist training required	3	2

(a) Ratings Scale: 1 = Low Utilization / Low Ease of Use; 5 = High Utilization/High Ease of Use

As mentioned earlier, a wide variety of engineering decision methods are available, varying in their complexity from moderately challenging to severely difficult. Most of these decision techniques support design alternative decisions. These include Decision Matrix, Pugh Method, Quality Function Deployment, and many more: all of these are not trivial and involve training and practice to truly become viable additions to your toolbox. Dependent upon your current and future positions, coaching in certain of these techniques will become mandatory, and it will be up to you to acquire that education at the right time.[4]

8.2.2 Organizationally Based Decision Models

As mentioned earlier, unlike engineering models, organizational decision methods can handle a much wider, more subjective, and complex information set. This capability means their fundamental methods will vary significantly from purely numeric engineering techniques.

Like the engineering methods, organizational models are classified by the complexity of the decision tool, from simple applications used by basically trained individuals or groups to complex applications, needing specialists to apply the technique. Again, a continuum links these different applications.

Of course, the major difference in organizationally-based models is the involvement of more individuals attempting to make a choice. By their very nature, group-based decision processes will be more complicated than individual, where most of these complications arise from an inability to control the beliefs, views, and certainties of the group members, a professional version of "herding cats."

This means the outcome of the group approach can be mixed. There is truth to the contention that "two (or more) heads are better than one," but that assumes all the heads are equal in organizational level and power. If not, the group can easily be hijacked by the most senior, longest-tenured, or most loquacious participant. Proceed cautiously.

Of the many, many organizational choice models available today, four are important to mention:

Management Science

The *Management Science* approach is the organizational analog of the purely rational approach in the engineering category. It is based on the use of statistical and mathematical tools to find an optimum solution to a particular organizational problem. Also known as the *operations research* approach, this method is best used for problems that are both analyzable and where the variables can be identified and measured. "Identified and measured" is the key phrase here. Unfortunately, these variables are the Achilles's heel of Management Science. Not all variables can be identified, and certainly not all variables can be measured with any degree of accuracy or precision. Dependent on the system's sensitivity, a small error in an input value can create a major error in decision output.

Carnegie

The *Carnegie Model* is the organizational version of the bounded rationality approach in the engineering classification. Carnegie emphasizes bounded rationality principles (such as limited time, information availability, and the mental capacity of the decision-makers), as its primary ground rule states that a purely rational solution often cannot be legitimately derived. As in many organizational situations, disagreement among managers may force the formation of coalitions who agree on decision goals or priorities. Thus, the Carnegie Model emphasizes the political process involved in decision making.

The fascinating thing about the Carnegie model has to do with how it is applied. Due to the factors mentioned above, managers tend to engage in "problemistic search," that is, looking for a quick solution in the immediate, local environment rather than trying to develop an optimal solution. Thus, a solution is often chosen to *satisfice* (satisfy + suffice) rather than optimize. This means management will select the first alternative that both satisfies the decision criteria and is "good enough" to be implemented. This is an important point, as the majority of decisions in engineering organizations are of this type. There is rarely, if ever, time to purely optimize. Satisficing is the name of the game.

Mintzberg Incremental

The Mintzberg Incremental process model is based on research into the actual decision-making processes of firms developing new products. This model emphasizes a step-by-step sequence of activities leading to a solution. In it, major decisions are broken down into smaller steps taking place in three major phases: identification, development, and selection, with each phase containing a specific series of activities.

One insight from the Mintzberg approach is that some organizations tend to make a series of small decisions in a certain order, ultimately resulting in a large, final choice. One disadvantage in this approach is that final decisions may take a very long time, making it unappealing for corporate use.[5]

Integrated Contingency

The Integrated Contingency philosophy is based on two dimensions. The first is goal consensus, that is, the degree of agreement on goals among decision-makers. The second is technical knowledge, i.e., the understanding of the cause and effect relationships in place that can lead to decisions that reach the group's goals.

This is much easier said than done. In this approach, complexity is high, multiple decision techniques must be correctly applied, solution accuracy is poor due to the multiple models used, and the speed to solution can be very long. Not a very appealing approach.

To précis these major models, the summary shown in Table 8.2 may be helpful.

So, what can we say about these decision-making tools? First, all these approaches recognize the purely rational approach is often inapplicable, as decisions are bounded by limits set by the external environment. These limiting factors are many: time available, the mental capability of the decision-makers, insufficient information, scarce resources, boundaries on personal relationships and on the social constraints of the individual. Second, due to their contingent nature, organizational decision-making models can easily fragment as the assumptions used for different situations change and multiply. Third, generally speaking, for organizational analysis the bounded rationality and Carnegie approaches appear to be the most-used choice models.

TABLE 8.2
Organizational Decision Method Comparison

Decision Characteristic	Management Science	Carnegie	Mintzberg Incremental	Integrated Contingency
Complexity	Low: simplest of all organizational decision models	Mid-level complexity: identifying boundaries is the challenge	High. Many steps involved increases the chances of error	Very high. Requires competency in multiple decision techniques
Solution accuracy/ precision	Accurate/precise if objective data quality is good	Moderate	Moderate to poor	Poor due to multiple decisions solutions employed
Problem consensus	Certain – Excellent	Somewhat Uncertain	Uncertain	Very Uncertain
Length of time decision is considered valid	Very short as input data can change rapidly	Short to mid-range dependent on quality of boundary knowledge	Short to mid-range due to many input variables	Very long, as no one wants to do it again
Speed to generate solution	Fast	Moderate to fast	Moderate to slow	Very slow, may be abandoned before completion

Finally, approaching decisions requires some basic analysis before jumping into doing the actual choice calculations. Contingency demands a thoughtful selection of method as a first step. And because of the substantial variability between decision model outcomes, for important decisions some practitioners may even repeat the decision analysis using two (or sometime even three) different models to triangulate in on a more valid outcome. In short, don't become a slave to a particular decision model, but take a moment and consider your decision-making options as a required first step.

8.2.3 Decision Risk and Failure

One important factor we haven't touched on yet: decision failure. There are dozens and dozens of conditions that cause either a failure to make a decision, or where a decision is made and found to be wrong.

The riskiness of every decision can be organized on a four-position scale according to the availability of quality information concerning the possibility of failure.

The first position is the certainty that all information for what the decision-maker needs is fully available. This position is unrealistic; all decisions must contain some level of uncertainty. The second level of risk means a decision has clear-cut goals and good information available but has associated alternatives that are subject to chance. The third is uncertainty, which means a manager knows which goals they wish to achieve, but information about alternatives and future events is incomplete. The final level is ambiguity, meaning goals can be achieved, yet the problem to be solved is unclear, the alternatives are difficult to define, and information about outcomes is unavailable. This is by far the most difficult decision situation, requiring the most robust decision techniques available, yet understanding the risk will still be exceedingly high. These levels are summarized in Table 8.3.

Many individuals or groups will try to assess riskiness through some form of numerical or quantitative technique, assigning values to certain conditions believed to be true at that time and then perhaps applying an algorithm to arrive at an answer. A plethora of methods are commonly available and can have varying degrees of effectiveness based on the method and the analyst's skill in using it. In short, most analysts at some point will employ a quantitative technique.

It may seem counterintuitive, but the importance of analyzing risk in engineering decision-making is its value in actually performing the quantitative assessment, not in the numerical result. Performing the steps in a risk analysis process gives the analyst a better visceral understanding of the risk to inform the current and future

TABLE 8.3
Risk-Level Characteristics

Relative Level of Risk	Description	Result	Comment
Level 1: Certainty	All information needed by the decision-maker is fully available and highly accurate	Essentially no risk provided correct decision tool is used	Unrealistic in practice
Level 2: Confidence	Generally good information is available, but alternatives are subject to chance	Low-risk/high-confidence situation normally judged to be "under control"	Overconfidence in the projected result is the risk
Level 3: Uncertainty	Goals are known but information about alternatives is incomplete	Analysis is widely known to be uncertain ahead of decision event	Unidentified alternative appears at last second to negate all prior analysis
Level 4: Ambiguity	Problem statement is unclear, alternatives are difficult to define, and information about outcomes is unavailable	Make decision anyway despite high discomfort with analysis. An important function of upper management	Very common situation in organizational decision-making

decisions. It's not about the destination; it's about the journey. Coping with uncertainty demands thoughtful analysis from all analytic approaches, coupled with making decisions when the time is right, such as before the future is clear to competitors.

8.2.4 A Different Type of Decision

Now, here comes something to think about. These decision models are all well and good. However, your local management, dealing with local questions, is more concerned with a different kind of decision. Let me ask you: in a normal, local engineering organization, what is the most desirable type of decision:

1) An optimized decision
2) An accurate decision
3) The right decision
4) A good enough decision
5) A fast and correct decision
6) A wrong but fast decision
7) A reversible decision

Table 8.4 may give a clue to the answer.

Management decisions tend to default to speed. But why speed above all else? The cynical view is there is an unspoken belief that fast results imply competency

TABLE 8.4
Ranking of Decision Importance Judged by Local Management

Decision Importance	Type of Decision	Rationale
1	A fast decision	Management continuously under pressure to answer questions from superiors immediately, no matter the quality of the answer
2	A good enough decision	Management likes low-cost, fast decisions, provided the answer is somewhere in the range of OK. Good enough fits the bill
3	A wrong but fast decision	Remember, there may never seem to be enough time to make a correct decision but there is always time to repeat the assignment
4	A reversible decision	See "wrong but fast decision" immediately above
5	The right decision	Depends on quality of the good enough decision
6	A highly accurate decision	Seen as wasteful of time and money. See "good enough" decision above
7	An optimized decision	Management normally unwilling to spend time and money on a luxurious solution

1 = most important; 7 = least important

and skill, that the decision-maker is "on their game." And fast decisions provide significant leeway to cover errors in the decision: the inherent uncertainty of the quick, incomplete information available at the time of the decision provides air cover. Voilà, a fast decision.

Perhaps. But there is a more legitimate reason for fast decisions. The compulsion for speedy decisions has to do with risk mitigation. Coping with uncertainty demands making decisions when the time is right, before the future is clear. Much of the art of decision-making is anticipating the right time to make it. This means the organization must focus on surfacing critical issues quickly, performing the necessary *approximate* analysis fast and then nimbly decide on the action in concert with critical timing requirements. Correct decision timing means less risk.

8.2.5 Implementing Decisions

In engineering projects of a certain size, we have a tendency to put total effort and energy into making the decision regarding "what we are going to do?" Endless scenarios, return on investment calculations, and risk assessments are launched to make sure the decision is "right." Unfortunately, many times the implementation phase, the "how do we intend to do this" portion of the project tends to get minimum attention and support. For decisions that are small to medium in size, overwhelming attention is sometimes placed on making the decision but not very much on its implementation.

Say your local group needs to upgrade their materials performance laboratory to speed up new product evaluations. Naturally, a user committee is formed, and after five months of weekly meetings generating equipment lists, specification arguments, and multiple visits to your director's office, your committee decides on the configuration of the new lab. Management approves the decision, and with smiles on their faces, everyone closes their laptops and makes for the door, happy the decision was made and relieved the effort is over. Yet very little is normally discussed about how this decision will be successfully implemented. An undeniable habit for many is, for small to modest projects, to immediately forget the decision just made and move on to the next topic or assignment. The decision itself is a fait accompli: implementation is left to flounder and, to management's surprise, six months later little (or nothing) has been done to implement it. The decision tends to be thought of as the totally of the assignment: nothing more needs to be done.

We can argue the reason for this. Perhaps it's the natural exhaustion that occurs after the crescendo of work on the decision. We take a break and somehow move onto the next urgent assignment, promising ourselves to come back to the implementation as soon as we finish this other quick task. Of course, we never come back. Another possibility is that implementation is just not considered a high-status assignment. In some organizations, implementation is perceived as being left to the "second stringers": it's old news and nobody wants to be part of yesterday's work. Or perhaps the implementation suffers from resource starvation, where implementation is viewed strictly as a cost so the necessary people, money, and equipment are not provided.

There are other reasons as well, but these reasons alone can be enough the scuttle a project. If management is smart it will lavish some of the same attention and funding on implementation as it does on the actual decision-making.

8.2.6 The Trap of Decision Making

The field of behavioral economics brings us to an unfortunate possibility: becoming snared in our premeditated beliefs about the decisions we are making.

Dr. Karl Weick has spent significant time examining this issue within decision-making, providing some essential insights and potential pitfalls regarding this temptation of choice.

Specifically, Weick suggests an alternative view of the bounded rationality model, where decision practice is actually different from what is believed. In 1995, Weick published the contention that much decision-making is not based on rational or even bounded rationality techniques, but instead is based upon the decision-maker's feeling about the topic, and, after a decision is made, will go back and search for reasons why the original choice was selected. This is surprisingly common, especially in complex decision-making environments. In other words, Weick argues that people justify their decisions only after they make them, and will try to convince others they "had decided" before they actually took the action.[6]

Early mentions of this trap go back several decades. The economist John Kenneth Galbraith mentions an early version in his thoughts on conventional wisdom:

> We associate truth with convenience, with what most closely accords with self-interest and personal well-being or promises best to avoid awkward effort or unwelcome dislocation of life. We also find highly acceptable what contributes most to self-esteem.
>
> Economic and social behavior are complex, and to comprehend their character is mentally tiring. Therefore, we adhere, as though to a raft, to those ideas which represent our understanding.[7]

Let's take the example of a classic suburban family in the United States, wishing to buy a new vehicle for two parents and their two young children. There are many, many vehicle choices available in the market, including sedans, coupes, crossovers, sport utility vehicles (SUVs), and pickup trucks, all with gas, diesel, electric, and hybrid propulsion systems. The family selects a plug-in hybrid electric SUV with a high-end trim package and excitedly brings it home. Their neighbor, spying their new purchase in the driveway, naturally congratulates them on their new vehicle. And inevitably, the next comment is a question: "Why did you choose that particular vehicle?" And just as inevitably, the new owner will list several functional and responsible features, such as the vehicle's safety rating, environmental responsibility credentials, command seating position, advanced communication electronics, and any number of sensible attributes.

But there's another reason, rarely, if ever, mentioned to the neighbor. Frankly, what is probably never shared is the defining decision point: that the vehicle chosen is a premium brand, and the vehicle makes a statement as to the status and success of the breadwinners in the family. The choice of this SUV becomes an emotional decision, and its features are used as a kind of "retrograde rationalization" supporting the purchase. This happens constantly, yet no one will hear somebody buy a premium branded product and actually say out loud: "I bought this as a status symbol." Yet it's stunning how many important decisions are made his way.

So, what can we generally say about some of the pitfalls that influence and ultimately drive decision-making in individuals and organizations?

First, reliable and accurate data on the conditions the decision will be made is rarely available. Cause-and-effect relationships are normally not well known, and generally, the more important the decision, the more complex and "softer" the information to make the decision will be. And the more complex the decision, the more gut feelings will be used in making a selection. And, of course, many decisions are sold to others based upon retrograde rationalization.

One subtle point regarding decision-making meetings. Many management decisions are not decided in meetings. Instead, management decides the result in advance, creating a "premade" outcome that is then "decided" in meetings that look more like theater than actual decision events. This is very common, and don't be surprised if you see this in your own organization.

These are only a few of the scores of pitfalls engineers need to understand and be aware of in decision-making situations. These snares are everywhere, and you certainly won't avoid them all. But even an awareness they exist can really help in understanding these traps.

8.2.7 Understanding the Data Driving Decisions

Carl Bergstrom and Jevin West, in their book *Calling B******t: The Art of Skepticism in a Data Driven World*, highlight an emerging trend in numerical analysis and presentation: that of data visualization overwhelming insight.

In the COVID-19 pandemic of 2020–2022, the U.S. State of Georgia tallied 87,709 cases of COVID on July 2, 2020, as shown in red on the state's official pandemic case map. On July 15, 2020, that number was 135,183. Yet the graphic on the Georgia map looked the same: there visually appeared to be no increase in COVID cases. What was causing this paradox?

Between July 2nd and July 15th, the map's threshold for cases to turn red increased from 2,961 to 3,769 cases, causing the map to now display an inaccurate visualization for a key indicator of public health. One single definition change in an important chart could have altered the trajectory of COVID care for an entire state.[8]

This example points to an emerging trend: as the amount of available numerical data grows, the need to show that data simply through visualization can result in data being massaged in the name of "clarity." This can create a world full of doubtful claims based on a series of spurious data points. After all, not all charts are created by data scientists. As a new hire, you must know how to spot visual nonsense. Correlation and causality are prime concepts that can be misinterpreted.

Contributing to the problem is that some scientists are not participating in the usual peer-reviewed publication process (see Chapter 3). Unfortunately, pressure to make exaggerated claims (both bad and good) is pushed by media, online outlets, and TED-like venues and expressed in poorly designed visualizations. The bottom line for you is that judging visualizations must become a more skeptical activity, and using data to drive important decisions has to be more cynically evaluated.[9]

8.2.8 Decision-Making Under Stress: The After Action Review

High-stakes decision-making is a special animal that can bite your ankle unless you treat it with some respect. While you probably won't be making critical decisions totally on your own at this early career stage, you will be working with people who will. And one side effect of high-stakes decision-making you may witness and will be obligated to share with your management.

The mental state of the decision-maker under stress has a huge role to play in the quality and appropriateness of an individual decision. Executives and other senior management are not infallible; each have their own personal filters and emotional reactions when weighing decisions, and the effectiveness of their filters is directly proportional to the importance of the decision.

In fact, the U.S. Army actively trains their commanders to confront this invisible possibility. Called the "After Action Review," or AAR, senior commanders repeatedly take part in major war-game exercises involving literally thousands of individuals and hundreds of vehicles against a simulated opposing force. In these multiday war games, commanders are subjected to combat-level decision scenarios. What is fascinating is that these commanders are individually video recorded for the entire game. Their entire set of conversations, directions, and orders are chronicled to give a complete picture of what happened in the command center during high-stress and high-consequence situations.

During one exercise, I witnessed a commander, in the middle of the night, give an erroneous order that resulted in the simulated loss of a major part of his force. During the AAR, this error was brought up. The commander stated (quite honestly and sincerely) that he did not give that order. And everyone believed that the commander sincerely thought that he had not given it. Yet when the recording made in the command center that night clearly showed the order had indeed been given, the commander was stunned. He had absolutely no memory of making that bad decision. Only when the recording was played did he realize that the "fog of war" impacted the safety of his soldiers. And that's the whole purpose of the AAR, to review with senior management that decisions made in the executive's mind may not translate into actual results.[10]

Of course, it's rare a business executive will be directing army divisions at 3:00 am, but consider that many senior executives look to the wisdom of military strategists such as Sun Tzu and Carl von Clausewitz for guidance in dynamic situations. In fact, many consultancies exist to provide this type of training to business organizations desiring to become more nimble in their decision-making. But the fundamental conclusion here is that executives, when pressurized, could likely say or do things that may not be in the organization's best interest.

What about less dramatic situations? A good (but somewhat rare) suggestion can be applied to your local organization. Whenever an action is required or decision made (which is about 99.7% of all assignments), you can apply the concept of AAR. What happened during the assignment? If positive, understand why it was a positive outcome. If not, why not? Was it a lack of clarity? Was it a misjudgment of the facts? Did the assumptions made about the situation prove false?

It's incredibly obvious this evaluation must be done: it's just "learning from your mistakes." Yet perhaps 98% or more of all assignments will never undergo an AAR. The reasons are simple. First, the perception is that an AAR is wasted time: why rehash recent history when there are new projects to start and new markets to conquer? And, second, an AAR could bring unwanted attention to those who might bear some responsibility for the outcome. Why ask uncomfortable, embarrassing questions which may require corrective action on someone's part when we can conveniently forget the whole thing? Fundamentally, it'll take a large and expensive project failure to trigger a true AAR. But for you personally, you should always take some time and study your outcomes, even if your group will not.

8.3 MEASURING SYSTEM RESULTS

Let's now change gears and look at something more concrete: measuring the results of a technology system. On the surface, measuring the results of the technical system you are operating would seem quite straightforward and simple. Each morning you turn on the system, apply some kind of input to it, a transformation occurs within the system and something of value pops out. You'll eventually come to know many of the technical details of the system: the voltage required, optical sensor sensitivity, temperature limits, and the like. And soon, you will be able to quickly identify what is a reasonable result and what isn't. Report that result into the balanced scorecard and you'll be satisfied that you are a contributing member of the team.

However, underneath there's a dirty little secret.

A technical system is designed to produce a valuable output built to a specification established to indicate quality. With production comes feedback information regarding that product's specifications. If the feedback information indicates an out-of-control condition, the operations engineer in charge (i.e. you) will normally need to take an action to correct the out-of-bounds condition. This is classic manufacturing engineering. Unfortunately, some problems are just not measurable, yet we as engineers attempt to make them measurable anyway; it's what we're trained to do. This results in many situations where we believe problems are moving toward solution, but in reality, we are just taking a guess that we are measuring something meaningful. Many, many times we assign as a metric something that's easy to measure, yet it may have nothing to do with true progress.

The quality guru Edward Deming shared a famous story about this. A factory wished to improve the acceptance rate of their product (ball bearings) by instituting a quality improvement program. This program consisted of an award given once a month to the "willing worker" who produced the highest percentage of bearings that met the quality control standards. Every month a ceremony was held in the factory to publicly award the fortunate winner. Yet after months of award ceremonies, quality remained at the same low level. Called in to solve the mystery, Deming's team discovered that the Statistical Process Control (SPC) charts showed the machinery producing the bearings was incapable of providing "in control" parts. This meant no matter which adjustments the worker made to the machinery; a random rejection rate would result. In other words, the program was not about quality control but instead became a lottery, where pure luck was the only factor for who was awarded. Only when the manufacturing equipment was redesigned to produce "in control" parts did the award program become legitimate.[11]

Measuring something for the sake of measuring something is not helpful. Make each and every one of your measurements mean something legitimate. Don't end up running a lottery.

8.4 TYPES OF TECHNICAL SYSTEMS

At this point, we need to touch on a couple of the different types of technical systems found within the Inner Core. How should a technical system be chosen to facilitate and optimize operational work processes? As you already suspect, there can be an infinite number depending on the conditions present. To keep things simple, we'll very briefly touch on two very different types of technical systems: *manufacturing* and *service.*

8.4.1 Manufacturing Systems

The seminal classification of manufacturing work was developed by Joan Woodward, who developed the work-type topology we still use today.[12]

Woodward defined three overall types of manufacturing:

1) Group I – Small-batch and unit production
2) Group II –Large-batch and mass production
3) Group III – Continuous process production

Each one of these groups has its own basic characteristics, as shown in Figure 8.3.

As you already know, the main focus in manufacturing today centers on forms of flexible manufacturing. Flexible manufacturing systems are very common and have been in place for many years now. These systems include computer-aided design (CAD), where digital systems assist in drafting, design, and engineering of new parts; and computer-aided manufacturing (CAM), where digitally controlled machines accomplish material handling, fabrication, production, and assembly tasks, increasing the speed of production for unique and intricate parts. Coupled with these machines are integrated information networks that link all aspects of the organization from design and manufacturing to supporting organizations such as accounting, purchasing, marketing, and inventory control.

Two relatively recent, game-changing developments in manufacturing systems are lean manufacturing (that implements the combination of just-in-time production,

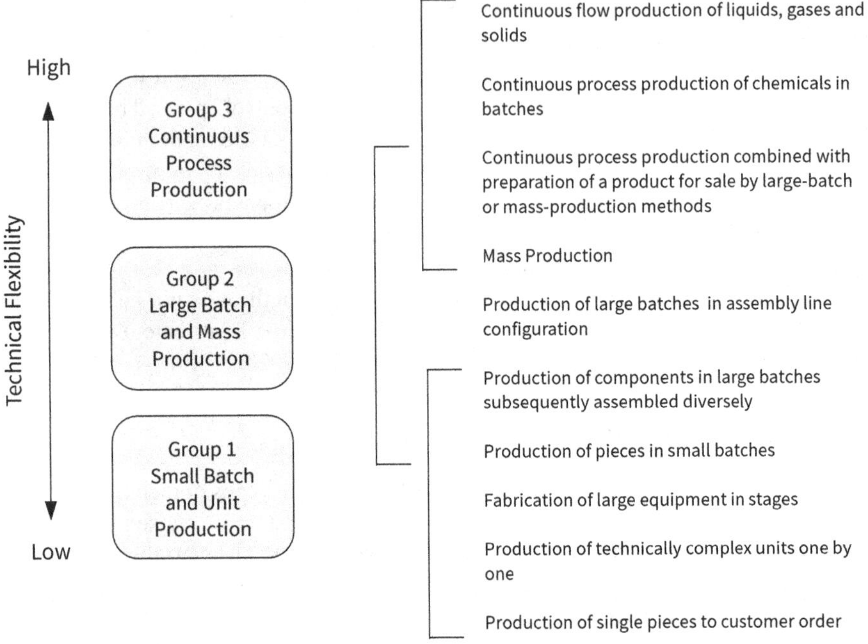

FIGURE 8.3 Woodward's manufacturing classification system.

quick changeover, and continuous improvement) and the revolutionary and rapidly expanding area of additive manufacturing. Additive manufacturing has already demonstrated its capability to radically expand its reach, through the use of new materials and techniques in an ever-accelerating cycle. This cycle benefits unique part production, to reducing assembly complexity, and to providing an infinite number of customized versions of the same base product. As of this writing, the scope and reach of additive manufacturing is still being discovered, yet we do know this technology has the strong potential to upend traditional thinking about mass customization and materials science. The current relationship between product manufacturing flexibility and production unit size is summarized in Figure 8.4.[13,]

8.4.2 Service Systems

Service is a special form of product which consists of activities, benefits, or satisfactions offered for sale. These are intangible and do not result in the ownership of anything at the end of the service. They are unusual in that services are produced and consumed at the same time, and that these services cannot be separated from their providers. This signifies that service systems are fundamentally different from traditional manufacturing systems, as the product is not a hard object but is an action perceived as satisfactory to a customer. This is noticeable in Figure 8.5.

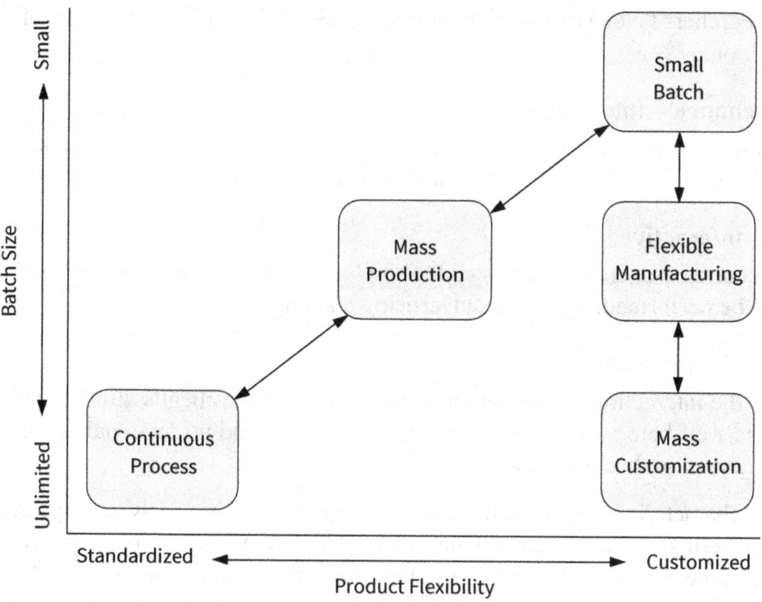

FIGURE 8.4 Comparison of classic flexible vs. traditional mass production manufacturing systems.

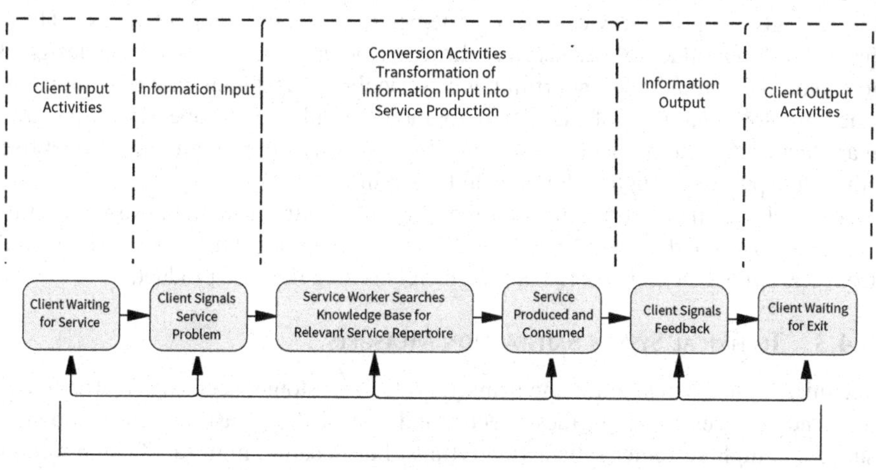

FIGURE 8.5 Traditional customer service process.

Researchers Peter Mills and Newton Margulies define three basic types of service interactions:

Maintenance – Interactive

A cosmetic relationship between an employee and client in order to maintain or sustain a relationship. Examples are banks and insurance providers.

Task – Interactive

A concentrated interaction between the employee and client where the focus is on the task to be performed, such as in advertising and engineering.

Personal – Interactive

Where the interaction focuses on the improvement of the client's direct intrinsic and intimate well-being. Common examples are teachers and professionals such as doctors, lawyers, and counselors.[14]

The characteristics of these three service types can vary widely, and the design of an engineering service system will depend on addressing unique service characteristics for each type. In terms of expense, training, support functions, and intangibility of the goal, these service systems are a substantial challenge to traditional engineering firms.

But this hasn't dampened the enthusiasm of engineering firms or manufacturers to enter this potentially lucrative adjacency. A relatively new hybrid form for manufacturing firms is the *industrial service transition company*. Manufacturers are now increasingly adopting service-based strategies to maintain competitiveness. These firms develop product-related services through a dedicated service division designed to exploit the commercial opportunities of servicing an installed base of equipment. They employ a strategy of integrated solutions to enhance competitiveness of the manufacturer's core product offering, as it is now difficult to maintain competitive advantage purely through technological leadership.

Examples of this hybrid are both the General Electric and Rolls-Royce aircraft engine divisions of their respective firms. Today, the majority of their income is derived from servicing their engines, rather than merely selling the base product.

8.4.3 Technical System Satisfaction Measures

Technical systems in engineering firms provide for customer needs by creating products and services fulfilling these needs, and one of the measures of a company's success is high customer satisfaction ratings. For decades, high satisfaction ratings have been the crowning achievement for all firms, including engineering firms specializing in product development. There is a cottage industry of companies that independently provide initial or time-in-service satisfaction measures such as J.D. Power, Ipsos, and Nielson. You often will see firms tout these measures in mass market outlets, all attempting to snag some conquest sales as a result.

In these surveys, customer satisfaction is commonly used as a synonym for quality, with a higher number indicating higher quality. Yet customer satisfaction is not quality: quality is a subset of customer satisfaction.

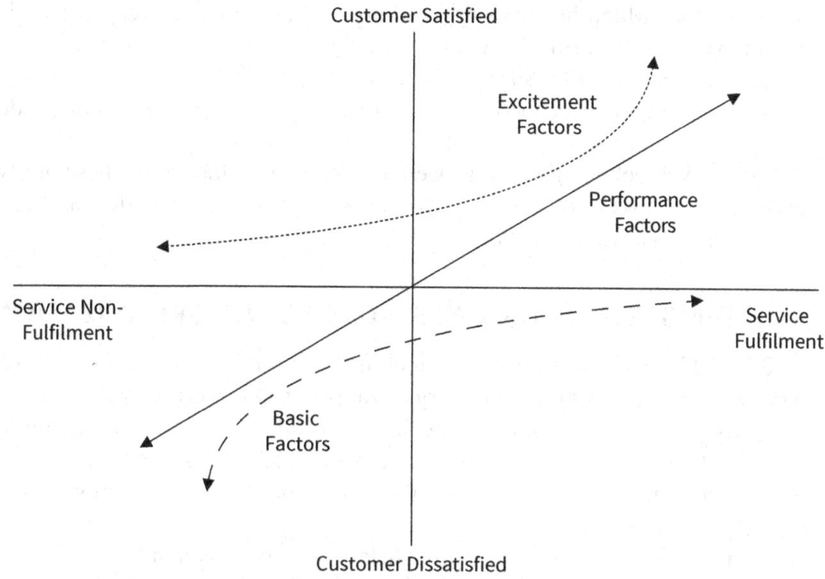

FIGURE 8.6 Kano customer satisfaction model.

And quality is much more than a numerical report. As we discussed in Chapter 3, quality measures really must partner qualitative data with the existing quantitative information. A good way to perform this trick is with the Kano customer satisfaction model. The Kano model, shown in Figure 8.6, is a marketing quality management technique used to establish client or customer satisfaction.

Three factors influence satisfaction. The first are *basic factors.* These are minimum requirements set by the customer which will cause dissatisfaction if they are not provided but do not create customer satisfaction if they are fulfilled or exceeded. These are prerequisites in the market and are taken for granted: these are expected and, if missing, are dissatisfiers. For example, if you purchase a new mobile telephone and it lacks a high-definition camera, this is a dissatisfier. If the camera is there, you'll take no notice of it.

The second factor are *performance factors.* These cause satisfaction if performance is high, and cause dissatisfaction if performance is low. These are directly connected to the client's explicit needs and desires, and are sometimes referred to as traditional quality measures. If something works, great, that's as expected. If it's not working and you must have it repaired, that's a dissatisfier.

The most interesting factors are the excitement factors. These are sometimes called "surprise and delight" features as these factors increase satisfaction if delivered but do not cause dissatisfaction if not delivered; the client feels they got something extra with their purchase or service. Premium luxury brand automobiles provide excitement factors continuously. The Tesla electric vehicle line makes it their specialty to provide surprises in their vehicle and support systems that cause huge satisfaction and "halo effects," which has helped Tesla move to the top of the electric vehicle business globally. Many firms realize the "excitement factor" quality measure and strive to provide these positive surprises whenever possible.[15]

To sum it up, service technologies are fundamentally different from product development or manufacturing technologies. You will probably or certainly spend time in the service technology area as well as product development. These are two distinctly different tasks but are designed to give you a broader understanding of what a technology system really is.

8.5 MELDING THE TECHNICAL SYSTEM TO THE ORGANIZATION

This one should be a no-brainer. The technical system delivers the core product, service, or desired output to the organization. It is the proverbial goose that lays the daily golden eggs. Yet there are situations when the connection from the Technical System to the organization becomes broken. This happens when attention is diverted from the core purpose of the company by political infighting, an external crisis, or both.

In the early 1990s, a global automotive firm began to develop a traditional four-door mainstream sedan (or saloon as our European friends call it). Just as the program got underway, a complete corporate reorganization commenced involving nearly all the senior management of the corporation. This reorganization took a very long time: about 24 months. During those 24 months the saloon program proceeded and the appearance and functionality were finalized, but without review or direction by senior management (who were too busy reorganizing to pay attention to the program). When the new management finally came around, they were presented with a less than exciting vehicle with even less than exciting features. That ship had sailed: it was too late to change either its appearance or competitive feature set. And it sank without a trace in the marketplace. The technical system and rest of the enterprise communication links broke, and there was no recovery possible.

8.6 REALIZATIONS

The technical system of the typical engineering organization can be immense, even when considering small firms. The number of technological tools available for your use, even in startups, is nearly infinite.

Yet the most useful tool that applies to the entire technical system is decision-making. A universe of decision methods abound: some technical and some organizational. Many are detailed and quantitative; fewer are broad-based and qualitative, and many, many are misused. Yet if correctly chosen, these tools can be applied with enthusiasm and presented with certainty and confidence. The step-by-step mechanics of applying these decision tools is not what we are about here. Instead, the aim has been to understand the conditional limitations of each method and the contingent foundation they rest upon. The major models outlined represent a wide range of viewpoints and assumptions regarding how and when to use certain decision tools, as all decision models fall cleanly into contingency theory. High-uncertainty situations tend to call for more complex, "organic" models, while low-uncertainty situations can use more rational, quantitative approaches.

Selecting and applying a combination of decision-making models to the same question is a potentially wise choice, especially for high-uncertainty, high-risk situations. And in both the engineering and organizational contexts, speed of decision-making tends to overshadow accuracy and optimization: urgency is the game.

Looking between the lines you may begin to see the limitations of our current decision-making processes. The main limitations are a lack of accurate data, uncertain cause and effect relationships, and even limitations on defining the problem to be solved. And future decision-making will only become more complex and urgent, and our models must be updated to contain this trend.

The After Action Review shows that decision-making under stress can result in bad decisions, including those that are not even remembered. And the act of decision-making can attract the majority of management attention while decision implementation (especially for certain smaller decisions) may be starved. As a professional, the many decisions you and others will make will have some weakness or flaw you must see, understand, and accept as you nonetheless use them to do your work.

The good news is that, for technical system design, the way forward is relatively straightforward. Woodward's fundamental studies in the typology of work remain valid. Service technologies vary from traditional product development technologies in that service is consumed immediately upon creation and generally cannot be stored. As a result, original equipment manufacturers are now increasingly adopting service-based strategies to maintain competitiveness and open up parallel revenue paths.

Finally, tools like the Kano Customer Satisfaction Model, with their three satisfaction factors of basic, performance, and excitement, allow product designers to better target their features to maximize customer satisfaction.

Now, with our survey of the firm's technical systems complete, let's address something a bit more squishy and ill-defined. Let's search for the Hidden Organization.

NOTES

1. Daft, Richard. 2010. *Organizational Theory and Design*. Mason, OH: Cengage Learning.10: 253–274.
2. Ibid., 10: 450–476.
3. National Research Council. 2001. *Theoretical Foundations for Decision Making in Engineering Design*. Washington, DC: The National Academies Press.
4. Herrmann, Jeffrey. 2015. *Engineering Decision Making and Risk Management*. Hoboken, NJ: Wiley.
5. Ibid., Daft 10: 466–470.
6. Weick, Karl. 1995. *Sensemaking in Organizations*. Thousand Oaks, CA: Sage Publications.
7. Galbreath, John Kenneth. 1958. *The Affluent Society*. Boston, MA: Mariner Books.
8. Bergstrom, Carl and West, Jevin. 2020. *Calling B******t: The Art of Skepticism in a Data-Driven World*. New York: Random House.
9. Ibid.

10. Townsend, Patrick. 1999. *Five-Star Leadership: The Art and Strategy of Creating Leaders at Every Level.* Hoboken, NJ: Wiley.
11. Deming, W. Edwards. 1988. *Out of the Crisis.* Cambridge, MA: MIT Press.
12. Woodward, Joan. 1958. *Management and Technology.* London: Her Majesty's Stationary Office.
13. Meredith, Jack. 1987. The Strategic Advantages of New Manufacturing Technologies for Small Firms. *Strategic Management Journal* 8: 249–258
14. Miles, Peter and Marguiles, Newton. 1980. *Toward a Core Typography of Service Organizations.* Academy of Management Review 5: 255-265.
15. Xu, Qianli, Jiao, Roger, Yanga, Xi et al. 2009. An Analytical Kano Model for Customer Need Analysis. *Design Studies* 30(1): 87–110.

9 Searching for the Hidden Organization

9.1 SEEING THE INVISIBLE: THE HIDDEN ORGANIZATION

An organization's hidden culture is one of the most interesting (and confusing) areas of understanding for any new company employee, not just for engineering organizations but for any collection of individuals in any line of work or play. Illogical, obscure, surprising, veiled, startling and sometimes upsetting, a hidden culture exists within all companies and firms, invisibly operating at all levels and at all times. While it can be located nicely on a diagram, as in Figure 9.1, in reality a company's hidden organization is a unique, one-of-a kind specter that inhabits every nook, corner, and cranny of your workplace. And it cannot be ignored.

Understanding and operating within your organization's hidden culture can be challenging for any of us, but doubly so for new graduates like yourself. Most of your mistakes and blunders will occur in this cultural zone, the only question being the extent of the damage. At the jump, I strongly urge you to, early on, focus your attention on this hidden world, where logical cause-and-effect ceases to exist and where Alice in Wonderland would feel quite at home.

Many of us may perceive that engineering organizations have one single culture. Yet on reflection, you'll find manufacturing plants are different than laboratories, product development is different than engineering sales, and technical service is different than R&D. Engineering cultures are uncounted and uncountable. And you need to know which one you're dealing with at any particular moment. In short, you must search for the hidden organization within your organization.

We need to expand on the introduction to culture presented in Chapter 5, to specify exactly what is the hidden organization (also known as *organizational culture*, *corporate culture*, *informal organization*, or *hidden culture*) and the characteristics that reflect its values and expected behaviors. We should also examine the importance of culture; that it is very real and can help you with your work or damage you greatly. Finally, we should perform some "training" to sensitize yourself to this hidden entity, to see it in all its overt and covert forms.

Let's start with some definitions. Nadler and Tushman define the informal organization as:

> Coexisting alongside the formal arrangements is a set of informal, unwritten guidelines that exert a powerful influence on the behavior of groups and individuals. Also referred to as organizational culture, the informal organization encompasses a pattern of processes, practices, and political relationships that embody the values, beliefs, and accepted behavioral norms of the individuals who work there.

DOI: 10.1201/9781003214397-12

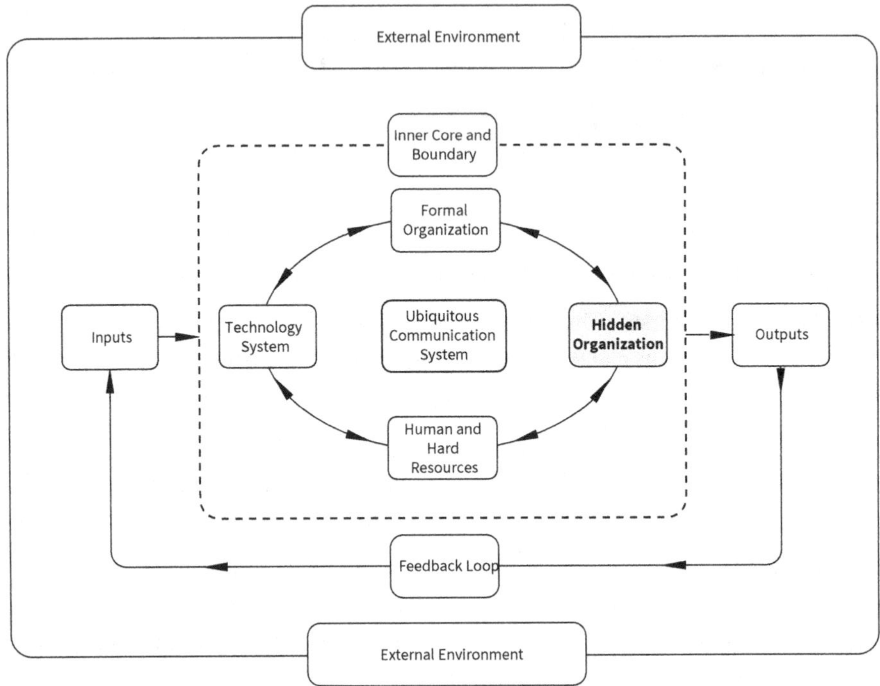

FIGURE 9.1 Hidden organization location within the inner core.

> It's not unusual for informal arrangements to actually supplant formal structures and processes that have been in place so long that they've lost their relevance to the realities of the current work environment.[1]

A more formal definition is given by Edward Schein in *Coming to a New Awareness of Organizational Culture*:

> Organizational culture is the pattern of basic assumptions that a given group has invented, discovered, or developed in learning to cope with its problems of external adaptation and internal integration, and that have worked well enough to be considered valid, and, therefore, to be taught to new members as the correct way to perceive, think, and feel in relation to those problems.[2]

Even though engineering organizations have special characteristics, for this book, the definition of corporate culture can be taken directly from the non-technical case as defined by Daft:

> [A technical or engineering culture is a] … set of values, norms, guiding beliefs, and understandings that is shared by members of the organization and taught to new members as correct. It reflects the common goals, problems and experiences of the group, conditioned by the technological tools and processes those members use.[3]

Today, while seeming like a vaporlike idea, corporate culture is actually a fairly ubiquitous term, and the phrase *engineering culture* is not that far behind. Major digital companies like Apple, Microsoft, and Google all tout their special culture that makes them unique and desirable places to work. Unfortunately, sometimes their description of their culture may emphasize golf simulators, ping-pong tables, and free cappuccino bars rather than a true description of the corporate culture itself.

What about a culture's purpose? The purpose of the hidden organization is twofold. The first is *external adaptation*, meaning how an organization as an entity meets its mission, goals, and tasks while dealing with its outside environment. This includes the specific goals pursued by the group, the means to accomplish those goals, the criteria for measuring those results and corrective actions if these goals are missed.

The second purpose is *internal integration*, where members develop a collective identity so to work together more effectively. This is where you will see a common language, a particular way of working, a certain distribution of status, power, and authority, and the criteria for the allocation of rewards and punishment for noncompliance.

Engineering groups arrive at their basic culture as they learn to use external adaptation and internal integration methods that are good enough to be considered legitimate and thus passed on to new members.[4]

Let's examine the importance of culture. As just mentioned, whatever definition you prefer, the concept of the organizational culture is very real, and certainly can have a direct, immediate, and potentially significant impact on your career and you personally. The blunders we spoke of earlier are real and can range from minor (where experienced employees might share a joke about a misstep at your expense) to much more serious, mistakes attracting the attention of senior management. Let's discuss a serious example first.

CASE EXAMPLE 9.1 THE SAGA OF LAURA KENDRIC

The McCallister Company was one of North America's largest retailers, renowned for providing rock-bottom pricing on all manner of goods, from groceries and housewares to sporting goods and every category in between. McCallister's was a beloved brand, developed over some 50 years by its founder Joseph McCallister, who promised the absolute lowest price on any item every day. "Low prices with no frills" was the McCallister credo.

By the early 2000s, McCallister's senior management felt an updated advertising presence was needed to replace their tired existing brand, consistent with the dawn of the new millennium. This meant fresh ideas were needed in the form of a new Vice President of Branding.

Enter Laura Kendric.

It's rare the hiring of a marketing vice president would grab national attention, but Laura was not an ordinary executive. Recruited to help modernize

McCallister's brand, Kendric was hired and fired in just ten short months, the victim of a serious mismatch of corporate culture.

Before joining McCallister, Kendric, 36, earned an edgy reputation as director of brand communications at a mid-level automobile firm, notable for attention-grabbing, buzz-producing TV commercials, and print ads featuring both provocative actors and ideas. Kendric was a star within the cloistered world of corporate advertising, winning multiple awards for her state-of-the-art concepts and entertaining quotes.

Given her colorful career, Kendric's hiring by one of North America's most low-profile companies struck her marketing colleagues and other insiders as strange. And no one was really surprised when Kendric was dismissed by McCallister after well short of a year.

Yet the reason for her firing ignited controversy. Two reasons were reported in the press, one overt and the other covert. The overt being that Kendric was fired for accepting gifts and gratuities from ad agencies she was doing business with; the covert was for refusing to integrate into an iron-clad corporate culture.

When Kendric was first approached by McCallister, Kendric was flattered by the exciting idea of transforming an entire company and its image. Yet from the first moment she arrived at McCallister in February, Kendric realized that fitting in would be harder than she had imagined. The core culture at McCallister was the concept of "Low Prices Every Single Day," deeply embedded for decades. The McCallister headquarters, with its windowless offices and gray walls, was hardly inspiring for a woman who liked color. One of the first things Kendric did was paint her office chartreuse with chocolate-brown trim. And that was her first mistake.

When it came to the executive suite, McCallister's senior management always worked in the background; there were no "stars" on the team. Individual attention was not the McCallister way, and a passive-aggressive subculture emerged to deal with any high-profile personalities.

Some new employees would keep a low profile and spend their first 90 days listening, but not Kendric. Interviews in national periodicals offer a glimpse into her thinking at the time.

> "I get overly excited," she [said], "I want to hit the ground running."
> "But we realized you don't know where the edge is unless you are willing to go over it once in a while."
> "Go, go, go."
> ".... if you don't ask, you don't get"
> "I think part of my persona is that I am an envelope pusher. The idea of change in general can be uncomfortable for many people, and my persona as an agent of change can prompt that feeling."

Meanwhile, Kendric's cultural mistakes came fast. One was the annual McCallister shareholder meeting, a no frills, cheaply run gathering focused strictly on financial matters. In one of her first assignments, Kendric transformed a dull shareholder meeting into a three-hour Broadway show, hiring a cast of New York Broadway-style actors who sang songs like "The Day That I Met Joe," the company's legendary founder. The show caused rolled eyes and shaking heads from the senior executives.

Another was the weekly senior executive "all-hands" meeting held at McCallister's headquarters every week. Each Friday morning, McCallister's CEO would gather together some 300 managers and executives for a wide-ranging meeting at McCallister's headquarters. Kendric, always on the road and unaware of how important it was to attend these meetings, missed several in a row.

No other executives clued her into the mistakes she was making. Even when Kendric heard from her own staff that "you shouldn't be doing" things like planning show tunes for the annual meeting, she ignored the advice.

The beginning of the end came with the decision to hire a new advertising agency for McCallister, a competitive bidding process worth approximately $640 million. For seven months, Kendric toured the country visiting the 30 advertising agencies that were bidding on the account. It was during these months of nonstop travel that Kendric made her fatal errors. Whispers circulated that she accepted gifts from these ad agencies and showed favoritism toward certain contenders.

During that time, her conduct surprised and alarmed other McCallister executives. She attended a dinner given by an agency at one of Manhattan's most fashionable restaurants, where she praised the agency and suggested it was favored in the bidding process. Her fellow executives were livid about the dinner because of the strict McCallister ethics rules that ban employees from accepting gifts of any kind from existing or potential suppliers.

Shortly after this dinner, that agency won the McCallister account.

After learning of these incidents like the dinner and other behaviors, McCallister conducted an investigation. Shortly afterward, Kendric was fired for violating company ethics policies. Three days later, McCallister announced the agency bidding process had been tainted and would be reopened without the winning agency's participation.

What happened? At the heart of the controversy was a culture clash that Kendric did not see. She later reflected on the experience. Regarding the weekly all-hands meetings: "Had I known," Kendric says, "I never would have been gone on Friday." Yet no one told her. And she acknowledged that her style and ideas did raise eyebrows at McCallister, and she was never at home within the painstakingly modest by-the-books culture of McCallister. Rather than adapting, she believed the rules didn't apply to her: she was "special."

What did she learn? "The importance of culture, it can't be underestimated."[5,6]

Kendric suffered from ignoring the key factor of corporate culture in organizational life. Unseen, lurking in the shadows, this "thing" has the power to damage, hurt, and even destroy careers and people, as it did to her. Of course, it can also enhance, improve, or advance an individual's career. This may sound a bit dramatic, but after witnessing a few compelling incidents about culture, you will see the importance of paying attention to it and working within its confines.

Corporate cultures are certainly not just the purview of North American firms. As many of you know, Asian business culture has a long-standing and strong set of collective, expected behaviors and beliefs. Today, many North American engineers, due to their increased interaction with their Chinese counterparts, struggle to make sense of these specific business and social cultures. For example, in his book *Barriers to Entry: Overcoming Challenges and Achieving Breakthroughs in a Chinese Workplace*, author Paul Ross lists challenges non-Chinese workers may face today. For example, increased respect for authority and reliance on informal communication for data transfer means working in China is ideal for someone who handles uncertainty, change, and taking the initiative in engineering or business. To date, immersing themselves in the culture by "doing" seems to be the only solution for non-Chinese to solve the cultural problem.

As an example, Ross mentions that taking a break by leaving the workplace, if only going to the coffee shop, may be regarded as impermissible at the office: it may seem like this is slacking even though it could be a useful device for maintaining productivity.[7]

The point here is, as much as it may cause discomfort, your performance within an engineering organization is more than a strict meritocracy. Yes, doing a good job will be rewarded. But that's only half the story. Remember the performance reviews we discussed? There were two axes of evaluation that everyone is judged: by performance in getting the job done (measured by a hopefully objective criteria), and then behaviors, those qualitative actions and interactions that contribute to comfort and confidence. And correct behaviors are driven by culture.

Learning to understand a corporate culture can be very difficult unless you have a little training in how to see it. Think of the concept of the electromagnetic spectrum. As we know, the full spectrum of electromagnetic energy extends from 10^2 Hertz up to 10^{24} Hertz, and is made up of both visible and invisible energy; the visible being a very small portion of the overall spectrum. Thus, the human eye can only see a small percentage of the existing electrical energy present. It's only when we have tools like spectrometers or other instruments can we then see the entire range of energy reaching us. This simile is shown in Figure 9.2.

Corporate culture is the same thing. You need tools that allow you to see this invisible spectrum of hidden beliefs, attitudes, and actions. And the best way to do this is through training. Not formal training, the type you would hire a consulting firm to provide (I don't think I've ever seen a company that offered specific cultural training). Instead, personal practice is the route that will help you see these invisible yet critical aspects of a technical organization. This practice is an activity that may not be efficient, but hopefully effective.

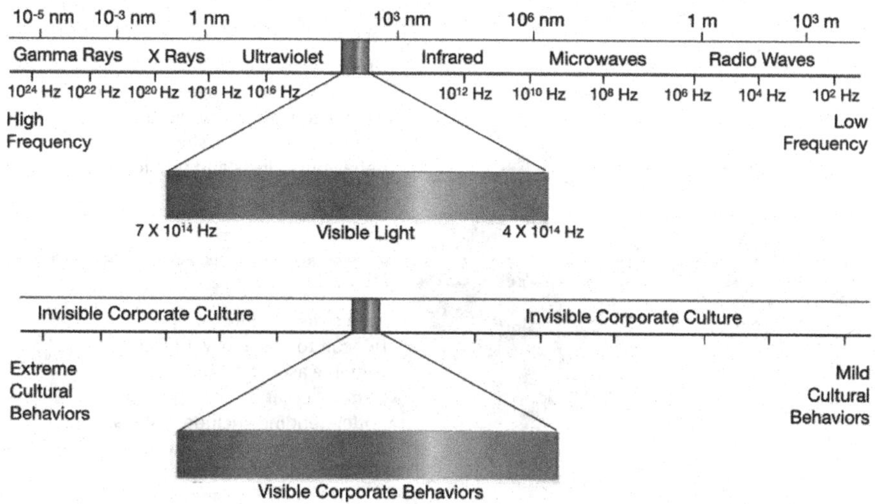

FIGURE 9.2 Comparing the electromagnetic spectrum to corporate culture.

9.2 REPRESENTATIONS OF CORPORATE CULTURE: THE SACKMANN AND SCHEIN MODELS

Several cultural models have been developed over the years. Two of them are particularly helpful in giving you the best training in understanding this squishy concept. These models, *Sackmann* and *Schein*, do a nice job of improving your sensitivity to perceive and understand various cultural situations.

Let's consider the Sackmann Model first, as shown in Figure 9.3. This model (also known as the *Iceberg Model*) is a vertical construct dividing the informal organization into two layers, "observable" and "underlying." The observable culture is superficial; evidence of the culture is easy to see yet is limited in what it tells you. Observable examples are ceremonies, stories, slogans, behaviors, dress and perhaps the group's physical setting. The underlying culture is more difficult to observe yet is much more valuable in defining and understanding what's going on. This underlying culture includes values, assumptions, beliefs, attitudes, and feelings expressed by members of that culture.[8]

An excellent example of observable culture as expressed through a physical setting is the Apple headquarters building in Cupertino, California. Apple spent a reported $5 billion to create a headquarters campus, designed as a seamless circle, nicknamed the "spaceship." The latest design thinking on enhancing workplace creativity and innovation is embedded in its blueprint, with minimal environmental impact and maximum energy efficiency deemed essential. Employee convenience such as seven different cafes, dry cleaning, dinners, free snacks, gaming, and all the rest are *de rigueur.*

It's not hard to see that the Cupertino headquarters is far beyond a mere building: it is a strong statement of Apple's corporate culture and a powerful recruiting tool.

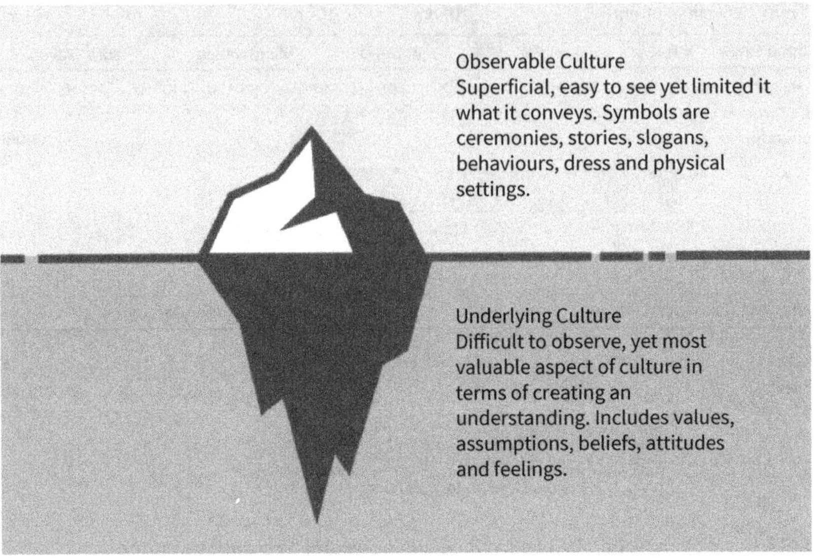

Observable Culture
Superficial, easy to see yet limited it what it conveys. Symbols are ceremonies, stories, slogans, behaviours, dress and physical settings.
Underlying Culture
Difficult to observe, yet most valuable aspect of culture in terms of creating an understanding. Includes values, assumptions, beliefs, attitudes and feelings.

FIGURE 9.3 The Sackmann (aka Iceberg) Model of corporate culture.

The second representation, called the Schein model, divides culture into three main layers arranged in a pyramidal configuration. Similar in some respects to Sackmann, the Schein model (shown in Figure 9.4) is a bit more detailed in parsing out layers of understanding. In this representation, the topmost layer denotes *artifacts and creations*. This is the portion of the culture that is seen most easily by its members. These include physical space and objects, written and spoken language such as slogans, acronyms and buzzwords, and the overt behavior of members remembered in various stories, legends, and myths. These are particularly good examples of common artifacts.

The middle layer, *values and beliefs*, gives a sense of what "ought" to be relative to the group. Usually tested over time and proven to work, values are group-based ideas that are overt, espoused (i.e., spoken) by group members and acted upon without hesitation.

The bottom layer and most difficult to see are the culture's *underlying assumptions*. These assumptions are validated beliefs that are taken for granted and treated as reality, often subconsciously. They are ultimate, non-debatable, foundational expectations and principles considered the underpinning of the local organizational society.[9]

A simple example of Schein's model is a comparison of two schools within the University of Michigan (U-M) campus at Ann Arbor. The U-M College of Engineering is large; some 10,800 students are enrolled, taught by 658 faculty in multiple facilities sited across most of the northern campus. The U-M School of Business is much smaller, educating 4,152 students (including MBAs) with 224 faculty located in a small cluster of buildings on the central campus.

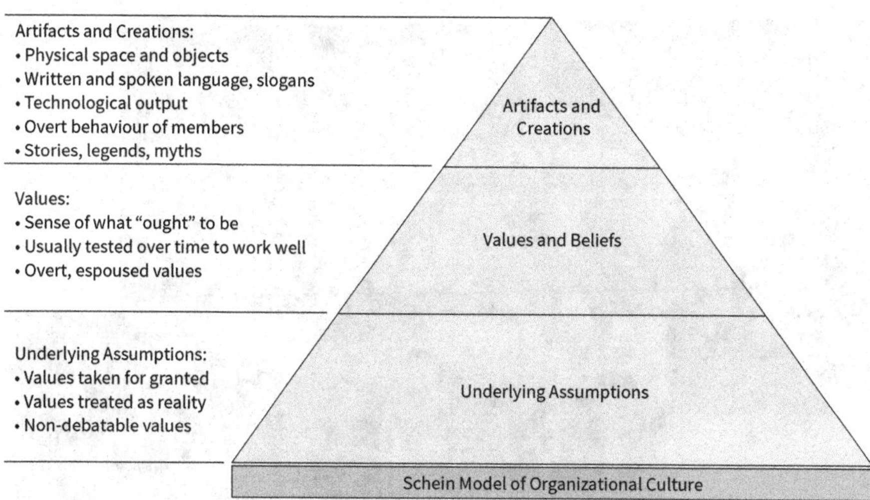
Artifacts and Creations:
• Physical space and objects
• Written and spoken language, slogans
• Technological output
• Overt behaviour of members
• Stories, legends, myths
Values:
• Sense of what "ought" to be
• Usually tested over time to work well
• Overt, espoused values
Underlying Assumptions:
• Values taken for granted
• Values treated as reality
• Non-debatable values
Artifacts and Creations
Values and Beliefs
Underlying Assumptions
Schein Model of Organizational Culture

FIGURE 9.4 The Schein Model of organizational culture.

An analysis of the two cultures shows some basic differences and similarities between them, as seen in Table 9.1.

Let's consider another simple example of group cultural identification and its resulting artifacts. Figure 9.5 shows photographs of two groups of technologists: one is a group of engineers taken in the 1950s, and the other a group of coders taken

TABLE 9.1
Comparison of Two College Cultures Using the Schein Model

Cultural Level	Engineering School Cultural Descriptor	Cultural Level	Business School Cultural Descriptor
Visual Artifacts	Pierpont Commons Student Center	Visual Artifacts	Ross Business School Building
	Duderstadt Center		Kresge Library
	Casual dress		Business casual dress
	U-M Engineering Student Government		Ross BBA Council
Values and Beliefs	Students do assignments so they can learn to solve problems	Values and Beliefs	Students do assignments so they can get a good grade
	Any problem can be solved using the scientific method		Any problem can be solved using Porter's Five Forces model
Underlying Assumptions	An engineering degree will give a person a strong chance at a getting a good job	Underlying Assumptions	A business degree will give a person a strong chance at a getting a good job

FIGURE 9.5 Artifact differences in technical groups: 1950s vs. 2010s.

around 2010. Examine the photos closely and detail the variations in the artifacts between the two groups. What are the differences?

You might be surprised to learn that the cultural differences between the two groups are .... nothing. That's correct: there is essentially no difference in the cultural markers between these two photos. The 1950s group is all dressed identically, with dark pants, white dress shirts, and narrow black ties. They are wearing a uniform, identifying them as members of this group of engineers, and their dress shows it.

There is no difference between these engineers and the coders. They identify themselves in the same way; the jeans and t-shirts are also uniform. The number of black t-shirts is striking: 12 out of 15 individuals are wearing this color. Nine out of ten wear blue jeans (of course, we can also point out another cultural artifact: the number of women in the photo as a marker for gender balance in the coding field at that time (4 out of 19, or 21%).

Cultural behavior can also extend to ignoring certain aspects of your firm's policies or initiatives. Discounting or failing to include culture in new initiatives can substantially reduce their success and effectiveness, or in some cases actually create a deleterious effect. In the mid-2000s, before the widespread adoption of desktop computing and streaming, the Ford Motor Company made a bid to improve its enterprise-level communication with its 300,000 employees by installing hundreds of large-screen televisions into all their global facilities. Placed in hallways, entryways, and commons areas, these screens ran company-created programming continuously,

24 hours a day, 7 days a week. The idea was to improve company communications and awareness through push broadcasting of corporate news stories, with a typical story on this private network lasting about four minutes. The result? To this day, no one can recall ever seeing a single employee standing and looking at one of these screens: the programming played to an empty house. The entire initiative ignored an important cultural fact: no one wanted to be seen standing and looking at a television screen when they believed they should be scurrying about doing the company's business. No one wanted to be seen as a slacker. Within five years, these orphaned television screens were quietly removed at great cost to the company, leaving nothing but bolt holes in the walls as witness to this failed idea.

What's important to note is not that the people didn't want the information, they just didn't want to be seen idly absorbing that information in public. Today, you might see a remnant of this behavior in your own company. If you visit a colleague at their desk unannounced, and they are reading a story on the company streaming site, the odds are even up they will immediately change the screen. Different technology, but the same culture.

Examples like this are where the Sackmann and Schein models can guide you in identifying your own organization's cultural markers, values, and assumptions. And developing those skills is a necessary prerequisite to truly understanding your own firm's culture. And this means practice.

9.3 GROUP CULTURE: THE DENISON MODEL

Even though corporate culture tends to be invisible, we can still categorize group cultures in a basic way. The Denison Model is a clear way of classifying broad-brush organizational cultures and links them to the external environment and resulting internal organizational strategies. It is powerful in that it is easy to understand and points to what a correct corporate culture for your company might be. This means it can be used as a diagnostic tool by your management to ensure your group culture is somewhere near the desired one.[10]

The Denison construct, shown in Figure 9.6, is a matrix linking an organization's level of control of its structure to its focus on the external environment. This continuum of structure (highly structured vs. loosely controlled) and company focus (high internal concentration vs. external environmental attention) results in four quadrants, described below:

Clan Culture

Consisting of a flexible structure with a focus on the internal workings of the group, the *clan culture* is based on a maximum human relations perspective, placing a high regard on employee involvement and participative management. Employees have substantial input on work methods and (to some extent) goals, and higher-than-average resources are allocated to employee well-being. Important exposed values of this model include taking care of employees, ensuring they are satisfied and fulfilled as much as practical, while the role of management is to integrate the culture across all

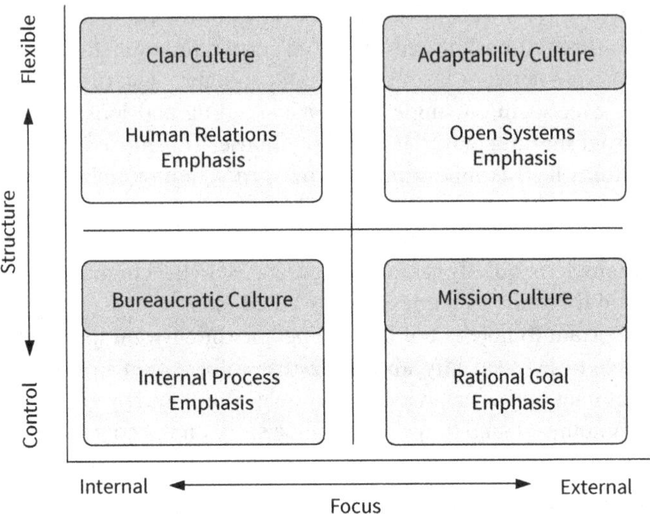

FIGURE 9.6 The Denison Model linking culture to organizational focus.

portions of the firm. The fundamental purpose of the clan model is to achieve high performance by focusing on the needs of employees, where employee investment in the firm creates a greater commitment to it. An excellent example is the Mary Kay Corporation, a U.S.-based beauty care product firm. Their strong commitment to the support of their sales representatives has made Mary Kay one of the largest direct sellers of skin care and cosmetics globally. Mary Kay products are sold in more than 40 markets around the world, enjoying \$2.8 billion in annual revenue at year-end 2020.[11]

Adaptability Culture

As the name implies, an *adaptability culture* emphasizes serving the external environment, using a flexible organizational structure to do this. This approach encourages "entrepreneurial" values (sometimes referred to as "intrapreneurial" values for large, legacy firms) and corresponding norms and beliefs. It encourages behavior that allows the organization to detect, interpret, and translate changing environmental signals into new responses, reacting quickly to environmental shifts in customer needs, wants, and desires. It values employee creativity and risk-taking with minimum penalties for failure. Management's role is to establish and reinforce a culture of continuous change to maintain relevancy moving into the future. An example of this is the 3M Company, where the 2020–2022 COVID pandemic changed their famous internal innovation model into one much more collaborative with external technical firms. The results were impressive: 3M designed and produced a new powered air-purifying respirator in just months, increasing the production of high-efficiency filters for COVID personal protective equipment. In 2020 alone, 3M increased respirator production from 20 million to 95 million per month in the United States and 2 billion respirators globally.[12]

Bureaucratic Culture

The bureaucratic culture tends to be the opposite of the adaptability approach. As its name implies, the bureaucratic structure is best for tightly controlled company structures with an internal focus, implying an unchanging external environment. This drives a methodical approach to their business with lower personal involvement coupled with a very high level of operational consistency and conformity. A highly integrated structure aims for best-in-class efficiency and a resulting hope for effectiveness.

Values in these bureaucratic cultures are cooperation and mutual support among employees through the strong and consistent application of established policies. Management's purpose is to continuously reinforce this culture of control, process improvement, efficiency, and a highly mechanized approach to work. A good example of this culture is the U.S. Internal Revenue Service with its annual deadlines, stable purpose, and easy-to-measure performance.

Mission Culture

The mission culture resides within organizations that value control of the business while keeping their attention on the external environment. The mission approach maintains a clear vision of organization's purpose and goals, having a significant dependence upon metrics for employee evaluation and performance, competitive firms, process improvement, and bottom-line financial results.

Mission cultures value beating the competition while using superior processes to help achieve this goal. Management's role is to strengthen this culture through training, education, and by example. An excellent illustration is the beverage company AB InBev, whose emphasis on professionalism, ambition, empowerment, and aggressiveness are key values. And for AB InBev, it works. Market share in 2020 was 25.7% vs. 12.2% for its nearest competitor. AB InBev owns the market.[13]

These four models offer a good initial understanding of how corporate culture directly impacts company competitiveness and performance.

Let's now move on to look at models focused on you as an individual to deal with culture on a day-to-day, local basis.

9.4 SELF-AWARENESS IN CORPORATE CULTURE

Let's say you've developed a reasonably good skill at reading your group's informal culture, and you begin to see some group behaviors and beliefs that are endemic to your local organization. For instance, say you perceive a tendency for your workmates to come in late and stay late, or perhaps tend to work at home four days a week and only come in on Monday afternoons. Or witness that "Taco Tuesdays" are actually an important group bonding experience that's not to be missed.

Whatever the specific behavior or event, once you see it begs the question: what are you going to do about it? What I mean is, what is your own reaction and decision regarding your participation in your local organization's culture? I'm not talking about grabbing a taco on Tuesday. I'm referring to your own decision as to how much you will embrace the prevalent culture, especially considering the power the culture can exert over your thinking and actions.

An example of this kind of decision happened to Reed Kitteridge a number of years ago. Reed was a new engineer, helping to design the Boeing 757 commercial airliner. It was a Saturday, and he was sitting at his desk trying to understand the internal arrangement of some avionics equipment in the fuselage. Reading his blueprints, he just could not visualize how the parts came together. In the assembly building a quarter mile away, the first prototype was coming together and one of Reed's colleagues suggested he walk over to the assembly area and look at the equipment bay in question. As Reed stood up and grabbed his coat, one of the older "gray heads" asked Reed what he was doing. Reed replied, "I'm headed over to see the prototype." The gray engineer's reply startled him: "So, going sightseeing on overtime, huh?" This stopped Reed in mid-stride. He immediately understood that leaving his desk to see the fuselage, while a big help to him, was seen by the others as a form of slacking. Reed took off his jacket and sat down. Now, the next regular workday, Reed went over and saw the prototype with absolutely no comment from anyone. But the message he received was clear: certain actions done during straight time were acceptable, but those same actions during high-cost overtime were considered sketchy. This cultural rule stayed with him for years to come. To the gray-headed engineers, overtime paid at a time and a half rate didn't include what was seen as nonproductive activity, even though it was.

Sometimes cultural self-awareness isn't left to be discovered; sometimes it's told to you explicitly and directly. In the United States Congress (a hidden cultural institution if there ever was one), one piece of advice is always given to rookie first-term congressional members: "Go along to get along." This means adopt the party leaders' beliefs and positions and become a loyal party member, voting for your leaders' policies even if it runs against what you promised your constituents back home. The same is true where you work. And a common cultural example that cuts both ways is the *quid pro quo.*

9.5 THE QUID PRO QUO

An early concept you'll encounter in the informal organization is the quid pro quo. The quid pro quo is a Latin phrase literally meaning "what for what" or "something for something," more commonly seen as a favor or advantage granted in return for something of roughly equal worth at a future time. It's an exchange of value between two individuals, parties, or organizations.

The quid pro quo is a staple of the informal organization. It is invariably covert and may be used immediately or separated by time and place. And while a quid pro quo is (on the surface) an exchange of value over time, it's real worth is that of an investment in a person or group; the development of rapport that becomes the basis for an ongoing relationship. This is particularly important in what's called *building capital,* which is a stockpile of a stockpile of favors done for another person, to be redeemed at a future time (sometimes called a "chit"). There is no way someone will not need the benefit that the quid pro quo provides. A technician solving an unforeseen circuitry problem, a graphic artist shifting your job to the head of the queue, or the all-important IT helpdesk expert who can resurrect your laptop from the dead. Your success depends on building and spending the capital the quid pro quo provides.

9.6 SOCIAL RITES AND CEREMONIALS

Social rites and ceremonials are very important components of the corporate culture, lending strength and viability to a given organization. At its core, a *social rite* or *ceremonial* is a public expression of an organization's culture. According to Trice and Beyer, a social rite "amalgamates a number of discrete cultural forms into an integrated, unified public performance; a ceremonial connects several rites into a single occasion or event."[14]

Rites and ceremonials combine different forms of cultural expression within group events, with well-defined beginnings and endings. In them, people use certain cultural markers: customary language, gestures, ritualized behaviors, artifacts, and impressive settings to emphasize these shared experiences and values. As mentioned above, often these values are shared through myths, sagas, legends, or other stories associated with the occasion. Sometimes, these events can even demonstrate certain irrational or unproductive behaviors that are actually functional and appropriate for some members of the organization.

You will find there are a surprising number of organizational rites and ceremonials routinely expressed in your new engineering group. Table 9.2 summarizes some of these common events and experiences you will undoubtedly witness firsthand.

TABLE 9.2
Social Rites and Purposes in Hidden Organizations

Type of Rite	Example	Social Purpose
Introduction	Boss personally walks you around the workplace; individually introduces you to each existing group member	Authority figure publicly certifies you are a group member in good standing, deserving of all group rights and privileges
Passage	Induction and basic training into a new technical specialist role	Facilitate transition of person into new social roles and statuses
Degradation	Written notice of the firing of a top engineering executive	Dissolve social identities and their power
Conflict Reduction	Collective bargaining	Reduce conflict and aggression
Enhancement	Annual awards night	Enhance social identities and increase status of members
Renewal	Organizational development activities (i.e., "Ropes Course" or group community service projects)	Refurbish social structures and improve organization functioning
Integration	Office holiday party	Encourage and revive common feelings that bind members together and commit them to the organization

CASE EXAMPLE 9.2 INTRODUCTION OF THE ORIGINAL IPHONE

An outstanding example of a dual enhancement and integration rite occurred during the legendary 2007 corporate launch of the original Apple iPhone. This presentation, given by Apple founder and CEO Steve Jobs, introduced the then-revolutionary idea of a single device combining a mobile phone, a digital music system, and internet access device, all packaged inside a minimalist design with a scrolling screen. This presentation has been repeatedly analyzed for its effectiveness, most notably by Nancy Duarte, author of *Slide:ology: The Art and Science of Creating Great Presentations.*[15] While the presentation delivery aspect is the most discussed, there is another message embedded in the presentation that truly gives it power. In it, Jobs repeatedly (and not so subtly) communicates to the crowd of faithful Mac developers the cultural message that they are the very special individuals: the smartest people around, working on the world's most advanced and disruptive technology that is changing human history while simultaneously defining what is "digital cool." Over and over, as he describes the function of this groundbreaking device, he reminds his audience (or perhaps more accurately, his "fans") that they are part of a unique and special organization and more importantly, are practitioners of a new way of thinking and being. The presentation is pure enhancement and integration, and well worth looking up on YouTube. Look for the cultural references: not in the product, but from the implied description of Apple.

9.7 THE POWER OF MYTH: THE FOUNDERS' STORY AND OTHERS

One of the most interesting cultural experiences you will witness has to do with "the power of myth": the impact of storytelling on a firm's culture. You may not think that storytelling has much place in an engineering organization, yet once sensitized you will see it nearly everywhere. Within just a few weeks of arrival, you will undoubtedly be exposed to various stories: some true, some not so much, and some merely exaggerations of prior events. These are the history of the organization, things that have happened to its members; always engaging and sometimes even funny. These tales give your organization its unique flavor and local rules for understanding and action. You will hear anecdotes that indirectly define the values of the group; the correct way to perceive, think, and feel in relation to your colleagues.

The fundamental corporate myth or story is a classic: nearly all engineering organizations tend to use the "founders' story" as a cultural indicator. These stories are carefully crafted parables that quickly become identified with the root of the firm's identity, purpose, and reason for being. They are also useful in large corporations to "humanize" their identity and are employed by corporate public relations departments to craft a belief of heroic (or perhaps just smart and well-meaning) employees trying to make the world a better place. We all know the famous story of the telephone inventor Alexander Graham Bell saying to his assistant: "Mr. Watson – come here – I want you." The Bell Telephone Company and its many, many descendants have used that story to great advantage for decades.

Another familiar example regards the Hewlett Packard (HP) Garage, the founding location of the HP company. Here's what HP says about their garage:

> Tucked away on a quiet, tree-lined residential street near Stanford University, the HP Garage stands today as the enduring symbol of innovation and the entrepreneurial spirit. It was in this humble 12x18-foot building that college friends Bill Hewlett and Dave Packard first pursued the dream of a company of their own. Guided by an unwavering desire to develop innovative and useful products, the two men went on to blaze a trail at the forefront of the electronics revolution.[16]

Marketing professionals even have a name for all this: *grand storytelling*. And for any entrepreneurs you may know, the web has dozens of easy-to-find articles telling them how to craft their own unique and engaging founder story to propel their brand forward.

The sheer number of founders' stories shows us just how effective these tales are in setting and maintaining a corporate culture. And from these stories can come your and your colleagues' ongoing behaviors and beliefs.

9.8 THE DYNAMICS OF CULTURE

While the founders' story implies the organization's culture is static and fixed in the past, in reality, culture is dynamic, and changing at an accelerating pace. Like an evolving organism, the informal organization continually morphs as new products are created, fresh methods are developed, and recent employees bring new experiences from the ever-changing external environment.

Let's consider the story of Marsha Fleming, an engineering analyst taking a routine business trip from Boston to St. Louis. One day Marsha's colleague John was in the snack room, looking for a Snickers bar when his coworker Darshan comes in. Darshan immediately says to John:

Darshan: "John, did you hear about Marsha's business trip yesterday? Unbelievable!"
John: "No, what happened?
Darshan: "Marsha was flying to St. Louis, and you know that big ice storm we had yesterday? Well, on takeoff Marsha's plane slid off the end of the runway. Thank God no one was hurt, but they had to take everyone off the plane on those rubber slides and close the airport. It's so lucky that Marsha didn't get hurt or anything worse."

"Now, here's the amazing part. Marsha went right back to the terminal, marched up to the counter, and immediately booked another flight to St. Louis for later that day. And she got on that next flight and flew right into St. Louis. I would never do that!"
John: "You're right, that is amazing! I wouldn't do it, either."

The next day Ahmed, Marsha's boss admiringly tells the same story during a weekly group meeting, citing Marsha's "get it done" attitude and perceived bravery.

The message is clear: management admired Marsha for placing her work above her personal concerns. That story was repeated for years afterwards, and a new cultural norm was born. The message? You, too, will be admired if you go the extra mile for the company, especially in some unusual way.

9.9 RESPECT THE COLLECTIVE

Tony Phillipson is a talented engineer with about 15 years' experience. He has been with his engineering firm through the tough times, when money was tight and there was barely enough cash to cover the payroll, let alone "frivolous" items like office improvements. For his entire career he had been working at a desk with a file cabinet that was bought during the mid-1960s, while Lyndon Johnson was the U.S. president and international phone calls had to be reserved in advance with the phone company. The company conference rooms had not been renovated since the late 1970s and lacked the basic amenities like projectors, speakerphones, and chairs without springs tearing your pants.

Tony was delighted when the company finally renovated his building's office space. His department's conference room was now a dream: new chairs in bright colors, broadband videoconferencing capability, and variable lighting. In the middle of the room was a beautiful conference table made from light oak and large enough to seat 12 people in comfort and style. Tony and his entire department agreed: the room was a knockout.

Soon after the renovation was complete, Tony went to the room to prepare the communications equipment for an upcoming meeting. Tony opened the door and was astonished: sitting cross-legged on the table, texting, was a new engineer, and judging by the condition of his beard, perhaps 24 years old. Not in a seat, but *actually sitting on the table*, dirty sneakers and all.

Tony was stunned but quickly became furious. Actually yelling at the engineer, Tony delivered a quick lesson in respecting the culture:

> What are you doing on that table? Don't you know that's brand-new? We haven't had a new conference table in here for probably 50 years and you're sitting on it with your dirty shoes. People work in here and sometimes eat lunch at that table. This isn't in your dorm room and it's certainly not your buddy's apartment. Get the hell off that table and don't you ever do that again. Get outta here.

The young engineer didn't need to be told twice. As Tony tells it, the engineer disappeared from view faster than lightning. So, what happened next? Tony repeated this "can you believe it?" story to anyone he ran across, including those at staff meetings, design reviews and especially in hallway conversations. The guilty party became the object of rolled eyes and whispered comments, and certainly the story got back to him. The lesson is clear: discover and respect the culture that others before you have provided. Remember, to Tony, the young man not only disrespected his current workmates, but also everyone who had come before him. With culture, it's not only about today: the past is still alive and relevant.

A favorite description of respecting the collective was from Peggy Noonan's book *What I Saw at the Revolution*, about her time as a speechwriter for U.S. Presidents Ronald Reagan and George H.W. Bush. In it, Noonan perfectly describes the special collective culture of the White House staff when traveling on Air Force One:

> I was going to go on Air Force One with the president of the United States to the University of Alabama, where he would give the stump speech to thousands of adoring students. I was so excited I dressed nicely, with an expensive sweater and a truly adult Norma Kamali black linen skirt, and I resolved to keep a friendly, open adult face on.
>
> We rode in limousines from the White House to the airfield. We rode right up to the plane. I said my name to a military aide who …. checked off names as we jogged with just the right amount of casual energy up the stairs. At the top I turned to see what the president sees when he stands and waves …. and turned into the coolness of Air Force One. There it was: the West Wing hum. And the West Wing aroma, flowers on the airplane.
>
> There was a name on each seat. There was no boisterousness, no loud talk. … I was too excited to read but I didn't want to look that way so I read the *Times*. People walked by. A steward. 'Would you like anything, ma'am?' Yah (with a yawn), think I'll take some coffee. (This isn't my first time here.) I …. looked around in a bored fashion. Intense men in suits.
>
> The famous smooth takeoff was smooth. Air Force One macho: No one wore seat belts. I unbuckled mine. Soon we were above the clouds and the president walked back to say hello. … this morning he wore his suit and walked from chair to chair. Ken reminded him who I was, and the president said of course of course and talked about the day ahead.
>
> Outside we scrambled for cars. … when you travel with a president, you get a little booklet that tells you not only what car you're assigned to but who is in it with you, and as soon as people get it they read it to see if they're being treated with respect.
>
> Back to the plane …. This time I saw that everyone was sitting in his own separate seat doing his own separate work, all of them together but really apart. They did not speak with comfort and familiarity to each other. I didn't feel I belonged. Now I wonder, did anyone?
>
> I remember Margaret Tutweiler, an aide to James Baker, knitting in the seat in front of me and saying to Senator Strom Thurmond as he boarded as a guest, "Senduh, you just sit right down here next to me and we'll catch up." She belonged, or must have felt she did.[17]

Notice that almost every scene of the trip had its own little cultural twist, with the travelers all observing the West Wing culture, showing their respect when traveling within this special community. Even though they were not really part of the group, they made sure to follow the culture anyway.

One last example of respecting the existing culture had to do with Amad Jamal, an 18-year veteran process engineer in an 80-year-old company named Brookdale, which specialized in manufacturing and packaging processed vegetables. Brookdale was rapidly moving into manufacturing plant-based meats, a major innovation certain to disrupt the food processing business. Amad's processing plant was being converted to support this new direction by installing two new continuous flow manufacturing lines, each to produce ground "beef" patties at a rate of over 1,600 per hour. Each line was estimated to cost $8.7 million, complete.

A notable feature of each line was a very large compressed nitrogen tank, some 38 feet high and 12 feet across. This tank provided critical pressure to quickly move the processed vegetable slurry into the patty extrusion machine. Each tank was cast steel and, due to its massive size, dominated the production line.

Amad was assigned total responsibility for installing one of two new lines. Assigned to lead the second line installation was a relatively new process engineer named Chuck Gregory. With just two years' experience, this would be Chuck's first installation. Though never said out loud, the installation would become a small competition between the "old dog" and the "young turk."

Both installations went well, with no job-stoppers or surprises, until the very end. With one day to go until their senior management was to visit and celebrate the new installations, Amad quietly told his crew to paint their compressor in a rich, gloss blue (Hawthorne's corporate color) and add a large company logo to the tank's side.

The next morning, Chuck arrived at the site and was horrified: Amad's installation was beautiful; clean, and professional. Chuck's installation looked like it wasn't finished; dull and grimy. It was instantly apparent whose installation was perceived as better, even though from a functional viewpoint each was absolutely equal. Both engineers were congratulated by their seniors, but most people present noticed a bit more warmth shown to Amad.

Amad knew his culture. He knew at Brookdale, all things being equal, excellent building and hardware appearance was seen as professional, demonstrating the extra effort supporting the company's desire to be the best. And a mere coat of paint made a huge difference.

A postscript. After the ceremony was over, Amad (with the smallest of grins on his face), offered Chuck his crew to help paint the unfinished tank.

Chuck accepted.

9.10 SUBCULTURES

As the name implies, subcultures are smaller, different cultures embedded within a larger organization's society. A subculture reflects the common problems, goals, and experiences that members of a team, department, or other small unit share that are different from the firm's culture as a whole. A subculture can successfully exist within a larger culture, even when the subculture holds conflicting values that can create inconsistent behavior as long as there is complete consensus on both cultures' underlying assumptions.

Let's discuss an example. A subculture exhibiting inconsistent subcultural behavior is the U.S. Navy Blue Angels flight demonstration team. The Blue Angels have, for decades, demonstrated excellence in formation flying of high-performance aircraft. The seven members of the team, drawn from existing fighter aircraft squadrons and rotated every two years, are a highly trained, small group-based organization consisting of the best pilots and crews the Navy has to offer.

As fighter pilots, you would expect this team to exhibit the behaviors and values of the stereotypical "Top Gun" fighter pilot: highly competitive, aggressive, rule-breaking Type A personalities on steroids. Yet if you look in-depth into the training of Blue Angels members, you'll see some unusual group dynamics and behaviors. One of the most interesting is the daily "séance," where each morning the entire team practices their demonstration at a conference table with their eyes closed, visualizing the speed and position of their aircraft relative to their teammates. Their right hand moves an invisible control stick while their left hand operates a nonexistent throttle. Watching these highly skilled pilots practice in an almost trancelike state is the antithesis of what we might think of as the typical supersonic warrior.

Another strong subculture is their heavy reliance on the After Action Review, as mentioned earlier. After each flight, every pilot performs a highly self-critical admission of each error they made in the exercise, followed by another hard-nosed evaluation by all the other members. This includes the team commander. And at the end of each individual self-examination, every pilot makes the same statement to the others: "I promise to correct my errors, and I am happy to be here."

This subculture, exhibited by a high-profile group with a very unusual and dangerous job, developed out of necessity. With potential death resulting from any mistake or miscommunication, the team uses any and all techniques they deem appropriate to avoid disaster. And if that involves daily, Zen-like training sessions, that's fine.

9.11 THE STUBBORNNESS OF CORPORATE CULTURE

Most long-established legacy companies eventually build up within their cultures a calcified view of themselves. Tending to focus on a successful past rather than an uncertain future, middle management develops a reluctance to embrace the changing external environment, preferring to enjoy the substantial perks of a past success. This spawns the organizational condition called *layers of clay*.

Layers of clay are a hidden cultural condition where those who have benefited from a past culture will actively erect barriers to establishing a new, required future culture.

Here's how it works. Think of an organization as a structure made up of three major layers: the very top layer where senior management lives, the very bottom where the rank-and-file workers reside, and an intermediate layer populated by middle and slightly higher management types. Say a major change to the company's product offerings is needed due to a surprise change in the external environment. As senior management is directly charged with ensuring the continued existence of the firm, they create and earnestly endorse the company's new direction. The firm's bottom rank-and-file population is open to this change, as they know declining sales mean layoffs and pay cuts. Supporting the new direction contains both an explicit and implicit promise of a better future for them.

The intermediate layer is a problem. As middle management, they have reached a moderate level of financial success, power, and influence. They may have just built a new home, bought that boat they've always wanted, or just completed their children's college savings accounts. To them, the status quo is very, very important. The reference to clay indicates their level can be exceedingly hard to penetrate. Without the support of the intermediate layer, executing an actual change becomes a significant challenge. At its core, senior management creates and directs a change, the bottom layer supports, but the mid-level attempts to stop the change because their "ox is getting gored." Only in a severe crisis does this resistance break down.

The idea of layers of clay can help answer the question "Why doesn't change within an organization happen quickly or doesn't even occur at all?" Logically and rationally, anything that is good for the organization and the people within it should be a no-brainer, it should happen promptly and without much discussion. Unfortunately, this is not a universal rule.

However, there is good news about the impact of layers of clay on an organization. Senior management and almost all of the rank-and-file know this exists and will do their best to reduce this layer. Fortunately, as we move through the third decade of the 21st century, the resistance to change is slowly fading as a younger generation of engineers achieves middle management rank. Time will eventually wash this layer of clay away, opening the way for a new culture to support the new business direction.

9.12 WHOSE CULTURE IS IT, ANYWAY?

Let's now look at the hidden organization from a different vantage point. Many large engineering corporations are multinational; spanning continents in pursuit of both market share and the talent necessary to gain it. Depending upon the type of organization structure, various countries will host local company facilities supporting research and development, production, tier-one suppliers, product development, and untold other groups required to conduct business in that country. On balance, it's very beneficial to have these "outposts," bringing the company closer to local customers and their needs. But a sticky situation emerges when we talk about corporate culture in these locations. In general, there will be two groups of employees: local workers from the host country and the expatriate employees, those individuals from the firm's headquarters country temporarily living there, from one to five years being typical. Which culture will be exhibited in the host facilities? What norms and expected behaviors will you see at these remote locations? Should these facilities follow local customs or attempt to emulate the headquarters culture?

The fact is, there is no universal answer: once again contingency theory raises its head. If an employee is in a foreign location, they will need to be on the lookout for a local solution to any cultural issue. In general, culture will default to a mix of lower-level, local cultural behaviors but also honor certain overarching, large cultural behaviors specified by the mothership. For instance, an overseas facility may require all employees (including executives) to remove the trash from their own work area each night (a local cultural behavior), yet senior management review meetings at that same location must conform to the headquarters pro forma and rules of presentation. Expatriate management is normally aware of and sensitive to these local cultural issues, but sometimes they are not.

A few years ago, a quintessential American multinational elevated a Lebanese-Australian executive to be the new CEO of the company. This company's identity was steeped in the American tradition. To many of its customers, its brand name was synonymous with America and what it stands for.

This new CEO made a telling cultural error from the get-go. Soon after taking charge, an email from the CEO's office was sent to each employee, stating that effective immediately, British language spellings were to be used on all company documents. This meant "color" was now "colour," "ton" was now "tonne," and "labor" was now "labour." Even the CEO's first name (previously spelled and pronounced "Jack") was now to be "Jacques." Obviously, the new European slant put the company's American workers into a bad humour.

The importance of this example was not in the pettiness of the directive but was seen as an attempt to reduce the identity and strength of the foundational culture, and was a preview of what was to come under the new regime. Immediately, Jacques lost much-needed goodwill from the North American employees of the company, not a good way to start. His errors only multiplied from there, and he had no reserve of goodwill to absorb his rookie mistakes. Finally, Jacques was ousted 34 months into his tenure as CEO and replaced with the great-grandson of the company's founder, a move as American as you can get.

You probably shouldn't be surprised when an organizational subculture pops up when you don't expect it. An interesting experience occurred when I was working as a design engineer on the U.S. Marine Corps AV-8B fighter. The schedule to complete the design phase was (as always) tight, and the design team was put on a standard 54-hour workweek (10 hours per day Monday through Thursday, 8 hours per day on Friday, and 6 hours each Saturday). After a solid year of this, company management announced that, to reduce burnout and improve productivity, every fourth Saturday would be a mandatory day off.

Several years later, I was taking an evening class on manufacturing practices when the topic of work schedules came up. I shared my experience about the mandatory fourth Saturday off. To my surprise, a fellow student with a military bearing (brush cut and tie inserted between the 2nd and 3rd buttons of his epaulette shirt) turned and told me that anyone who needed a fourth Saturday off "obviously wasn't tough enough."

I stopped in mid-sentence, not sure what I just heard. Not until I thought a bit did I realize this was a classic subcultural example. This student was what we called a "plant rat," a manufacturing employee who has a strong personal pride in their commitment to work long hours at their job. You can identify this type by many clues, such as when they refer to "getting the product out" or "good SPC numbers" or especially "in control." While others might prefer to spend their non-working hours outdoors or with family and friends, the subculture of the plant rats' work ethic can be strong and compelling, and they own the culture with pride.

9.13 IN PURSUIT OF CULTURE

At this point, we know that cultures exist, that all are different depending on the company in question, and at first, almost all cultures and subcultures are invisible to you. We also know that new employees should try to integrate with the culture as fast as they possibly can, even if it may not make sense at that moment. There are many ways of quickly integrating, including the obvious artifacts (style or ways of dress, language, etc.) and the more hidden markers (behavior patterns and expectations).

Obviously, a good first place to look for your group's culture is when you are completing any task or assignment involving several people from your unit. Analyze their activity. For example, in larger meetings, seemingly small things (such as seating position) can influence how a meeting proceeds and how people interact. Let's take an old-school example that you might think no longer applies. Consider a formal, planned meeting around a U-shaped conference table. Traditionally, the ranking member will tend to take the position at the bottom of the U. Other participants will spread out on either side of the ranking member using both sides of the table. The lowest members will sit at the top of the U, and any overrun will take seats against the wall on either side. This simple matter of seating shows the explicit and implicit hierarchy of the group; you can identify the ranking member easily. If you suspect this is old-school behavior that isn't relevant today, now consider this: let's say your director asks everyone to meet outside on the lawn by a big oak tree. Who gets to lean back against the tree, who sits in the shade, and who sits in the hot sun? A good bet is the ranking member leans against the tree: the seating position hierarchy still exists. And now think about a Zoom, WebEx, or other virtual meeting. Who gets their "picture box" placed at the top of the screen? Is anyone's box bigger than the rest? Who gets to black out their video? Who's microphone always stays on while everyone else is muted? The point here is that no matter the surface technique used in a group interaction (a conference room, tree, or virtual), there is still a need for the participants to implicitly identify the power structure.

Corporate culture expressed through dress has been talked in every "how to" article since the mid-1950s, with the rise of *The Man in the Gray Flannel Suit*.[18] Other than the comparison of a group's dress discussed earlier in this chapter, I'll just add two additional points:

1) Believe it or not, dress is always important. As mentioned earlier, it's important to know that every group adheres to the same consistency of dress; everyone uses dress to signify they are part of a particular group. Their cultural identity with the group is the key point, no matter the style of dress.
2) Don't wear a costume.

A point you'll hear constantly repeated is to make friends with people who can help you. There is always an invisible communication network present in your group and department: it is informal, private, and valuable. Do not mistake a company website or a formal, new employee communication network as all the information you need. The good stuff doesn't reside there; it's inside people's heads. And that means tapping into the hidden culture within your group and department. This means drinking a Monster at someone's office or workstation can be very valuable time spent.

In a different realm, it's better than even odds that the subculture already in place will require you to "pay your dues," which is the idea that you must invest in the group before appropriate membership will be granted to you. Dues is not education; it is actual work and a real demonstration of commitment and investment in your colleagues. In some engineering subcultures or jobs, the "new one" will be kidded or picked on by older, more experienced coworkers. This is not necessarily a bad sign. It may be the normal part of the dues-paying process. You will find out soon enough.

9.14 REALIZATIONS

There is a lot of "soft" information in this chapter, and because contingency theory is so strong, distilling culture down into a few pithy "do and don't" checklist items is doomed to failure. Instead, let's briefly review the main ideas of what we've discovered.

First, an engineering company is a social organization, an entity made up of engineers who must work together. The company mirrors the values of its founders, owners, and hired managers, and the resulting corporate culture is a composite reflection of the values of its inhabitants. It has a founding story, but that story also changes over time. And one of the responsibilities of senior management is to ensure that culture changes in an appropriate and positive manner in response to a changing environment.

For success, it is essential to understand how a company's culture relates to its employees and the outside world. Recognizing the existing culture is important from the standpoint of its breath and most importantly, it's strength. And yes, that strength can cause you to be hypnotized by the existing culture.

Your company, department, or group culture has an importance far beyond what one might think. There is an old saying that "culture eats strategy for breakfast," meaning the impact of the current culture can be more powerful than an organization's strategy and plans. There is some truth in this.

A professional has a good understanding of the informal organization in which they reside. They understand its power and honor its place in the company's environment. They work within it, and only in the rare instances where the culture is negative, destructive, or illegal do they knowingly negate the culture, usually by self-selecting out. Professionals are not frightened by culture, nor are they subject to the temptation of being subsumed by it. They see culture can enrich their work life and even act as a teaching tool. The more experience they have, the better their ability to use the culture positively. And they respect the culture's strength, and unlike Laura Kendric, know that the informal organization can damage them deeply.

Finally, subcultures give "running room" for smaller groups to establish customized cultures that enhance the local group's performance and success. Subcultures should be allowed to prosper and grow under the protection of the main culture. While your firm has its culture, you will spend essentially all your time in your subculture. Learn it well.

NOTES

1. Nadler, D. A. and Tushman, M. L. 1997. A Congruence Model for Organizational Problem Solving. In *Managing Strategic Innovation and Change*, ed. M. L. Tushman and P. Anderson, 159–171. Oxford: Oxford University Press.

2. Schein, Edgar. 1984. Coming to a New Awareness of Organizational Culture. *Sloan Management Review* 25(2): 3–16.
3. Daft, Richard. 2010. *Organizational Theory and Design*. Mason, OH: Cengage Learning.
4. Ibid.
5. Barbaro, Michael and Elliott, Stuart. 2006. [McCallister] Fires Marketing Star and Ad Agency. *New York Times*. December 8, 2006.
6. Berner, Robert. 2007. My Year at [McCallister]: How Marketing Whiz [Laura Kendric] Suffered a Spectacular Fall in 10 Short Months. *BusinessWeek*. February 12, 2007.
7. Ross, Paul. 2020. *Barriers to Entry: Overcoming Challenges and Achieving Breakthroughs in a Chinese Workplace*. London: Palgrave Macmillan.
8. Sackmann, Sonja. 1991. Uncovering Culture in Organizations. *Journal of Applied Behavioral Science* 27(3): 295–317.
9. Schein, Edgar. 1990. Organizational Culture. *American Psychologist* 45(2): 109–119.
10. Denison, Daniel and Mishra, Aneil. 1995. Toward a Theory of Organizational Culture and Effectiveness. *Organization Science* 6(2): 204–223.
11. Forbes. 2021, Mary Kay Corporate Summary. https://www.forbes.com/companies/mary-kay/?sh=39a4f1da2bc9.
12. Benjamin, David and Komols, David. 2020. How the Pandemic Changed 3M's Approach to Innovation. https://www.forbes.com/sites/benjaminkomlos/2021/12/29/how-the-pandemic-changed-3ms-approach-to-innovation/?sh=11210cd91ae0.
13. AB InBev. 2022. Our Culture. ABInBex Corporate Site. https://www.ab-inbev.com/careers/working-with-us/our-culture/
14. Trice, Harrison and Beyer, Janice. 1984. Studying Organizational Cultures Taught Through Rites and Ceremonials. *Academy of Management Review* 9: 653–659.
15. Duarte, Nancy. 2008. *Slide:ology: The Art and Science of Creating Great Presentations*. Sebastopol, CA: O'Reilly Media.
16. Hewlett Packard Corporation. 2022. A Home for Innovation. Hewlett Packard Corporate Site. https://www.hp.com/hpinfo/abouthp/histnfacts/publications/garage/innovation.pdf.
17. Noonan, Peggy. 2003. *What I Saw at the Revolution: A Political Life in the Reagan Era*. New York: Random House Trade.
18. Wilson, Sloan. 1955. *The Man in the Gray Flannel Suit*. Lebanon, IN: Da Capo Press.

10 Ubiquitous Communication

10.1 ORGANIZATIONAL COMMUNICATION: IT'S NOTHING PERSONAL

Ask any engineering manager, director, or other executive what the most valued organizational talent is, and they will probably give you one word: communication. Communication is a skill, not a science. Articles and books galore talk about communication: your personal communication, communication between groups, high fidelity communication, communicating with technology, communicating without technology, and on and on. An entire army of communication consultants, presentation gurus, and management experts all promise they have the secret to good communication. The IABC (the International Association of Business Communicators) has thousands of members in over 100 chapters in North America alone. And for good reason. Communication is "sticky." It's the glue that holds an organization together, a cement that sticks to everything. Technical divisions, departments, groups, and individuals all depend on communication to effectively and efficiently get something done. Exchanging information through communication is how it's accomplished.

Multi-modal communication is critical in all organizations: big or small, old or new, complex or simple. Communication conveys information, and organizations use information strategically to make sense of ongoing changes in its environment, acquire new knowledge for future innovation creation and make decisions about future courses of action.[1]

Communication in firms is not defined by you or other individuals; it is defined and governed by the organization itself. The organization instructs you as an individual engineer how to communicate, not the reverse. There are rules and boundaries, conditions and contingencies that govern organizational communications, and everyone must get to know them.

Looking again at the entire Essential Engineering Framework (Figure 10.1), the Inner Core shows the communication component as the entity holding the four other components together. Here, communication binds these four entities tightly together. A technical organization needs this construct to create and support an integrated, holistic information transfer system. Without this cement, the Inner Core collapses. This is the nature of organizational communication and why we will spend some substantial time on it.

It's only right to emphasize this communication. After all, communication, be it written, spoken, presented, texted, emailed (or merely whispered in the hallway), is

DOI: 10.1201/9781003214397-13

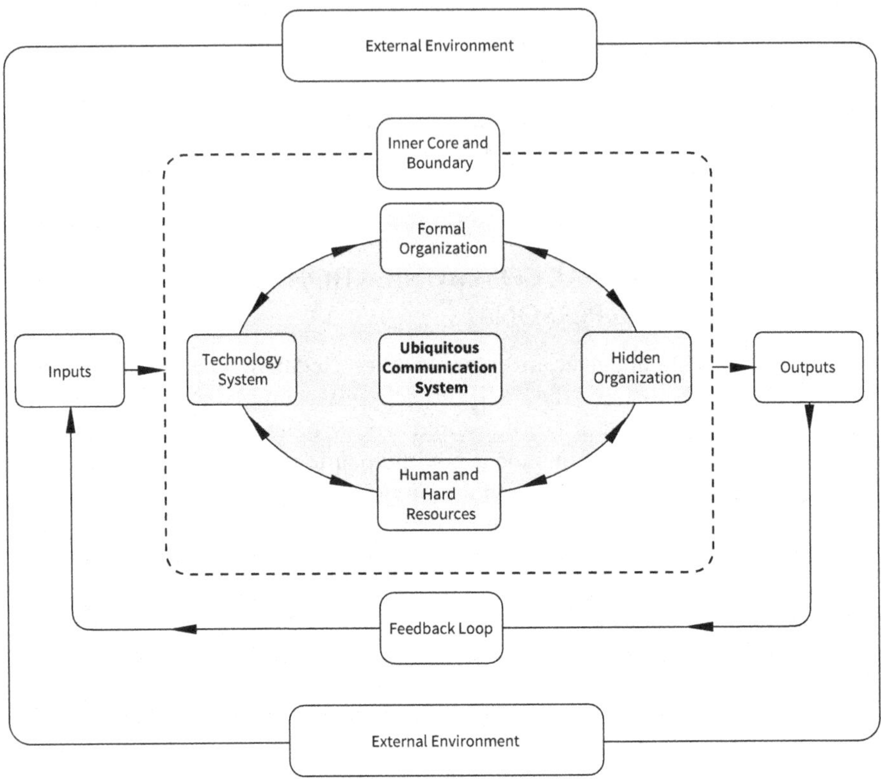

FIGURE 10.1 Ubiquitous communication system.

one of the most important activities an engineer performs each day. Not only is the calculation or analysis of results important, but equally important is the dissemination of that knowledge. Communication is the critical enabler of the organization; an organization does not exist without it. With it, the group can move forward and deliver amazing technology; without it, the organization will fail no matter how astounding the actual technology or product is.

Communication is also a paradox. For something that we do hundreds of times per day, communication can be breathtakingly confused, unclear, mistaken or ill-advised, or perhaps just plain wrong. It can be complex or simple, clear or chaotic, timely or irrelevant, passive or active. Not surprisingly, individuals communicate to others in ways far beyond what they think they do. Communicating is considerably more than speaking, or writing, or tapping on a mobile phone. Individuals communicate to others with every gesture, every facial expression, the gait of their walk, and tone of their voice. The timing of an email and how one sits in a seat: all are part of ubiquitous communications. And whatever forms it takes, we can't emphasize communication enough.

Simply put, communicating correctly (meaning clear, concise, accurate, and memorable information transfer) is hard. A quote attributed to the Irish playwright George Bernard Shaw summarizes the issue:

> The single biggest problem in communication is the illusion that it has taken place.

Unfortunately, despite continuous pleas by management to "communicate well or often or clearly, or all three," the reality of new engineer training has been to place scant emphasis on the "how" of communicating well, leaving true communication skill to be gained through hard-won experience in the workplace. A simple count of available communications courses offered at major engineering schools tells the tale that this skill is rarely if ever emphasized and even with communication training and assistance from a university engineering program, new hires begin their career at a deficit.

Let's say a manager has hired two new engineers. Both are doing similar assignments and the manager believes both are technically competent. It's no surprise that the better communicator will be perceived as the more capable individual. In fact, even if one engineer's technical answer is slightly deficient vs. the other, yet that engineer's communication skills are better, the superior presenter could be seen as the better engineer. High-quality communication is that important.

But here comes another paradoxical situation. Experience shows there may be a predisposition in some engineers' minds against strongly communicating or advocating for their new technology or other work product. While personally proud of their work, they may feel that strong advocacy may smack of "salesmanship" which frequently is anathema to the technical mind. A common attitude is "My invention (or innovation or design or work product) is so good, it speaks for itself." No, your invention is deaf and dumb. You must advocate for it through communication because no one else will.

So, in this chapter, we talk a lot about communication. Not as a cookie-cutter, checklist-driven guide to communication theory, nor as an attempt to replicate all those communication books mentioned earlier. Rather, we'll talk a little bit about a communications framework and then discuss the common "holes" in organizational information transfer that are generally missed: those messaging exchanges and usages you will encounter every day.

With that said, let's get a better handle on this whole notion of communication.

10.2 COMMUNICATING IN ENGINEERING ORGANIZATIONS: DEFINITIONS, MODELS, AND DESIGNS

Five minutes on Google is enough to convince you that once again, the definition of communication is strictly governed by contingency theory. Dozens of definitions are offered: the number of communication types and ways of organizing and structuring communication can be scores or more. The first conclusion you may reach is that the study of communication is really a minestrone soup, with too many chefs contributing too many "special" ingredients to the pot.

10.2.1 A Working Definition

With so many definitions to choose from (meaning, of course, there is no single, agreed-upon definition of communication at all), let's just avoid attempting a "formal" definition of organizational information transfer and instead create a definition that makes sense within the context of this book. And rather than force you to read a number of definitions only to be told what's wrong with them, let's just agree that our definition is this:

> Communication involves the transfer of an idea from one mind to another. The communicative process consists of a sender, a message, a medium, and a receiver. First, the sender carefully selects the symbols (usually words) that form the message. Then, the sender must determine which method, or medium, of sending the message will work best. The receiver hears or reads the message and makes an interpretation. If the receiver's interpretation matches the sender's intended meaning, true communication has occurred.[2]

This is a good, simple, and accurate definition. But what about the next step, establishing a model that describes the basic communication process? The answer comes from Corey and Pierce in their Transactional Model, shown in Figure 10.2.

The transactional model is the most robust of the common communication models. For example, people are not labeled as "senders" or "receivers" of messages in this model but as *communicators*, implying that communication is achieved when people both send and receive messages nearly simultaneously. Corey and Pierce advocate that communication is a cooperative action in which the communicators

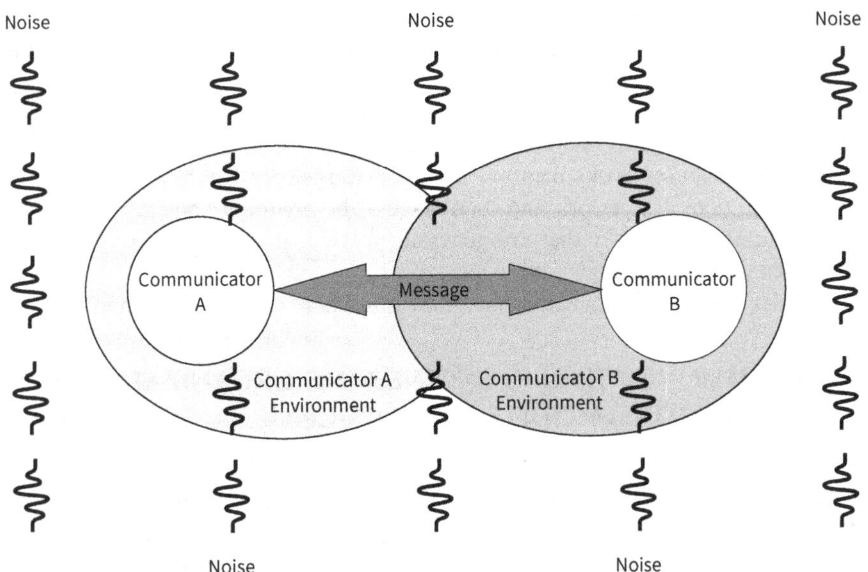

FIGURE 10.2 The Transactional Model of Communication.

actually create the process, outcome, and effectiveness of the interaction. They contend people create shared meaning in a more dynamic process in this transactional model.[3]

This model is good, but the simple processes implied in it are much more complex in practice than first imagined. In an ideal world, the message is clear, and the medium is optimum; one commutator's part of the process is successful, and if well-executed, the other communicator should be able to clearly interpret the intended meaning. Of course, this rarely happens.

10.2.2 Designing a Communication System

This definition and model are helpful, but now let's use them to create an engineering communication system. In daily use, there are really only two ways of *how* to communicate: spoken (through speech and hearing) and written (through vision). Both are underpinned by language and thinking. Yet when it comes to how we actually *execute* the communication, we find ourselves in a thicket of endless methods and techniques. So let's make it simple. Let's use the transactional model to maximize the potential for a holistic and successful engineering communication by using this model to *design* a technical communication system. Let's examine the steps in this design process.

Emphasis Area

The first step in designing a successful communication is to establish the emphasis area the communication will use. An *emphasis area* is the major way a communication will be transmitted. It parses the proposed communication by the way the information transfer can occur, recognizing that communication has the best chance of being successful by stressing one of these emphasis areas. Researchers such as Corey and Pierce contend that all organizational communication falls into one of five emphasis areas:

1) By Communicator: Here, communication methods and techniques are based on the number of people involved in the information exchange. This is usually described as pairings: individual to individual, group to group, individual to group, group to enterprise, etc. This results in the communication system being designed around serving the needs of any number of participants.
2) By Efficiency: An approach where the best pathway of a message is designed and optimized to ensure minimum loss of signal, maximum understanding, and best efficiency of the transmission. Note we are not talking exclusively about electronic or digital transmission technologies here, but instead the complete pathway traversed by a given message. For example, an efficiency approach could be described as an optium design "from brain to voice to a chosen transfer medium to ear to brain."
3) By Psychology of Understanding: Here, the prime concern is the quality of understanding of a message in the context of the sender's and recipient's

feelings and mental filters. The reality that the meaning of a message by different individuals or groups is highly contingent on the prior understandings, experiences, and expectations of either the sending or receiving parties or both.

4) By Method of Transmission: Some communications are organized by the method of communication chosen (verbal, written, listening, visual, etc.). This method is surprisingly common, perhaps as it is straightforward, simple to apply, and avoids the more complex and uncertain aspects of psychology and drives more toward maximum efficiency.
5) By Menu: Simply put, this is an approach to communication where solutions to different aspects of the communication problem are selected from several of the previously listed techniques. Highly flexible and satisfying, this menu approach presents a quick, plausible method of designing a holistically successful message. It fits well with contingency theory. For these reasons, the menu tends to be selected as the most preferred process for quality communication.

Opportunity for Message Transfer

The next step focuses on the *communication opportunity*. A communication opportunity is defined as an information transfer event or occasion; an opening or chance to execute the transfer; the time and place a given communication can occur. It is a contingent approach in that the exact method to best communicate depends on the location, time, and environment of its transmission. It is predicated on the belief that no communication can be successfully accomplished unless an appropriate opportunity is presented. Table 10.1 summarizes this stage.

There are seven main opportunities: in traditional meetings, in 1:1 meetings, through traditional discussions, in hallways or elevators, auditorium presentations, and written communications. Each opportunity has many execution conditions: planned or "snap" (i.e., instantaneous), formal or informal, with prepared material, or ad hoc. Each opportunity (hopefully) has an appropriate duration to complete the transmission and yield a successful outcome. Finally, the communicators can be singular or multiple, and the transfer venue can be any number of spaces or locations, including virtual.

Table 10.1 summaries how each communication opportunity has a set of characteristics associated with it that can provide guidance. For example, if you need the best opportunity to discuss a technical analysis with a fellow engineer, the traditional discussion opportunity is appropriate. To discuss your career, a 1:1 is essential, and to brief the chief engineer on your latest test result, a hallway is good if the timing is urgent and the message is simple. If not, a written communication or presentation is appropriate. The goal is to select the message opportunity that best fills the need.

Information Transfer Method

The next step addresses the design's transfer method. Since organizational communications are highly contingent, selecting the appropriate method (or methods) of

TABLE 10.1
Basic Opportunities of Organizational Communications

Communication Opportunity	Planned or Snap Information Transfer	Formal or Informal	Prepared Material or Ad Hoc	Opportunity Duration	Designed Outcome: Awareness/ Discussion/ Decision	Receiver: Individual or Group	Typical Venue
Meeting	Normally planned, a few are snap	Both formal and informal	Active contributor: Prepared Passive participant: Ad Hoc	Only long enough to achieve meeting goal, "refund" unused time	Any: Awareness, Discussion or Decision	Group	Conference Room or other enclosed, formal or semi-formal space, and/or virtual
1:1	Normally planned	Normally informal	Prepared is highly recommended	Only long enough to achieve meeting goal, "refund" unused time	All	Individual	Individual Office/ Small CR. Virtual not recommended
Traditional Discussions	Both planned and snap	Normally Informal	Can be either	Only long enough to achieve desired outcome	All	Either	Tends to be in an informal location: open air collaboration space
Hallways	Snap	Informal	Ad lib	Under 5 minutes	Awareness	Individual	Everywhere
Elevators	Snap	Informal	Ad lib	Approximately 1 minute	Awareness	Individual	Everywhere
Auditorium Presentations	Normally planned, rarely snap	Normally formal	Prepared	Aim for 30 minutes, maximum 45. Include questions	All	Group	Large conference room/auditorium/ virtual
Written Only	Planned	Both	Prepared	Variable	Awareness/ Decision	Individual	Anywhere and everywhere

information transfer involves a fair number of considerations. These contingencies and resulting questions include:

1) The urgency of the message: how soon must it be received?
2) The importance of the communication: what is the critically of the message being sent? What will happen if the message is not received?
3) The half-life of the information being communicated: how long is the information relevant? Is the "freshness" of the information short or long?
4) The need for clarity (i.e., quality) of the communication: will it be understood once received?
5) The advocacy of the message: is the message passive or active? Is it a one-way-only message without feedback, or is two-way communication required?
6) The message complexity: is the message's content simple or convoluted? Is the message nuanced or straightforward?
7) The communication's audience: is the message intended for many recipients, a few, or only one?
8) The message's sensitivity: is the communication confidential or public domain? If confidential, what level of secrecy is required?
9) It's historical requirements: can a record of the communication be easily obtained, either immediately or at some point in the future? Will it be metadata only or must content be included?
10) It's content classification: is the communication routine (a weekly status report, for example), or is it a one-time special message (say, a product safety issue has just emerged)?
11) The message's context pairing: is the context of the communication identical between the sender and recipient?

Analyzing these 11 requirements will help determine the best communication method or mode to entrust the message. Generally, you will always have several methods available to you, some rarely used and others quite common. A summary of these methods is shown in Table 10.2 and is meant to provide a starting point in choosing the specifics of the particular communication design you need to accomplish. Many of these are obvious, but some may be new.

Like other concepts, the contingency of communication is both good and bad. It prevents you from using a comfortable, single communication channel and just hoping it works. Instead, it forces you to create a specific, customized approach for each situation, resulting in a hopefully optimized information movement. But it takes time, energy, and some experimentation to ensure it works, which is counter to the routine urgency your message may demand.

Let's walk through a simple example. Say you have developed a set of technical specifications for a new high-efficiency centrifugal pump. You now need to transfer these specifications to a group of five engineers located in an adjoining department. This downstream group will access the feasibility of your specifications for a manufacturing setting and pass judgment on your work. They assume

TABLE 10.2
Contingency Approach to Organizational Communication

Communication	Delivery Urgency	Information Importance	Half-Life of Relevancy	Clarity / Fidelity	Passive or Active Delivery	Complexity of Message	Number of Recipients	Ease of Archive	Routine or Special
Book	Very Low	Low to Very High	Very Long	High	Passive	High to Low	Many	High	Routine
Magazine/Journal	Very Low	Very Low to Moderately High	Months	Medium to High	Passive	High to Low	Many	High	Routine
Website/Digital format	Moderate	Very Low to High	Very Short to Moderately Long	Moderate to High	Passive	Low to Moderate	Many	Moderate	Routine to Mildly Special
Telephone/ Mobile	Mildly Urgent	Low to Very High	Days to Weeks	Moderate	Passive to Active	Moderate	Single or Few	Very Low	Both
Email	Mildly urgent	Very Low to High	Weeks	Moderate	Moderately Active	Low to Moderate	Single to Many	Moderate	Both
Text	Urgent	Low to High	Minutes to Hours	Low	Passive to Active	Very Low	Single to Many	Low to Moderate	Both
Multi-Participant, In Person Meeting	Mildly Urgent	Moderate to Very High	Days	Moderate to Very High	Active	High	Few to Moderate	Low to Moderate	Both
One on One, In Person Private Meeting	Urgent to Very Urgent	Moderate to Very High	Hours to Weeks	Very High	Active	High	Single	Low to Moderate	Special
PowerPoint / Keynote Presentation	Moderate to Urgent	Low to Very High	Very Short to Very Long	Moderate to Somewhat High	Passive	Moderate	Single to Many	Moderate to High	Both
Video Conference	Moderate to Urgent	Moderate	Days to Weeks	Moderate to High	Moderate to Active	Moderate to Somewhat High	Single to Many	Low to Moderate	Both

you are professional and competent. What is the best way to perform this information transfer?

You think about this communication task. The specification is contained in a 26-page PDF document with three drawings detailing the pump configuration. Two areas of the specification have a nuanced meaning, not easily understood from either the document or drawings. You do not know these five engineers.

The framework asks to first select an emphasis area; how the transfer will take place. The first four options all seem to have some impact on the design of the transmission, but not exclusively. Therefore, option five (the menu) is a good bet.

Next, thinking about the transfer, you decide to alert the group about the two nuanced areas of the specification that are prone to misinterpretation. You resolve to speak with the group personally to highlight these potential misunderstandings, and you want all five engineers to equally understand the same two areas. You also desire an opportunity for verbal question and answers, as you are new and feel your work may have missed some important points.

The message transfer selection then provides you with opportunities to achieve your transmission requirements. From Table 10.1, a meeting, a discussion, or an auditorium presentation look promising. Yet the auditorium is much too formal, plus it does not allow many two-way question-and-answer opportunities. So, either a meeting or discussion is good. Since you wish to promote two-way, sincere, and informal communication, the traditional discussion opportunity is your choice.

The information transfer method is easy. The communication will be multichannel: a face-to-face conversation with question and answer, a discussion or summary file for use during the meeting, then leave a document, file, or other evidence of your work on a website or as a digital record. The key is to provide both a verbal and written status as of that date.

Now here comes the most important part of this example. Remember, the real purpose of the meeting is to officially transfer your work for the feasibility check. This will probably be a formal or documented step in your firm's engineering development process. As such, you must keep a formal record of the transfer: that it took place on a certain date, that adequate documentation was provided, that it was officially accepted by the downstream group, and any other evidence that showing a complete review and transfer occurred, including any nonconformance items and their disposition. Why? Being a process step, your work is subject to review at any time by any outside entity. If your documentation is not complete, you are open to an out-of-process correction, which can be painful and a reputation killer. While you might believe after-the-fact documentation doesn't matter, especially under the pressure of moving on to the next assignment, you must complete all your process steps, and you must take the time to keep a record of the successful transfer.

There it is: a discussion in a conference room with a written file of the transfer material, coupled with questions and answers, including a file location or website, plus certification that the transfer occurred successfully. Your communication is complete.

This framework may seem trivial, yet it's important new engineers learn the discipline and train to automatically use this process. It's very true that even moderately experienced engineers already know these steps: it is probably second nature to them. Your aim is to to arrive at that level of knowledge as well.

10.3 TOWARD THE GREAT DIVIDE: NAVIGATING THE COMMUNICATION NETWORK

At this point, let's take stock of where we're at. We know that organizational communication is highly contingent. That a universally agreed-upon definition of communication doesn't exist, but we have a workable approximation we can build upon.

We know that technical communication must be clear and highly accurate in spite of the difficulties caused by the fuzzy boundaries that surround it. And there are dozens upon dozens of information transfer modes, methods, and paths available to choose from, including hybrids.

That said, let's now examine a selection of the most common communications situations all engineers face in technical organizations and share some suggestions on how to proceed.

10.3.1 Communication Opportunities and Situations

We need to discuss in greater detail the communication opportunities introduced in Table 10.1. Essentially all your work-based communications will fall into one or more of these categories, and each category will have some noteworthy dependencies we should recognize and address. We'll discuss several of the most common situations, all brought on by the dictates of contingency theory.

The Meeting

A meeting is a bounded, real-time gathering of two or more people for the purpose of achieving a common goal through conversation and interaction. It can be formal or informal, planned or spontaneous. Its purpose is to achieve an outcome of awareness, discussion, or decision, or some combination of these.[4]

The motive for your attendance in a meeting can vary, but most reasons fall into four bins: attaining knowledge or education around a topic, exchanging information or ideas with others, deciding or being part of a decision, or portions of all three.

Attaining knowledge or education is passive, involving sitting there, listening, and (hopefully) absorbing. Discussion is an active exchange of knowledge between individuals, and decisions are a conclusion or resolution reached after consideration. This means that for a participant, communication in meetings is all about the discussion, the give and take, and the contribution you personally make to the meeting. How you communicate in meeting discussions is our focus.

The Formal, Planned Meeting

At this early point in your career, a planned, formal meeting is one you normally might not attend. If invited, your presence is as a guest; summoned to either present information, convey results, or add to a discussion regarding a single topic. Typically, the meeting participants are of a higher rank, and you probably haven't had prior interaction or experience with them; they are as unknown to you as you are to them. The meeting takes place in a formal manner, with exact participant lists, written agendas sent to participants in advance and rigid meeting conduct rules. Probably, your topic is one of several to be discussed, with an exact start and finish time specified on the agenda.

As mentioned earlier, formal meetings have three different types of outcomes:

1) Awareness: Providing participants necessary information and background on a topic without requesting a specific discussion or action.
2) Discussion: Presenting the status of a topic to promote discussion and guidance among participants without requesting a decision at that time.
3) Decision: Sharing summary information on the topic to receive necessary direction and take decisions.

Preparing for these occasional meetings requires unfamiliar work. When invited, your preparation should include the following questions. What are the meeting organizers looking for you to provide? What do *you* wish to accomplish? And how do you achieve this to create a "win-win" outcome for you, the decision-makers and the audience members?

The first question is easy: ask. Surprisingly, there can be reluctance to ask clarifying questions, as the questioner is loath to potentially look foolish or be seen as lacking. Don't get caught in this trap.

Next, explicitly decide what *you* are trying to accomplish. What do you wish to gain from this opportunity? Is it experience? Visibility? Learning the concerns of senior management? Survival? Whatever the reason, keep it in the front of your mind.

The third question is more complex. Determine what "rules" are in play. This can be tough, as many of these rules are not written down, and you may not even realize there are rules at all. Be assured there are. The meeting organizer, if you speak to them personally (no email or text) can give you helpful hints on what these rules are. Without this guidance, management may think "This person hasn't gotten the word yet" and discount your participation.

One final note about formal meetings. It should be obvious, but please do not arrive at executive meetings late. Irrespective of the amount of espoused formality or informality in your company, in a meeting, it is still imperative that you are there early and ready to go the minute the executive comes in. Distracting a meeting already begun through a Zoom or Webex chime (or opening the conference room door) is an irritant and bad look.

The Informal, Routine Meeting

The *routine meeting* is an informal work-related gathering that is local, periodic, and standardized, where everyone's attendance is expected at each session, and where repetitive business is discussed. Called by many other names (staff meeting, weekly topics meeting, team huddle, operating committee meeting, and the like), a group of employees gather around a table, either virtually or physically, sipping their coffee as the boss pontificates on whatever is the topic *du jour.* After 30 or 60 or even 90 minutes (always an even increment), the group disbands and then attempts to catch up for the 30 or 60 or 90 minutes of work time lost. Uncounted trees have been sacrificed by consultants sharing how to improve the outcomes and reduce the waste of these routine meetings, to no avail. It is by far the most common type of meeting you will encounter.

There's not much to say about these meetings. They can be primarily one-way information transfer meetings (with perhaps some discussion thrown in) and your contribution is normally limited. Occasionally, you may be asked to prepare something for the group, (an unexpected test result, a piece of competitive information, or a process improvement), but in general, these are low-pressure, passive gatherings. Enjoy.

The 1:1

The 1:1 is a form of informal, planned meeting between two specific people: you and your boss. It's informal in that it can take place in a variety of locations (the boss's office, the building's commons area or even over lunch at a local restaurant) and more of a conversation. It's planned in that it occurs regularly, perhaps once a week. Its purpose is different for each person participating. For you, it's to inform your boss on the status of your work, and to gain information on company or department topics you need to know. The boss's purpose is somewhat different: to review the condition of your work and to judge (over time) your technical and personal qualities as a department employee.

The rules of this meeting depend heavily upon and are dictated by the relationship you have with that boss. A 1:1 between a new employee and their boss will tend to be more reserved, formal, and more circumspect. Obviously, a boss and engineer who know each other well will be more informal and easier in their conversation.

Keep in mind this is your meeting, not the boss's. They may have scheduled it, but you run it. Many styles prevail; some engineers will conduct a strictly verbal exchange with no written notes or record kept. Others prefer drafting a simple agenda and providing some sort of document or file to the boss for each item as a record of the meeting. Either way, the style and details of these 1:1s will (and should) be at the boss's preference. Independent of the particular details, the overall thrust of a 1:1 is to enhance your bosses' confidence in your technical capability, their comfort with you as a person, and grow the trust between both of you going forward.

As mentioned previously, the 1:1 is also the corporate version of a quiz, performed verbally and given weekly. These ongoing presentations to your supervisor are ubiquitous; it behooves you to practice these as you be giving them often.

The Traditional Discussion

Many people do not see much difference between a meeting and a discussion. After all, both are normally planned events, involve multiple participants, achieve a give and take in the form of questions and answers, and can produce outcomes ranging from simple awareness to major decisions. However, there are actually substantial differences between the two. Meetings tend to be more formal than discussions, with additional prior planning involved, increased use of formal agendas, extra upper management attending (perhaps managers and directors or higher), occur in formal or semi-formal spaces and are more presentation-like (i.e., one-way information transfer) in their style. Generally discussions, while planned, are informal in language and pattern, rarely use formal agendas, occur between working-level engineers and supervisors, exclude upper management, and use relaxed or collaborative settings to promote free information exchange.

A major benefit of discussions that is normally lacking in meetings is establishing confidence and comfort with engineers in other departments and organizations necessary to achieve success. And like the ultimate goal of 1:1's, an additional idea behind discussions is to eventually establish trust and create a professional working relationship. Discussions build bridges between individuals, and thus connections to departments.

Hallways and Elevators

A hallway talk is literally that; a brief discussion between two (or no more than three) individuals in a public space, designed to convey information of various levels of importance. The author Michael Lopp in his book *Managing Humans: Biting and Humorous Tales of a Software Engineering Manager* stresses the importance of hallway conversations. Hallway conversations are short, informal conversations about the status of projects, goals and similar topics. They are great for bouncing ideas off people or just letting them know what you are up to. The important insight is the hallway is a brief conversation to gain insight, perspective or guidance. But please note: it is *not* about decision-making on any topic of importance. In fact, one of the major downsides of the hallway is it can be a major source of confusion and miscommunication.[5]

This situation is easy to understand. Say you are assigned to a project involving three other engineering groups, all led by a business office director. You see that director in the hallway one day and decide to ask them for a decision that you were going to ask the entire project team at their meeting the following week. After all, getting a decision now saves a week of waiting. You ask the director who, not knowing any better, makes a snap decision and continues to walk down the hall. Having a clear decision, you proceed for the next week guided by that decision, happy in the knowledge that you are working efficiently. Unfortunately, for that week no one else knows a decision has been made, and even worse, it happened to be the wrong decision. Without all sides of a technical issue being heard simultaneously, a bad decision can easily result. Meanwhile, there is now an added problem: there is a good chance the director forgot he or she made the decision. That's right. With so much information to process, many executives' memories are overloaded and one-minute conversations in a hallway are easily lost.

From this, a good approach to the hallway interaction is to use it for decisions *only if the decision is very minor and involves matters that are 100% in your control.* For any other situation, wave off.

One other point about the hallway you need to know. Executives use hallways as an effective "trolling" method to uncover the truth about what's happening within their projects and overall organizations. Because the hallway is usually between two people, not overheard by others and by their nature informal, people are willing (sometimes very willing) to share information with little or no filtering. This is a strong and important information sharing device. For example, a former Secretary of the U.S. Navy (an organization of some 300,000 people) believes in the rule that "the quality of communications and information exchange is inversely proportional to the formality of the meeting. Much more information is exchanged in a hallway than in a formal meeting room."[6] It won't take you long to see this in action in your own organization. If you elect to share this way, it's certain that your information will be used at the pleasure of management. But don't forget, you may regret it.

An elevator pitch is slightly different. It is a brief (perhaps 30 to 60 seconds) episode of one-way communication. It can be a method of introducing yourself, communicating a key status or result, or merely making a connection with someone. Obviously, it's called an elevator talk because the simile implies it's completed in the time spent taking an elevator ride with someone.

Some career consultants view elevator talks as great tools to use any time you want to establish a network connection or meet a prospective executive looking for a new person to hire. That's fine, but think of it more as rehearsal for organizational life in general: it's a practice session in effective verbal communication. Whatever the motivation, continuously hone and update your elevator skill, as you will eventually use it.[7]

Auditorium Presentations

You will be called on to perform occasional auditorium-style presentations. These are different than meeting presentations. As Table 10.1 outlines, these are a very specific form of communication with certain common characteristics. They are almost always a formal, scheduled in advance, one-way "broadcast", as a single presenter speaks out to many recipients for a set duration of time. Obviously, these occur in auditoriums but also large conference rooms, and of course can be virtual. With a limited (if any) opportunity for two-way question and answer, these presentations are tightly controlled and can be high-anxiety experiences.

Uncounted books already exist that will give you presentation guidance and tips for a variety of audiences. I suggest you pick one up, as it will ground you in the fundamentals of the all-purpose presentation. For the engineering or technical audience, let me share just a few suggestions specific to those individuals.

Technical talks always involve a multistep process that is easily divided into predelivery, delivery, and post-delivery stages:

Predelivery

This is the thoughtful preparation used in answering some fundamental, anticipated questions before you even begin.

1) First, analyze your audience. Who are they? What technical level are they? What is their technical "language"? Why should they listen to you? What is your message? This may seem obvious, yet in reality, these self-directed questions are rarely asked, let alone answered.

2) Establish your logistics. Where and when will the presentation be given? How will it be given? What is the medium? And most importantly, is the technology compatible and reliable? Any experienced presenter will tell you over 50% of presentations fail because the laptop couldn't talk to the projector, the zoom link wasn't tested, or the remote "advance" button lacked a battery. Don't ever assume someone else will handle this: they won't.
3) Follow the "rule of four" in presentations that states that at most only four ideas can be presented at one time (in fact, only three ideas are optimal for a single presentation). Make it clear that you are communicating just those three or four ideas, and do not overload the presentation with information.
4) Develop your material using a storyboard. Veteran presenters recommend using a storyboard approach (a paper and pen, drawing technique) to establish a narrative "arc" to their talk. An excellent reference is the previously mentioned *Slide:ology* by Karen Duarte.
5) Remember, a picture is worth a thousand words and hardware is worth a thousand pictures. Consider using hardware and other hands-on exhibits that attendees can examine and handle before or after the talk. This is old school but still highly effective.
6) Borrow with pride. An early step is to search for any previously developed material that you may repurpose for your presentation. Not only is this efficient, as being previously used and approved, but your management will probably expect you to use it as a timesaving move.

Delivery

This is the actual execution of your presentation. Those "how-to" books I mentioned earlier are your best guide for delivery, but I need to emphasize these points:

1) Maintain time discipline and take a shorter time to complete your talk than allotted.
2) Never, never read the talk.
3) Avoid distracting mannerisms.
4) Maintain eye contact with various members of your audience.
5) Memorize the introduction and the conclusion, but not the main body.
6) For a new presentation, practice the talk at least three times, timing yourself each time.
7) As I mentioned above, it is crucial that you familiarize yourself beforehand with your electronic aids. Physically handle the controls and personally verify each component is working and ready.
8) Adjust the level of technical communications to the level of the audience, but still keep your actual message simple and clear.
9) If practical, engage in a brief question-and-answer session.
10) Your talk should be snappy.
11) Remember, the main point in preparing a presentation is deciding what to leave out.

Post-delivery

Post-delivery is where additional, unexpected benefits to you can occur. You should plan to:

1) Continue discussions with interested persons outside the venue or at a later time.
2) Follow-up on actions you promised, either during the presentation or in discussions after the event. Be sure to *actually do* what you committed to do, otherwise, you lose credibility.
3) Analyze your talk regarding how it was received. Was your audience analysis correct? Did you stay within your time budget? What did you do well, and what would you do differently next time? Again, this may seem obvious yet less than 10% of presenters actually conduct an After-Action Review and learn from it.

Finally, there is an old joke regarding presentations that still has relevance today:

> "The ideal speech should have a grabby beginning, a powerful ending, and they should be very close to each other."

The Written Opportunity

Hopefully, you have already taken a writing or prose course at university or used another opportunity to train in the fundamentals of good, written idea transfer. Even better is a technical writing class especially designed for an engineering or technical practitioner. Coming into a new position means having a reasonable foundation in technical writing already in your skills portfolio. If not, I would sincerely suggest you find a suitable writing class or training opportunity locally to enhance your writing ability: clear and concise technical writing is that important as a professional skill.

Like in the presentation section above, I am not going to cover the fundamentals of technical writing here, as you should already know or move to acquire them. But I want to cover some organization-specific points.

The first is your home organization's or department's standard writing expectation. Every engineering company will have an expected writing style or approach and a certain look and feel for their written communications, be it on paper, slide, email, text, website, or discussion board. This approach may be explicit, using a commercially available formatting and composition method (*Write to the Top: Writing for Corporate Success* was the required approach to business writing at one of my employers) or implicit (someone wrote a particularly clear and clean document that was then copied by everyone). Whatever the reality in your firm, you need to discover it, learn it and ultimately embrace it.[8]

The second is the future of the "document." Prior to the coming of desktop computing, written communications (test reports, recommendations, decisions, methodologies, meeting notes, and the like) were detailed on paper, with this information and other supporting content included, written in a form that could be archived for later historical use. Paper was king, and copies were mechanically created to fill endless file cabinets.

Obviously, digital communications changed all that. Initially, written documents maintained this paper-based form and were merely created and stored electronically. Then, a fundamental change occurred: engineering writing moved from Word and its competitors to PowerPoint-style presentation decks. Documents that might have been 400 words per page now were reduced to bullet points containing 25 words or less. Written communications were pared down to essential wording only, and efficiency of transmission (minimum words, always in bullet form) was placed ahead of holistic understanding. This is essentially the nature of written technical communications today.

What matters is how you approach this reality. With documents now containing just scores of words vs. the hundreds previously used, you now must develop an organized, easy to access, personal archive of those background test methods, analysis, decisions and assumptions used in creating the final deck. Again, this sounds trivial, yet many, many engineers experience deep trouble when asked to produce their development material sometime after their deck is submitted, especially when asked to produce the material immediately. Without a legitimate, personally maintained reference system, your proof of the validity of your deck will disappear like smoke. And it is your personal responsibility to do this.

10.3.2 Not All Ground Rules Will Be Communicated

In communication, you will encounter (with distressing regularity) surprises having to do with ground rules, specifically with those that are communicated to you and those that are not. A good assumption in company communication (and in fact in all aspects of company life) is that not all ground rules will be told to you. In fact, take as a matter of faith that very few ground rules will be purposely shared. Or, go for the full Monty and just assume that *no* ground rules will ever be actually told to you. Now, they may be hinted at, or mentioned in passing which you might take as a pleasant conversational remark but fail to realize it was meant for you. Remember, your workmates speak in code and the rules for you are no different. Either your supervisor or coworkers will think the rules are too trivial to mention explicitly, or they will forget to share them until after you've made a mistake. So, that leaves you to uncover these little nuggets of guidance yourself.

For example, a classic ground rule that's mostly forgotten to be communicated has to do with your first business trip. When assigned to your first trip, your supervision may or may not share with you the nuts and bolts of the travel policy. Travel policy is the rules and regulations imposed by the organization that you must follow on any business expedition. Classic travel policy examples include booking the cheapest flight possible, only using company-approved hotels, spending only so many dollars per day on food, and so on. These rules can be quite extensive, having been developed over decades as the result of financial audits and previous mistakes. Since these ground rules are normally not communicated to you, it's a good idea to specifically ask for verbal or written guidance on all travel rules and regulations. Travel is a case where small mistakes can have outsized visability for you as a new engineer.

Overall, for any company situation, the only true way to uncover these tidbits is to ask: ask your supervisor, ask your experienced colleagues and for meetings, ask the meeting organizer. Active engagement with the people and "society" of your local organization is the true way to minimize unforced errors. Don't be shy.

10.3.3 Calling an Audible

A snap presentation or talk at any venue usually involves *calling an audible.* Fans of American football know the audible as a situation where a play is planned by the offense but is changed at the very last moment in response to the defensive alignment. The quarterback shouts a coded word or phrase and the plan is changed mere seconds before play begins. A snap presentation is essentially the same thing: your presentation is changed moments before it is to begin in response to changes with the audience or environment. It happens quite often.

Pete Shaw was a test engineer with about two years' experience in digital temperature sensor evaluation and related connectivity systems design. Pete's job was twofold: test the sensor functionality of new designs and use the test results to analyze the relative performance of the new system vis-a-vis existing benchmarks.

Enter Helmut Gretzell, Chief Engineer for Connectivity Design. Gretzell was highly respected for his engineering skill and feared for his very blunt, public takedowns of any engineer not meeting his exacting standards. Presenting to Gretzell was always in adventure in trepidation and anxiety.

Pete was assigned as the sensor test analyst for one of Gretzell's programs that was behind schedule and at risk. One day Pete was asked to provide an extremely fast analysis of an unusual sensor proposal. There would be no time to test; only a predictive study would meet the timing for a formal review meeting with Gretzell and this team of some 30 managers. Pete developed a quick analysis based on his experience and hoped for the best.

Pete's topic was last on a packed agenda, and sure enough, his topic was announced about three minutes before adjournment. Instead of his prepared remarks, Pete calls an audible:

Pete: "I was asked to provide an evaluation of the proposed sensor design. There was no time to test, so this is a judgement evaluation only. This evaluation was performed by breaking the design into six subcomponents (labeled one through six as shown on the slide), quantitively analyze them separately, and add up the results of each. Subcomponents 1, 3 and 4 exceed the reference benchmark, 5 and 6 are equivalent and 2 is below the referent design. This analysis shows the new design to be 7% better than the reference. Thank you."

Dead silence followed. Then Gretzell spoke:

Gretzell: "Do you work for this company?"

Pete: "Yes."

Gretzell: "How long have you worked for us?"

Pete: "Two years."

Gretzell: "That explains it. You gave a complete briefing, with everything I needed to know, in under 60 seconds. But by next year you'll probably be talking as much as the other gasbags at this table. Don't ever forget what you just did here today.

Meeting adjourned."

Ninety-two words in under 60 seconds. Pete gave a masterful performance that day due to his ability to call an audible and pull it off, an excellent lesson for us all.

In short, always be prepared for a surprise presentation or discussion about your project. I'm not saying you must have a PowerPoint stashed somewhere in your laptop, but I am saying to always be able to do a five-minute talk on your project, as a hallway, discussion, presentation, or other format as we discussed. Remember, you must be able to speak intelligently on your work no matter the circumstances.

10.3.4 Control Issues in Presentation

It's distressing but true: your presentation will not be in your control. As we just discussed, no matter what venue or gathering, surprise is the norm. Your allotted time may be cut in half, or the time slot changed, or the digital equipment has failed, or you get a constant interrupter, or unexpected strangers are present. As you are not in control, you must be nimble. What is your backup plan for each situation? For example, if your presentation software has failed, be prepared to simply state the conclusion of each slide without visuals. Proceed with your presentation verbally and don't attempt to fix the electronics for more than 10 seconds or so.

One important thing to ponder regarding control: you may not have a choice in making a presentation. You might think to yourself, why are you making the pitch? Why isn't your boss, coworker, or someone else presenting your material? It could be one of three reasons. First, perhaps you are the only expert on the topic or the only one available at that moment to give the brief. Secondly, your management may decide that you need training in presentation, and this opportunity is an excellent chance for you to gain experience in this unfamiliar skill. The third reason has to do with positive exposure. If your management is invested in advocating for you in a positive light, they may specifically ask or require you to make the presentation so you will become known to upper management or give you an opportunity to otherwise "shine." Note that all three reasons are good, which link to an old saying in professional settings: "never turn down an opportunity to speak in public," but with the caveat "just have something to say." Even though it may be an uncomfortable, or even terrifying experience, grab it and run.

10.3.5 Standardizing Communication

In more formal engineering organizations, there is a strong desire to standardize communication methods, especially for routine, periodic meetings or reviews. And for good reason. Standardized communication aids in understanding; reducing the time to communicate a given thought while also delivering higher accuracy. The simple fact is that engineering executives prefer routine reports that look the same and are highly standardized for ease and speed of understanding. This led to the rise of the *communication proforma*, a pre-existing template where standardized information is routinely placed and presented. With the proforma, the recipients do not have to relearn the type and location of information for each report; it's comfortable and reassuring. It's also much faster to create the report. This is another advantage of the BPR method described previously. Do not be surprised if you will be standardizing your reoccurring assignments into one of these platforms.

But the key word here is *routine*. As covered later in this chapter, there is a critical exception to the proforma approach you must be aware of.

10.3.6 Context, Leveling, and Bias

Qianyan Shea had a really interesting job. Working for a large technical multinational, she was assigned the daunting task of becoming the organization's technology "futurist." In Qianyan's firm, fresh information regarding new technologies from outside her company's industry was traditionally ignored, at the risk of missing key technological developments that could impact her company's future. Qianyan's job was to correct this weakness; to look outside her firm's specific environment, searching for breakthrough technologies that in the long term could change her firm's products or services. As a futurist, she used many different techniques to ferret out these long-term, "faint signal" technical developments.

There are many methods to do this, one of which was a highly effective technique called "genius futuring." As its name suggests, individuals who are considered "geniuses" in a specialty technology are interviewed to mine their thoughts and predictions. Typical candidates would be luminaries such as Freeman Dyson (quantum field theory and astrophysics), Richard Dawkins (selfish genes), Bert Sakmann (cellular signaling), and Norman Packard (chaos theory).

As part of her research, Qianyan's management asked her to specifically interview a number of these brilliant individuals. Naturally, she approached this assignment with (not a little) trepidation. How in the world was she supposed to interview a "genius" when she wasn't one? She didn't consider herself anywhere near qualified to talk to these stars of technology, let alone understand what they were saying. She needed help. Fortunately, she found the author Thomas Bass, who had recently written a book titled *Reinventing the Future: Conversations with the World's Leading Scientists*, in which he interviewed paradigm-changing thought leaders. Qianyan shares the story:

> I finally sat down with Tom [Bass], and he gave me some excellent insight about communications. "You don't need to worry about communicating with these people." he said. 'They are smart enough to know very quickly your level of communication ability and the context of the discussion. The worthwhile ones will automatically adjust their own speech, vocabulary and way of explaining to match your own level of

understanding. In a way, "they will come to you"; you will not have to go to them. Frankly, any person you are interviewing who doesn't adjust their own communication level just wants to feel superior and is not really interested in sharing their knowledge. Stay away from these people. Search for the experts who want to truly reach you.

The crux of Tom's message was that in communication, ensure all parties have the same context and level of understanding and adjust the exchange level to match that context. As discussed before, a simple but important principle.[9]

A few words about a related concern: bias. *Bias* is the tendency for us to hear or understand only information that we accept or agree with. Bias is a filter preventing clear communication and cannot be completely negated; we can only reduce it. This means we must continuously combat bias through clear, precise, interesting, well-organized writing and speaking, especially in technical communications where the transfer of exact meaning is especially important. Not only must the communication be clearly conveyed, it must be highly accurate and with minimum bias so there is little chance for misinterpretation. This is very hard to do.[10]

10.3.7 Simple, Declarative Statements in Presentation

Clarity, brevity, and efficiency are essential in communicating, especially when presenting to senior management. This calls for moderately paced, simple, declarative statements as applied to presentations. A surprisingly good example had to do with a large North American manufacturer planning a presentation to a delegation of executives from a smaller technology company located in Japan. At the time, this Japanese firm and the North American manufacturer were in negotiations for a potential joint venture. As part of their due diligence, a Japanese business and engineering team visited the North American firm to receive a briefing on the design of the proposed joint project. To the best of anyone's knowledge, none of the Japanese delegation knew English, and the North American attendees certainly did not know Japanese at that time. Thus, a professional translator was engaged to facilitate technical communication between the two groups. Don Rustori was assigned to make the presentation on the project's design feasibility to the Japanese, and was very concerned about the language differences.

To his surprise, rather than having a difficult talk, Don discovered the briefing using a translator was exceedingly easy to do. Unlike a normal presentation, where the audience expects a steady flow of words and ideas at a brisk pace, in a translated meeting, the presenter makes one simple, declarative statement and then waits 5 to 7 seconds for that statement to be translated. To Don, those extra seconds were a blessing as he could use the time to clearly organize his thoughts and prepare his next statement while keeping the message as simple as possible. The speed of the presentation slowed and surprisingly, Don began to feel more in control of his talk. And because non-English speakers were listening, he was forced to simplify each statement. The lesson here is that a moderately paced presentation using simple, declarative statements leads to a highly effective, highly satisfying presentation. And your audience will appreciate it.

10.3.8 Silos, Jargon, and Acronyms

There are a few additional concepts in organizational communication that you should be aware of and understand as a new employee.

The first is *silos*. The term silo (or *chimney*) has become synonymous with narrow-mindedness, extreme control, and institutional bias. Silos are a metaphor for formal or informal business structures that prevent the corporate whole from becoming greater than the sum of its individual parts. They are vertical organizations that are surrounded by mental "fortifications" or "brick walls," implying little goes in or out. For years, reducing silos has been a popular topic in management literature.

Silos extend to the transfer of knowledge through communication. In communication, the silo effect is a purposeful blockage of information moving from one organization or person to another. Communication silos are based on the false premise that information can be controlled for advantage, even though the digital era proves each day that information control is ultimately a fool's errand. And it is your job as a new engineer to identify the communication silos in place around you, know why they are in place, and establish your own appropriate boundaries and ground rules to guide your relationship to them.

For example, your department management may insist on silos due to trust issues with other departments, or perhaps even with other employees within your own department. You as a new engineer must individually decide how a communication silo (or any other silo for that matter) may impact your own personal responsibilities and develop your own response plan. Then, have a full, honest discussion with your management to fully understand your management's ground rules and reach a resolution, either formally or informally. Getting silos out on the table will pay dividends, as you will then know where both you and your management stand and avoid conflict.

Another interesting facet of our communication discussion is *jargon* and *acronyms*. Jargon is special words or expressions that are used by a particular profession or group. For example, the word *stat* in a hospital setting is jargon for "quickly" or "emergency." A *wrap* in movie or TV production is the completion of a unit of work; in American football a *dime* is a forward pass that travels at least 30 yards in the air and fits into a catch window of one yard or less. All are jargon.

We all know acronyms. An acronym is a subset of jargon, a word or name formed from the components of a longer name or phrase, usually using individual initial letters. Examples include NATO (North Atlantic Treaty Organization), EU (European Union), and NASA (National Aeronautics and Space Administration). While an acronym is a particular form of jargon, both are essentially the same and have similar characteristics.

Jargon and acronyms are something that every professional, academic, scientist, and engineer will come to know and sometimes even love. Jargon is a way for groups in a local or closed society to have their own specific, special language that communicates knowledge quickly and accurately. Technical occupations today almost demand their own jargon. As with most concepts, there are advantages and disadvantages to using it. It can be an efficient way to communicate with coworkers in the same group, provided they also know the definitions, and it can give a sense of belonging to that specific organization. Jargon also implies deep knowledge of a particular field, and its use can demonstrate mastery of the technical content of the subject. Yet there are also some negatives to using it. These include being seen as a way for the speaker to "show off," or leave an outsider feeling excluded from a conversation, or even leave an impression that the speaker has a hidden agenda. That said, on balance, jargon has a positive place in communication if used carefully and appropriately. Table 10.3 shares some recent examples of jargon from the technical community.

TABLE 10.3
Technical Organization Jargon Examples

Jargon Phrase	Meaning
"Groundhog Day"	A subject that has been discussed in meetings multiple times in the past without resolution or movement
"Snowplow"	Assignments come in and are accepted faster than they can be completed, creating an ever-increasing backlog of unfinished business
"Download"	Provide a briefing to a boss or higher executive
"Quarterback"	A local task force leader (noun); taking charge of a local task / assignment (verb)
"Tee It Up"	Introduce a topic at a meeting
"Deep Dive"	A detailed review
"Log off'	Remove your company badge when leaving the building (for lunch, going home, etc.)
"Loading Lips"	Writing a speech or presentation for an executive
"Secret Sauce"	The thing that makes your work special and unique
"Go Native"	Local corporate culture overwhelms an outsider embedded in a team
"Call the Question"	To make a final decision or finish a project or effort
"Open the Kimono"	Share detailed, uncensored corporate information with another company or individual, as in a joint venture
"Air Cover"	A boss providing support (budget, manpower, or political) to people working underneath them
"Mission Creep"	Starting a project with one goal, then adding more goals during the project without additional time or resources until project fails
"Drink the Kool-Aid"	Believe totally and completely in a position, even though everyone else believes it wrong, nonsensical or dangerous
"Heavy Breathers"	Senior management
"Calendarize"	To schedule or plan an action for a specific day or time
"Uplift"	A budget increase
"Put on Your Big Boy Pants"	Step up and do your job
"Ankle Biters"	Coworker constantly reminding you to do something you promised
"Mouth Breathers"	Not too bright a person

10.3.9 Listening and Voice in Personal Communication

Let's touch on your personal communication for a moment. Personal communication is just as the term implies: the communication you personally bestow upon others, be it your boss, your peers, upper management, people in adjoining organizations, or colleagues outside your group. This used to be most commonly done by verbal means, but text messaging is now the main alternative. Either way, you are totally responsible for the message's content and the clarity in expressing it. Luckily, this is one area where you have total control. Of course, the ground rules and expectations of personal communication within your local group or wider organization are unique and specific to that location and need to be learned quickly.

Listening

There is a simple, single rule for new engineers when it comes to personal communications. All those thousands of "how to" books on communications mentioned earlier tend to give similar advice when coming into a new organization, and this book would be no different: *listen.*

In her article *Just be Quiet and Listen*, Laura O'Connor simply divides listening skills as either bad or good. Bad listening means waiting for the other person to stop talking simply so you can start. Ironically, some people don't even wait until the other person has finished. There are some people who think of themselves as "assertive communicators" who are often the worst listeners. Eager to get their point across, they interrupt so much that another person can't complete their sentence. Another characteristic is pretending to listen but not really listening at all.[11]

Greg Harris was a 10-year veteran engineering manager in a civil engineering firm named Kueller-Brace (KB). KB was a global construction engineering firm specializing in urban commercial buildings having an average height of over 35 floors. KB had a presence in over 130 countries and was known as an aggressive, no-nonsense firm.

Greg's was on the hook for a status report to KB's Board of Directors regarding their latest construction project: a 47-story commercial office tower in Singapore. Things were not going well, but Greg had a solid recovery plan and was comfortable that it would be approved in short order. Gregg was scheduled to enter the meeting while it was already in session, which was standard procedure.

At the appointed time, Greg entered the boardroom and froze. Sitting around a large oak conference table were the 12 members of the board, each speaking in turn while passing around a three-foot-long, highly decorated wooden stick.

Yes, a stick.

Greg gave his presentation, and despite the distraction caused by this wooden object, it went well. Then as it came time for the question-and-answer period, Greg suddenly realized the purpose of the stick.

It seemed the aggressive nature of the company was reflected in their board meetings. Members consistently interrupted each other, cutting off discussion with snide comments, asides, and constantly making argumentative comments designed to score debating points but nothing else. It had gotten so bad that the CEO brought in a communications consultant who, witnessing the listening problem, installed the talking stick method.

The idea is incredibly simple: if you have the stick, you talk; if not, you don't.

Adopted from tribal culture, talking sticks are a powerful symbol, a communication tool used to foster an atmosphere of active listening and respect. As only the bearer of the stick has the right to speak, and those present must listen quietly to what is being said. And because the power is vested in the stick, not in any single individual, it's easier for the rule to be enforced.

As for Greg? After being initially embarrassed that the board needed such a child-like method to listen, Greg realized that at least they were trying. And he vowed he would never need to use a talking stick at work or at home.

Listening completely is hard, especially when you have a significant investment in the conversation. A common technique you may have been told is when listening to nod your head and keep eye contact. Experts argue this is not really listening. Good listening is looking for facts, emotions, and indications of the speaker's values. The fundamental aim of listening is to build a relationship so the other person trusts you. And remember that there isn't complete communication until both of you understand and agree on what was said.[12]

Voice

Another important point has to do with you speaking in meetings, be they formal management reviews or routine departmental meetings. There are three general types of verbal comments you can make: questions, facts, and opinions. Everything you say in a meeting will tend to fall into one of these categories. Here's a framework for your first few months or so:

1) Listen and don't speak. You don't know the answer.
2) When ready to launch your first comments, they should follow a distinct hierarchy. First, whenever you decide to finally add to a conversation, your first comments should be in the form of questions. And I mean good questions. A bad question to ask your supervisor is, "where is the bathroom?" (Don't laugh, I witnessed this happen). Rather, a good question in a meeting would be, "When is this information needed?" or "Is there anything special I should know about this problem?" A question is useless unless it makes progress towards an end goal.
3) Once gaining some experience in the form of questions, you can then move on to the second category, which are statements of fact. When ready, you should be willing to state facts relative to the conversation, but just facts. And obviously, your facts must be right. The rule of thumb is when you share a fact for the first time, it must be right. When you share a fact for the second time, that also must be right. When you share a fact for the third time, it certainly must be right. Only after three times have you established a small reputation for having your facts straight. And then you can risk perhaps having a fact found incorrect. Remember, you are establishing your

bona fides by having your facts correct. During this time, if you don't know for sure, don't speak.

4) The final category are opinions. You already know this, but I'll state it anyway: as a new engineer you are assumed to know nothing, and the people around you believe this, even if you indeed know something. Early on, no amount of talking by you will change this perception. Only after establishing your bona fides will you be able to voice an opinion based solely on your past reputation for speaking with truth or accuracy. There is a quaint old saying attributed to Maurice Switzer that goes something like this:

"Better to remain silent and be thought a fool than to speak and to remove all doubt."

10.3.10 Touchstones and Mavens

Your department or local organization probably has either a *touchstone* or a *maven*. The touchstone (who used to be the group secretary back in the old days, now replaced by today's office coordinator, department controller, or perhaps the director's administrative assistant) who occupies the communication and social "crossroads" of the department. Almost everything going on within the department flows through their desk. And I mean everything. Key reports, status notices, expense reports, personnel evaluations, annual raises and bonuses, future plans for high-potential employees; all these move under the watchful eye of the touchstone. If you don't understand something, you go to the touchstone. If you feel something is unfair, you get the touchstone's opinion. If you need a specific piece of information, the touchstone probably knows where it is. And if you are behind or forgotten something to do, the touchstone might give you a gentle reminder before your supervisor notices. Everyone depends on the touchstone.

The touchstone is also surprisingly influential. In addition to handling the group's daily tasks and information flow (acting as "oil between the gears"), they are also an important *de facto* advisor to the manager or director. And their viewpoint counts; they have cache with the boss. This means you don't want to be on the wrong side of a touchstone. And frankly, it has always been ironic that one of the most valuable individuals in an organization invariability receives some of the lowest pay or status in the group.

In the early 2000s, the people who were normally the touchstones, the administrative assistants, disappeared from the lower levels of the organization. Those who remained served the higher echelons of the engineering group, the directors and chief engineers rather than the supervisors and managers. While financially advantageous, what was lost was the natural location of the touchstone, leaving regular engineers and others devoid of their friends and advisors. One of the challenges resulting from this change is there is now no natural place for a touchstone to reside, no single place where so much information and knowledge can be had so easily.

Enter the maven. Mavens and touchstones are similar; people who acquire and share information within an organization. The maven's role is knowing the information and the connections to promote understanding of what's going on. According to Malcom Gladwell (who first coined the label) mavens are "information specialists," or "people we rely upon to connect us with new information." They accumulate knowledge, especially about the inner organization and know how to share it with others. According to Gladwell, mavens start "word-of-mouth epidemics" due to their knowledge, social skills, and ability to communicate. As Gladwell states: "Mavens are really information brokers, sharing and trading what they know."[13]

Mavens are incredibly useful, and you need to find them. There is no magic formula to finding a group's maven; just engage with colleagues and coworkers until they become visible. Be curious and observe.

10.4 SPECIAL CASES

Communication, being such a contingent subject, presents countless special cases that are guided by unique and powerful environmental conditions and ground rules. These are one-of-a-kind cases, situations you'll share as stories with colleagues in the hallway for years afterwards. They tend to be more important than normal communications and in extreme situations, can have lasting effects on your firm and perhaps yourself.

10.4.1 Communicating in the Senior Management Meeting

Decades ago, engineers did the daily work and the engineer's supervisor or manager presented the results of that work to the director or vice president. If the engineer was lucky, perhaps they could sit in on the senior management presentation, as long as they sat at the back along the wall and remained mute. In fact, sometimes the supervisor or manager would actually remove the engineer's name from a presentation or document and substitute their own name. Such was the corporate world in the dark ages.

Luckily, those days are essentially gone, replaced by the expectation that, no matter what rank you are, you will be presenting your own work to the seniors. And you need to know how to do it. Let's look at the two main categories of formal meeting, the planned meeting, and the snap meeting.

The Formal, Planned Senior Management Meeting Presentation

What about presenting to a formal, planned meeting of a group of senior management types? Obviously, you should first return to our first communication principles and start there. But due to the audience, you must accomplish additional preparation.

First, think about the makeup of the seniors you are presenting to. They are undoubtedly time and energy starved. They are probably tired, distracted, and balancing multiple worries simultaneously. Knowing this, attempt to make their life easier through limiting your presentation as follows:

1) Announce what your purpose is.
2) Be extremely brief unless asked to expand. Use only as much time as is necessary to achieve your communication goal. This means using a

minimum of their time and energy, which is their most valuable commodity. Remember, nobody complains if you "refund" some of their time back to them.

3) Answer the questions being asked, using simple, declarative sentences that are clear (more on this later). There is an old joke that when you ask an engineer what time it is, they'll tell you how they built the watch. Resist the temptation.
4) Do not go into background on test methods and process unless it is a critical new method, or you are asked to specifically discuss it. Unlike at university, where the test method is always described and defended, in industry your management assumes you are competent and as a professional are using the correct method. Employing the appropriate method is understood without being mentioned.
5) At the end, explicitly ask if there are any questions, comments, or concerns.
6) If requesting a decision and one is given, your last statement should be a clear repeat of the decision. If a discussion topic, summarize your understanding of the outcome simply.

Finally, as simple as it sounds, you must speak up. Speaking up has two advantages. For one, people can then actually hear what you have to say (which is your whole point in being there), and two, it shows confidence in what you are presenting and who you are. Just project your voice and let your knowledge fill the room.

An aside: Occasionally your presentation will go terribly, terribly wrong. Despite your best effort, your audience strongly disagrees, or is highly skeptical, constantly interrupts, or is openly disrespectful. It's obvious that you've stumbled badly. When this happens, be prepared for *shunning behavior*, where walking out of the meeting, the other participants will not even look at you, let alone speak. You have instantly become a ghost, and will remain one for the next day or two. It's easy to suggest you laugh it off, but it will still sting. Just learn whatever lesson there is to learn and move on.

The Formal, Snap Meeting Presentation

One of the common situations you can experience is the formal, snap meeting presentation. As we've mentioned, a snap presentation is a very challenging version of a talk in front of a group or audience, where you are given extremely little or no opportunity to prepare. This is unnerving enough when given to an informal group of peers or colleagues; it can become terrifying when attempted in front of upper management or other important audience.

From the 1940s to the 1980s, Henry Ford II was the president and CEO of the Ford Motor Company. During this time, Ford was consistently ranked within the top five corporations in the U.S. Fortune 500 rankings. Mr. Ford (nicknamed within the automotive world as "Hank the Deuce") was the grandson of Henry Ford and internationally famous, the equivalent of a Bill Gates, Larry Ellis, or Jeff Bezos today.

One of Mr. Ford's closest company and personal confidants was the Group Vice President for Public Affairs, a gentleman by the name of Walter Hayes. Hayes was considered to be one of the three most powerful men within the company at that time. An elegant Englishman with an impeccable accent and dress, his job was to know everything about each upcoming vehicle, truck, and concept car being developed within the company (and everything else for that matter). An impressive and intimidating figure.

At this time, Greg Scribner was the aerodynamicist for the Ford Probe IV concept vehicle. The Probe IV was a major effort to create the world's most aerodynamic four passenger, fully functional road vehicle. An area of major global competition, aerodynamics defined the phrase "Advanced Technology" in the automotive sector for an entire decade.

One morning a designer came breathlessly into Greg's office. "Greg," he said, "Walter Hayes showed up in the design studio just now, and wants a briefing on the aerodynamics of the Probe. You need to get down there right away."

Sometimes a snap presentation is a good thing. Since there is no time to prepare (or overprepare) your pitch, you can relax as you have the built-in excuse that if it doesn't go well, you can say you didn't have time to prepare.

Entering the studio, there was Hayes in a Savile Row suit, elegant white hair and matching tie and pocket square. When top executives like Hayes arrive, they always attract a crowd and this was no different: all of Greg's bosses plus many other management types gravitated to the studio. Greg was commanded to begin his briefing.

As he tells it, Greg never forgot the first line of his presentation: "The purpose of the Probe IV concept car is to develop the most aerodynamic efficient, fully functional four passenger vehicle in the world." Greg was immediately stopped by Hayes:

> "Young man, thank you for telling me that. In almost all the presentations I ever hear from engineers, no one ever simply tells me what the purpose of the project is. They always launch immediately into the details of the whatever they're doing and I never have any context. So, thank you and please proceed."

Greg's management beamed and Greg, at least for a day, was a hero.

The lesson learned is simple. In any presentation make a clean, short statement on the purpose of your talk. Again, always use simple, declarative sentences stripped of any unnecessary words. Strunk and White's seminal book on writing, *The Elements of Style*, sums it up neatly, "Omit needless words! Omit needless words! Omit needles words!".[14]

I guess we should omit needless words.

10.4.2 Bad News: Reporting Serious Problems to Management

This much is certain: you will run into problems in your work. Most will be trivial, requiring no extra help in solving. Many will be moderately difficult, perhaps requiring a coworker or colleague to assist. A smaller number will involve your supervisor, who will probably collaborate and advise you before settling on a path forward. But a critical few will require the attention and direction of your manager, director, or even executive director to solve; an experience prone to stress, worry, and acid indigestion dependent on the cause of the problem and who is responsible for the solution. These problems require a specific type of briefing.

These are situations where the communication you provide must be successfully received and understood immediately, independent of the normal, routine mode of transmission. In a true crisis or emergency, none of the routine communication modes described previously are appropriate. Simply put, for extremely urgent and critical messaging, you must ensure the timeliness of the message and its clarity. For these messages, it is your responsibility to cut through the normal communication barriers when any form of safety, reputational, legal, or other negative impact comes to your attention. In short, *do not* use the usual, routine channels for communications here.

Regarding timeliness, there is always a reluctance to share bad news, famously expressed by Shakespeare in the 16th century. His was the phrase, "don't kill the messenger."

Yet bad or negative news needs to be shared with management as soon as possible, so they may begin countermeasures and avoid or reduce the negative consequences.

A more serious example of failing to tell senior management bad news happened about five years ago at a major digital products company. Mike Whittan, the executive director of the company's Advanced Technology organization, was charged with performing an annual technology review for the senior management of the entire organization. Mike's annual technology review was a key report card on how his large, technically savvy (but expensive) sub-organization was providing future-looking technologies for the general consumer. Mike's goal was to emphasize their latest customer-facing products.

For this review, Mike's engineers had developed a suite of seven related technologies, all designed to work in unison to provide an enhanced customer experience. Each individual technology enhanced the adjoining 6 other features, providing an overall best-in-class functionality.

To demonstrate these technologies, instead of showing each technology separately, Mike decided to place all seven of them together in specially constructed, enclosed display area so that each executive could, with one visit, experience the entire suite. The interior of this enclosure would be designed with a futuristic theme, giving each executive an exciting and memorable experience. This is a very common way of demonstrating concept technologies by immersing decision-makers in a special environment, and from that experience, these top executives would hopefully approve those seven technologies to begin mass production.

With five months to go before the review, Mike assigned a four-year supervisor and two new engineers to have the enclosure designed and a specialty interior design house execute the futuristic vision. The supervisor was given the five months to complete the project with a budget of $315,000.

Early on, the supervisor and two engineers experienced delays in hiring the design house and getting the enclosure begun. Additional delays happened due to the designers optimizing their styling rather than beginning actual construction. Three months into the project, the supervisor realized the enclosure was going to be late, very late. Rather than immediately report this bad news to Mike, the supervisor elected to keep quiet and hope the time could be made up with overtime without anyone noticing. Yet as the remaining weeks flew by, the schedule only got worse.

With three weeks to go, the timing issue getting more acute and substantial overtime now being spent without approval, the build was in serious trouble. Unable to hide the status any longer, the supervisor went to Mike and reported that to deliver the vehicle on time would take an additional $252,000 to pay for the overtime already spent, plus the additional premium time needed to finish the build. Mike was asked to approve the additional funding.

Mike was livid. Here, he was presented with a huge bill without warning. As Mike said: "This is not a decision, it's a gun to the head." Mike was forced to spend over $250,000 in unbudgeted funds to avoid a major embarrassment with senior management.

While the technology suite showed well with senior management (and in fact did get approved as a new product line), the upset caused by the schedule and funding overruns had lasting effects. Within three months, the supervisor was transferred to a different organization within the larger company, and while the two engineers remained in the organization, they were sidelined on secondary projects. The supervisor was not fired from the company per se, but was released from the home organization and the two engineers banished to an outpost within the department. The key mistake was made early on by not reporting impending trouble, eliminating options Mike could have taken to correct the situation. Closing out options for management is a sin.

Messaging clarity is equally important when reporting bad news. A well-known example is the Columbia Space Shuttle disaster. On February 1, 2003, the Columbia broke apart on reentry into the earth's atmosphere, killing its crew of seven and dealing a severe blow to America's technical prestige. Foam insulation from the shuttle's external fuel tank separated during launch, impacting and criticality damaging the shuttle's wing leading edge. This damage allowed hot gases to enter the wing on reentry, melting the interior structure and ultimately causing the vehicle's disintegration. While neither the crew nor ground controllers knew of the damage, certain structural engineers suspected that damage may had occurred, it was merely a hunch. While the investigation board assigned primary blame on the foam strike, it also identified another culprit: a communication failure rooted in the routine of formal management meetings, specifically the use of presentation slide proformas.

In *Response Paper ESD.85J: Columbia Accident Investigation Board*, Former Secretary of the U.S. Air Force Sheila Widnall cited NASA's use of Microsoft's PowerPoint slideware and its deleterious effect on communicating complexity and decision-making. To quote Widnall:

> One week before Columbia's scheduled re-entry … engineers … formally presented the results of their numerical analysis [on the potentially damaged wing] … They projected a typically crude PowerPoint summary … with which they attempted to explain a nuanced position: first, that if the tile had been damaged, it had probably endured well enough to allow the Columbia to come home; and second, that for lack of information they had needed to make assumptions to reach that conclusion, and

> that troubling unknowns therefore limited the meaning of the results. The latter message seems to have been lost. This particular PowerPoint presentation became a case study for Edward Tufte, Yale's brilliant communications specialist, who subsequently tore into [PowerPoint] … for its dampening effect on clear expression and thought. The [Accident Investigation Board] later joined in, describing the widespread use of PowerPoint within NASA as one of the obstacles to internal communication.[15]

The board concluded that shuttle management had become too reliant on presenting complex information via PowerPoint templates instead of traditional technical reports. When NASA engineers assessed possible wing damage during the mission, they presented the findings in a confusing PowerPoint slide, so crammed with nested bullet points and acronyms that it was nearly impossible to understand. "It is easy to understand how a senior manager might read this PowerPoint slide and not realize that it addresses a life-threatening situation," the board stated.

This communications Achilles heel is seconded by former U.S. Navy Secretary Dr. Donald C. Winter, who during his tenure, pushed to "ban" PowerPoint within the Navy, stating that the software retards clarity and insight in policy-level decisions.[16]

The takeaway here is that for routine reporting of uncontroversial data in standard meetings, communication proformas can help speed management understanding. The exception is when reporting major problems, presenting unusual, critical or nuanced information, where misunderstanding will have strong consequences. In this case, the presenter must advocate for a unique communication mode that fully imparts the urgency, criticality, and consequences consistent with the importance of the message.

Knowing this, an engineer has many paths to choose from when deciding how to proceed. While each boss has a different preference in how to absorb information, a good bet is to approach the briefing in a way that maximizes information quality while taking ample time to ensure the message is correctly received. This is one time not to minimize presentation duration.

When briefing the organization's leadership, the presenter needs to communicate three things:

1) What the briefer knows about the situation.
2) What the briefer doesn't know about the situation. They must honestly tell the seniors what the presenter doesn't know. This is as important as the information the briefer does know about the issue.
3) What is the recommendation or solution regarding the situation? This is critical, as if no recommendation or solution is brought forward, management will feel obligated to impose one, and odds are no one will like their solution.

One important caveat will condition a presentation's preparations. Remember, each boss has a worldview, a certain set of beliefs, facts, and experiences that impacts how they see their environment and their reaction to it. For example, some bosses may have had their formative experiences in a highly formal or bureaucratic organization, where T's are crossed and I's are dotted. Or they may be informal in word and deed. No matter their style, the key is to try to match your message to their best method of absorbing information. Otherwise, their stress may block the accurate reception of the information.

So, how do you pull off this little trick? After all, communications can be like a boxing match: engineering vs. finance, engineering vs. marketing, engineering vs. industrial design; sometimes one engineering specialty vs. another. You will run into this surprisingly often. The key to solving this problem is to understand the others' language: the language of design, the language of finance, the language of marketing. You must be prepared to adjust your language and meaning to match your audience. Only then do you have a chance of legitimately changing minds, especially of any individual or group dealing with important problems.

10.5 REALIZATIONS

Organizational communication is complex. There are many failure states, some within your control and some not. The proper design of the organization's communications system is to identify and address the unseen, critical elements of a successful information transfer system. Conversely, an improper communications system can hurt, or even destroy, the ability of an engineering organization to function.

Communication is a series of linkages between a starting and ending point. Communication is all about the quality of these linkages, directly dependent on type, strength, language, and distance between those endpoints. This is obvious; however, one point isn't. The context of the communication is absolutely critical if a message is to be successfully transmitted and received. All parties taking part in communication must experience the identical context at the identical time.

We've spent a substantial amount of detailed effort discussing communication in meetings, presentations, hallways; in both formal and informal venues, and when sharing bad news or in crisis. These can be helpful, but please remember these are tactics in communications, not a strategy: the strategic purpose is to achieve that goal of clear and accurate transfer of thought between communicators.

As of this writing, the global COVID pandemic is beginning to recede, and new learnings in digital distance communication abound. Initial discussions regarding the enhanced role of electronic communication due to social distancing are starting. New "rules of the road" have been proposed and certain innovative protocols have been informally agreed to. A trend that started before the pandemic and has accelerated since is the pending disappearance of printed paper, replaced with new forms of digital documentation. And many people are waiting for the next breakthrough in digital communications, the next step beyond WebEx, Zoom, and the like. Whatever the specifics, the intent of communications is not to use *a* particular technology but use *any* appropriate technology that maximizes the quality of communication.

We're getting close to finally designing your personal roadmap. Just two penultimate steps remain: to complete some final preparations by integrating the components of the Inner Core, and then top it off with the large role personal ethics will make in the entire roadmap creation. Our next chapter tackles these two steps.

NOTES

1. Choo, C. W. 1996. The Knowing Organization: How Organizations use Information to Construct Meaning, Create Knowledge and Make Decisions. *International Journal of Information Management* 16(5):329-340.
2. U.S Department of Education. Introduction to Technical Communications. In *Technical Communications.* Waco, TX: Technical Communications Center for Occupational Research and Development.
3. Corey, Dr. Amy and Pierce, Tess. 2009. *The Evolution of Human Communication: From Theory to Practice.* Open Library: Pressbooks. https://ecampusontario.pressbooks.pub/evolutionhumancommunication/chapter/chapter-1/.
4. Lucid. 2022. *Glossary of Meeting Terms.* Portland, OR: Lucid Corporation. https://www.lucidmeetings.com/glossary/meeting.
5. Lopp, Michael. 2012. *Managing Humans: Biting and Humorous Tales of a Software Engineering Manager.* New York: Apress.
6. Winter, Donald. 2012. 74th United States Secretary of the Navy. Interview by Robert M. Santer, September 28, 2012. Transcript.
7. Center for Career Development. 2022. *What is an Elevator and Why Do I Need One?* Princeton University. https://careerdevelopment.princeton.edu/sites/careerdevelopment/files/media/elevator_pitch.pdf.
8. Deborah Dumaine. 2004. *Write to the Top: Writing for Corporate Success.* New York: Random House Trade.
9. Bass, Thomas. 2015. Author. Personal Recollection of Interview by Qianyan Shea. About 2002.
10. Ibid., U.S Department of Education.
11. O'Connor, Laura. 1995. Just be Quiet and Listen. *Detroit Free Press.* May 22, 1995.
12. Ibid.
13. Gladwell, Malcolm. 2002. *The Tipping Point: How Little Things Can Make a Big Difference.* New York: Little, Brown and Company.
14. Strunk, William and White, E. B. 2000. *The Elements of Style.* Needham, MA: Allyn & Bacon.
15. Falkenthal, Dietrich. 2005. Response Paper: Columbia Accident Investigation Board. ESD. 85J. MIT Engineering Systems Division. https://ocw.mit.edu/courses/ids-900-integrating-doctoral-seminar-on-emerging-technologies-fall-2005/c96e1f866d0e6a81df8922acadbfc486_columbia_df.pdf.
16. Ibid., Winter.

Part Four

Creating a Personal Roadmap

11 Final Preparations
Integration and Ethics

11.1 FINAL PREPARATIONS: REACHING THE PENULTIMATE STEPS

Blessedly, we're almost done. So far, we've identified the fundamental organizational framework that all engineering organizations use. We've parsed that model into its elements and individually examined those pieces. We've then looked deeply into the five key components of the framework's Inner Core and discussed the factors that can help operate within that core, hopefully preventing the mistakes and miscues that lurk in its dark corners.

Our next step is twofold: connecting the components of the core into a strong, dependable, holistic model of mutual understanding and dependency and then applying a capstone of personal ethical considerations to the model. These final steps will complete the preparations developing your personal guide, the navigation tool you will apply each day as you transition into the professional ranks. The good news is that neither of these final two tasks is particularly difficult. Establishing the obvious linkages between components is straightforward, and developing your personal ethical understanding is fundamentally simple once the main points of ethical belief and behavior are understood. The only challenge is applying some mindfulness to both these activities.

Let's start with connecting the components.

11.2 MELDING THE COMPONENTS TOGETHER

Let's go back to Chapter 4 and look again at the original Essential Engineering Framework, specifically the Inner Core. As in Figure 11.1, the five components embedded in the external environment are each different by virtue of their specific purpose, function, and rules of operation: each component is unique. Yet if this is completely true, how can we reasonably expect to meld these disparate pieces into an integrated whole, something that makes sense and is useful to an aspiring engineering professional?

The trick to integrating these components is *dependency*. Each of the five components is dependent upon the others, perhaps not on the surface, but will become visible when we perform a component-to-component analysis. All these components are adjacent: they lean on each other like a Roman arch where one missing stone brings the entire edifice to the ground. In short, the five components "need" each other to function and allow you to drive toward your goals.

DOI: 10.1201/9781003214397-15

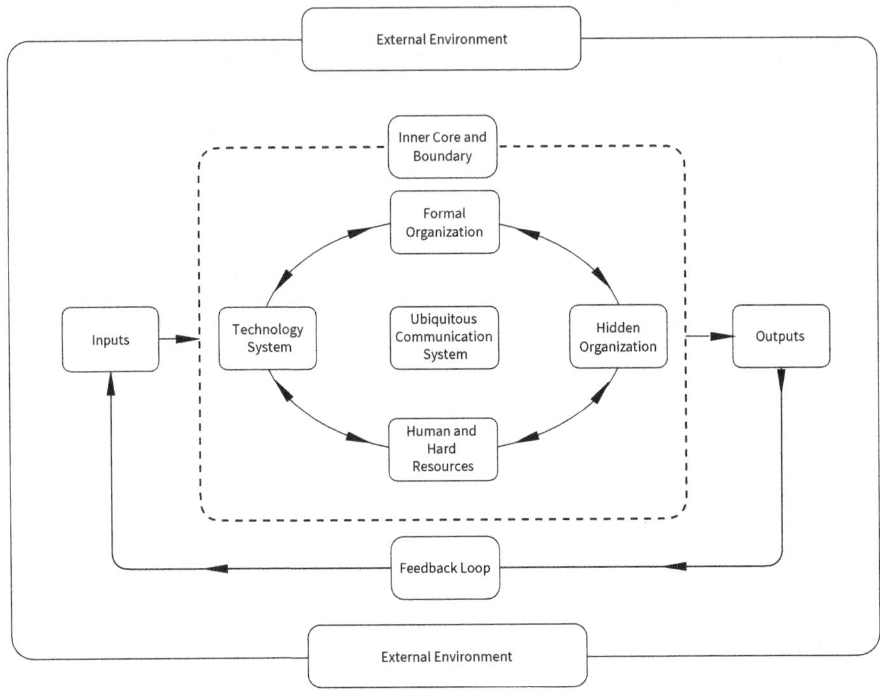

FIGURE 11.1 Essential Engineering Framework.

This has practical implications. You need to understand the basic purpose, characteristics, and limitations of each component well enough to naturally and quickly see the dependencies present. And not just the obvious ones, but the hidden connections as well. That's what melding the components means.

Let's take the simple example of the hiring and onboarding process you experienced as you joined your new engineering company. You probably view this as a linear process per Figure 11.2.

Say you receive an offer to join the firm as a design engineer. You may negotiate the conditions of your employment, such as pay, benefits, bonus (if fortunate), work location, or home department. You and the company agree to terms, and you begin employment on the appointed day. After a morning of paperwork and routine business, you are escorted down to your work location where, under the eye of your new supervisor, you officially begin your career.

You may or may not undergo training immediately, but after preparation of some sort, the daily work begins. And after a length of time, you are accepted as a group member in good standing and your work becomes a legitimate contribution to the organization.

This is most likely a universal process, and for you it is indeed linear. Yet for your organization your hiring is a networked, interdependent combination of components and subcomponents that come and go over time. To see this, let's rack up each

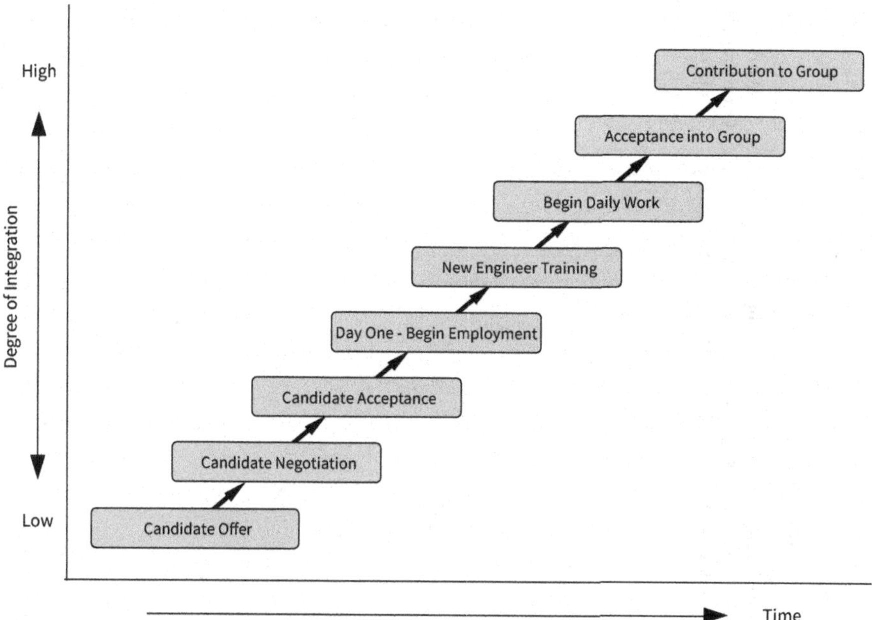

FIGURE 11.2 Generalized hiring and onboarding process.

process step against our five core components and establish which step is dependent upon which component. We can then step down a level and judge which subgroup in each component impacts your onboarding, as shown in Table 11.1. Breaking the process down this way yields eight steps involving seven connections to a variety of the five components. These are:

Human and Hard Resources: Human Resources Department Subcomponent

In this role, HR is responsible for two tasks: identifying, engaging, assessing, and deciding if you are a potential engineering employee, and if hired, shepherd you through the formal onboarding process (as much as there is one). A single HR employee is assigned as your central contact throughout this hiring phase. They are your point of contact through the first six steps and then essentially disappear. In large groups, this is a temporary relationship; in smaller organizations this person subsequently becomes your long-term connection to the world of HR. Even if they become your ongoing connection, their active interaction tends to end after some training is accomplished, except for normal career or personnel questions.

Human and Hard Resources: Finance and Legal Subcomponents

In the context of your onboarding, your connection to the Finance and Legal subcomponents is short and hidden from you. Finance approves the negotiated compensation

TABLE 11.1
Generalized New Engineer Onboarding Process

		Process Step							
Component Impacted	Subcomponent Impacted	Candidate Offer	Candidate Negotiation	Candidate Acceptance	Day One Begin Employment	New Engineer Training	Begin Daily Work	Acceptance into Group	Work Contribution to Group
Human and Hard Resources	Human Resources Department	X	X	X	X	X			
Human and Hard Resources	Finance and Legal Departments	X	X	X					
Human and Hard Resources	Facilities Department				X				
Technology System	Information Technology Department; Functional Tools Group				X	X	X		
Formal Organization	Local Group Management. Overall Department Organization. Policy Deployment; Standard Operating Procedures				X	X	X	X	X
Hidden Organization	Corporate Culture; Department Subculture; Social Rites						X	X	X
Communication	Internal, Local Group Connectivity; Level Setting				X	X	X	X	X

of your offer, and Legal primarily preforms the routine checks on your legal status (citizen, green card holder, conditional resident, foreign national, resident alien, and so on) and ensures you are allowed to be hired.

Human and Hard Resources: Facilities Subcomponent

The Facilities department connection is also quick and hidden. Their job is to arrange for and deliver your personal infrastructure: desks, chairs, cabinets, cubicles, open-air work locations; all are their responsibility. This might seem like nothing: you show up and all your equipment is (hopefully) there. But this infrastructure doesn't arrive on its own; it appears through the intervention of a number of facility staff. Personal infrastructure like this is normally in short supply, and you may unknowingly be part of a competition with other employees to score that newer desk or on what day your special ergonometric chair might arrive.

Technology System: Information Technology and Functional Tools Subcomponent

The Information Technology (IT) Department and Functional Tools subgroup in the context of your onboarding is a very short but visible connection to you. You need a laptop, a phone, a workstation, and any other unique hardware or software, either as an external purchase or sourced inhouse. Your initial involvement with IT is hopefully short but it doesn't disappear completely: you will have ongoing, low-frequency interactions with IT throughout your tenure. The same goes with whatever subgroup handles any special technology you require. You will be talking with these people again.

Formal Organization: Local Group and Departmental Subcomponents

Obviously, the formal organization is where most of your integration will take place. You will be focused on your local group (engineering peers, colleagues, supervisor, and manager) and your overall department organization (your director, executive director, and adjoining departments). Clearly, you will be highly dependent on and most closely integrated with your local group, and with the systems and environment immediately surrounding you. This means you will continuously immerse yourself with policy deployment, standardized operating procedures, scores of ground rules, and all the other considerations listed in Chapter 6. Your interactions are immediate and ongoing with many individual relationships. This is home.

Hidden Organization: Corporate Culture and Departmental Subculture Subcomponents

As you might anticipate, this is probably one of the most difficult components you will integrate into. As we've discussed, this is where most hidden connections reside, and they can last for a very long time. On arrival, you have no relationships, no personal investments, no common history with others, no group knowledge, no first-hand reputation with them, and no trust established. In short, in the eyes of the group, you have nothing to contribute. Each of these characteristics must be purposely established and demonstrated, and soon. Your first action or interaction must be right; your second interaction must be right; your third better be right.

After the formal organization, the hidden organization is the priority and needs your full attention. Because one day within the first month or so, a hallway conversation will occur between any two of your group's members. It will go basically like this:

Question: "How's the new kid doing?"
Answer: "They seem to be OK",or
"They just don't seem to get it".

That's it: short, small interactions in the hidden organization can convey an outsized message about you as a group member. For something so short, that exchange has a powerful impact.

Communication: Internal Connectivity Subcomponent

It's no surprise that a required part of integration is communication, especially communication at the personal, group, and department levels. As we learned in Chapter 10, communication is the glue holding the other four components together and is essential to you starting from Day One. Communication may seem a bit overwhelming so early on, especially if your group members like to use a lot of jargon and acronyms. But communication will come naturally enough if you are curious and show a willingness to engage.

As you see, for the simple task of onboarding, there are a surprising number of connections that are made before, during, and soon after your arrival. The good news in an exercise like this is that making the connections is easy once you become sensitized to it. Your first few months will, in large part, be spent learning to make those connections: understanding how these components interrelate is one of the main goals of your initial employment. Just realize that integration of these five pieces is multifaceted, involving connections with many people and entities. Complete understanding will not come quickly, but a working level knowledge can be attained quickly enough if it's kept in mind.

In our onboarding example, it's easy to see the connections within the Inner Core. But let's discuss a situation where the protagonist does not meld the components together into a unified whole.

Say you are a new engineer at a national, high-end sporting goods manufacturer located in Boulder, Colorado. This company makes equipment in all areas of sports: baseball bats, basketballs, hockey sticks, football helmets, soccer shoes, and snow sports are all part of its equipment portfolio. They have an excellent reputation for providing high-performance sporting goods at a reasonable price.

You are the first mechanical engineer ever hired by this manufacturer, as they believe more advanced technology in their products will give them an edge in their existing market space. As a new member of their advanced product group, your specialty is investigating and developing advanced materials to improve their offering's performance, including unique "aspirational" equipment. Today, you are asked to perform your first major assignment: an evaluation of a new pair of downhill racing skis developed by a second-tier startup supplier.

High-performance skis are an interesting product. For maximum speed in downhill racing, the skier must continuously apply forward pressure to the ski. All modern performance skis need to engage the tip at the top of each turn to start the ski carving and to prevent chatter, a speed killer. If you transfer all your weight into the middle of the ski, you'll lose precious speed. The idea is to drive pressure onto the tip of your downside ski by flexing hard at the ankle against the inside corner of your ski boot. After you feel the tip engage, you can then allow pressure to build under your foot. A tough trick, something that only the most expert skiers can accomplish. But what if a dilettante skier (as 99.98% of us are) could do the same thing? Faster speed and greater satisfaction.

This is where a first-tier supplier comes in. They are offering your company a revolutionary new ski, one based on the piezoelectric effect. The piezoelectric effect is the ability of certain materials to generate an electric charge in response to an applied mechanical force or stress. One of the unique characteristics of this effect is that it is reversible, meaning that materials exhibiting the effect (the generation of electricity when stress is applied) also exhibit the converse (the generation of stress when an electric field is applied). Since piezoelectric materials can change shape when an electric current is present, these materials can bend, stretch, or contract depending upon the amount of electricity applied. The idea here is that the ski could bend and flex independent of the terrain, continually changing shape to adhere closer to the ideal performance contour. A terrific, game-changing product, but only if it works.

Your job is simple. Evaluate the performance of the new ski and recommend if your company should take out an exclusive contract with the first-tier company for $24.3 million. An interesting assignment and important too.

Evaluating something as sophisticated as a piezoelectric ski is not easy, and you envision measuring an electric charge versus deflection per unit time, yielding a set of characteristic performance graphs.

Eager to begin and being told time is of the essence (it always is), your first step is to figure out just how to measure the deflection of the ski under a range of electrical and environmental conditions. Finding no suitable test apparatus available, you investigate and submit a purchase order to your boss for $112,000 for the design and purchase a new testing stand. Your supervisor, while uneasy, approves the request and incurs a sunk cost of $112,000, with delivery in six weeks' time.

After the six weeks, the test stand arrives, and you perform the test. The result is a nice, tidy set of deflection versus electric charge graphs that indeed shows significant deflection when power is applied. This claim made by the first-tier supplier is proven true and you recommend the purchase at the original investment level of $24.3 million.

And your management's reaction is … unenthusiastic.

The issue here is not measuring deflection but *judging performance*. It's really about evaluating the entire ski package, not a single mechanical characteristic. While you accomplished the engineering task of proving the piezoelectric ski does work, a true performance analysis was not done. This type of analysis is meant to answer higher-level questions. Does the piezoelectric ski actually change shape fast enough to be used in downhill or mogul racing? If so, does it really result in a faster time? Can the average customer perceive any performance difference compared to a normal ski? Overall, even if the ski works as advertised, will the customer be satisfied with it or just view it as an advertising ploy?

At its root, the word *evaluate* has a certain meaning for you and a totally different meaning for your management and the company as a whole. You viewed evaluation as strictly an engineering function; your supervisor viewed evaluation as judging a total product system. What was needed was a holistic evaluation plan, incorporating all five of the Inner Core components. Your plan addressed only the Technical System. No total cost analysis was performed, no communication occurred except with the supervisor; no real-use testing occurred. Using an integrated, component-based approach would have avoided this embarrassing result.

A few points to consider. First, no matter how quick the evaluation deadline is, you need to step back, have a Red Bull or Monster, and think about it. Was the assignment clear enough? What does performance truly mean? Is your boss asking for help on a holistic assessment or a simple mechanical engineering evaluation? Each of these questions will identify which of the five components you need to employ. For example, to completely evaluate the ski, you need help from most of your department subgroups, plus organizations outside your group: the business office, the reliability and durability offices, the warranty group, the finance department plus several others. If the ski truly delivers the benefit as touted by the supplier, you have no worries. If the piezoelectric effect is working but did not translate to improved performance by either an advanced or average skier, management would make a judgment whether to produce it or not. If the piezoelectric affect was not working at all, would the company still go ahead and offer it for sale, even if it was a false claim? In short, this evaluation involves judgments from the other core components. This is the difference between a savvy engineer and a brand new "one trick pony."

But how about expanding this concept even wider by establishing an understanding between your engineering organization and a completely unrelated group, say the company's sales organization? Let's look at a case example using these same components to link unrelated divisions together.

CASE EXAMPLE 11.1 THE DDS CORPORATION FLEET CONQUEST

DDS is a global provider of home and business security systems, operating in 34 global markets with annual revenue of nearly $10.9 billion. Their field technicians use a fleet of some 4,500 light utility vehicles to install and service their products. Prior to 2019, DDS purchased these vehicles exclusively from the Legacy Mobility Corporation. However, in 2019 DDS elected to open their fleet procurement process to other automotive manufacturers, including Legacy's chief competitor Vibrant Motors. On hearing this, Vibrant Fleet

Manager Mike Garcia identified an exciting opportunity to seize substantial conquest sales away from Legacy.

DDS did not know very much about Vibrant's vehicle capabilities, as DDS's perception was that Vibrant "just made cars and trucks." Discovering that DDS had no knowledge of Vibrant's strong leadership in vehicle occupant safety and technology, Garcia contacted Tony Lewis in Vibrant's Research and Advanced Technology Division (R&AT). Together they quickly determined that targeting DDS through education on safety technology, plus hands-on visits to witness actual safety crash tests, would radically change DDS's perception of Vibrant and thus, the competitive landscape.

Mike and Tony quietly arranged multiple visits by DDS to the Vibrant Research Laboratories, witnessing testing and having Vibrant's leading experts in automotive safety present their latest work.

The visits and safety education went well. The DDS management delegation was very surprised at the depth and breadth of expertise and resources devoted by Vibrant toward safety. Improved safety had a double benefit for DDS. First, it reduced injuries to DDS drivers which substantially cut insurance costs. Secondly, it could demonstrate to DDS drivers they were driving the most advanced, safety-equipped vehicles on the market, improving employee morale and job satisfaction ratings. Coupled with additional technology systems, Vibrant built a compelling case.

It worked. In early 2020, DDS placed a conquest order for 4,500 Vibrant "Connector" light utility vehicles, replacing their entire fleet of Legacy vehicles. Net revenue to Vibrant was $114 million dollars. In their purchase letter to Vibrant, DDS expressly stated that their briefings on vehicle safety research, and the passion of the safety scientists they met, was the key factor in their decision to leave Legacy and move to Vibrant.

The lessons of the Vibrant case are not hard to see. This example is all about the integration of components from separate organizations, demonstrated by:

1) Mutual communication linkages, where safety department members acted as technology "brokers" to attain mutually beneficial outcomes for the both DSS and Vibrant.
2) The hidden organization where, by quietly acting as an ad-hoc group avoided the barriers and "bureaucratic drag" of traditional command and control, demonstrated by not asking for vice presidential approval of the project in advance.
3) Human resources in the form of talent; those experts who educated and convinced the DSS executives on the importance of safety.
4) The technology system which created the unique, advanced safety products that DSS desired.

In essence, Mike and Tony melded a one-of-a-kind set of components in the communication, culture, talent, and technology spaces, creating a "win" for both organizations.

One final point. We know *how* this example worked for Vibrant, but *why* did it work? For one, the groundwork was prepared several years before the DDS opportunity. Certain R&AT engineers had previously established strong, personal relationships across the many silos existing between R&AT and other Vibrant organizations. Vibrant R&AT was now connected personally to many other internal organizations, including the Vibrant Fleet Sales activity. By establishing *in advance* these relationships at the personal level, the traditional, formal collaboration process was "short circuited" and major opportunities could then be quickly exploited. The fact that Mike Garcia knew specifically to call Tony Lewis, in a company of over 72,000 employees, proves the point. Garcia knew the connection existed and used it: it was part of his prior, personal integrated roadmap.

In your own search for connection and integration, what if the connections you see just don't make any logical sense to you? What if, despite your best efforts, you can't resolve the links (or lack of linkages) between certain components? Frankly, there is a good chance some of them won't. We need to spend a few minutes on why this is and how to resolve this cognitive dissonance within your own mind.

11.3 RESOLVING THE CONTRADICTIONS: ORGANIZATIONAL SENSEMAKING

Integrating the components depends on making sense of each interaction between them. *Sensemaking* is an attempt to reduce multiple meanings and handle the complex information used by people in organizations. In Chapter 8, we discussed the work of Karl Weick, who contends that the practice of day-to-day work is actually quite different from what we might believe.

Weick maintains organizing is a continuous, evolutionary process with the goal of reducing confusion, that employees place limitations upon the organization to avoid issues and that groups evolve as they make sense of themselves and their environment. Weick's framework for understanding organizations means making sense of the often overwhelming complexity and contradictions people create.[1] Communication becomes key because of its role in this sensemaking process.

Now here's a controversial statement. From this perspective, strategies, plans, rules, and goals are not things that exist in an objective sense; rather their source is in the group member's way of thinking. From a sensemaking perspective, the question of whether someone's view of the world is correct is not meaningful. Instead, the correctness of a position is contingent on the point of view that is being used for evaluation. The basic idea of sensemaking is that reality is an ongoing realization that emerges from an individual's efforts to create order and understanding from complex environments. In short, sensemaking allows people to deal with uncertainty and ambiguity by creating rational accounts of the world that can cause them to enable some sort of action.[2]

This is certainly counterintuitive, but let's play out this idea with a simple case. What if making a connection between you and a core component doesn't make any sense? What if, for whatever reason, the connection between you and a person in the financial office is totally illogical and just not working? For example, let's say you absolutely need to purchase some carbon nanotube material for a new innovation, but your request is repeatedly denied by the financial controller. You know your innovation is a game-changer, that it will fill a current white space in the market, and you're certain you cannot proceed with your project without this critical material. But your pleas to approve the funding have absolutely no effect on the controller. In short, this lack of integration with finance makes absolutely no sense. Why is this, and how can a successful connection be established?

The answer is in organizational sensemaking. You see your innovation as changing the competitive landscape: it changes the external environment. You see what the future will be. The controller only sees a request for a piece of expensive material, with no context as to what it is for, why it is important, and what its benefit to the enterprise is. You see a new world; the controller sees an unnecessary expense.

Lack of context and the resulting view of reality helps explain the controller's position. The simple solution to this problem is to sit down with your controller (face-to-face, as dynamic message delivery is best) and explain the new environment this innovation will create; to bring the controller along on this journey to a better world. It means demonstrating the benefit of the new material; it means active communication between both parties. (An aside: here we run into our old problem of not advocating for our work; the "my technology is so good, it speaks for itself" issue. Don't fall into this mindset: spread the word.)

From Weick's idea, researchers Chris Argyris and Donald Schön went further to argue that people have predetermined "mental maps" regarding how to act in various situations. These involve the way they plan, execute, and judge their actions. It is these maps that guide people's actions, rather than the beliefs they explicitly state, which is very similar to the retrospective sensemaking theory. In other words, in complex organizations what people say and do are highly inconsistent; "the words and the music don't match." A common example of this thinking is the case where organization members respond to changes in the environment by detecting errors which they then correct to maintain the status quo, rather than resolving incompatible situations by setting new priorities and norms, or by restructuring the norms themselves together with new strategies and assumptions. In other words, maintaining an existing paradigm rather than investigating if a new paradigm is now in place.[3]

In trying to resolve sensemaking, a paradigm shift is a very good starting place to start investigating. You will be very surprised how often this takes place. Look for the incompatible assumptions and mindsets in place: this is where you may find the key to making sense of a situation.

11.4 THE ONE CRITICAL EXCEPTION: YOUR PERSONAL CODE OF ETHICS

Throughout this book, we've been talking (somewhat insistently) about contingency theory. So much that your eyes may glaze over when you see those words and you'll probably never utter that phrase again. But let me share just one last point about contingency. Contingency theory has been the one touchstone that has been a constant in all the topics we've discussed. Now, it's time to reveal the one exception to the contingency theory model. I'm talking about personal ethics.

Ethics can be a slippery idea, and like most important ideas is subject to confusion and misinterpretation. Here, a *moral* or *ethical principle* is simply the belief that what is correct or incorrect is determined and accepted by an individual or social unit, i.e., how they should behave. This is based on what a person's conscience (integrity, principle) says is correct or incorrect, rather than what the law says should be done and is good or right when judged by the standards of a typical person or society at large.

That sounds simple, but there's much more to it than a single definition. Let's deconstruct this concept and clarify what ethics is and the applications it entails.

In this book, we establish three categories of ethics based on where they are applied. The first is corporate or managerial ethics, which are behavioral guidance rules aimed at steering groups of corporate individuals in decision-making. Social media and other news outlets delight in reporting corporate ethical lapses of this kind. You don't have to think very hard to recall such major corporate ethics disasters as the Exxon Valdez oil spill, the British Petroleum Deepwater Horizon explosion, the Dow Chemical Bhopal catastrophe, the Volkswagen diesel scandal, and on and on, ad infinitum. You are probably most familiar with these when you think of a firm's moral behavior.

The second type is critical: personal ethics. As we've repeatedly established in this book, there are many definitions of the same phrase, and personal ethics is no different. Here, *personal ethics* are moral principles governing the appropriate conduct of an individual or very small group.

The third type is a subset of personal ethics called *professional ethics.* Every profession has a code of ethics, a series of statements defining acceptable and unacceptable professional behavior meant to provide guidance and comfort to those professionals facing ethical dilemmas. For us, *The Engineering Professional Code of Ethics* developed by the National Society of Professional Engineers (NSPE) is our guidepost. The current code is exhibited in Table 11.2.[4]

These codes are good and can certainly provide initial guidance when faced with a professionally based, ethical choice. Where things get tough are when the ethical choices presented to you are not addressed in the code. These can be something as big as the construction of a high-rise apartment building or something more insidious like bumping the amount of money claimed on a travel expense account.

What makes engineering ethics slippery is that all three categories can exhibit varying amounts of overlap. This is an important realization: the amount and nature of corporate, personal, and professional ethical smearing can make what normally are clear-cut choices muddled and the engineer befuddled.

To gain some clarity, Richard Daft has helpfully defined a topology of five ethics approaches or belief systems applicable to an engineering organization.[5] Briefly, they are:

The Virtue Approach

This approach says that moral behavior "stems from personal virtues, given in an external law, which is revealed in religious scripture or apparent in nature and then

TABLE 11.2
National Society of Professional Engineers Professional Code of Ethics

Summary – The Engineering Professional Code of Ethics for Engineers	
Preamble	Engineering is an important and learned profession. As members of this profession, engineers are expected to exhibit the highest standards of honesty and integrity. Engineering has a direct and vital impact on the quality of life for all people. Accordingly, the services provided by engineers require honesty, impartiality, fairness, and equity, and must be dedicated to the protection of the public health, safety, and welfare. Engineers must perform under a standard of professional behavior that requires adherence to the highest principles of ethical conduct.
I. Fundamental Canons	Engineers, in the fulfillment of their professional duties, shall: 1. Hold paramount the safety, health, and welfare of the public. 2. Perform services only in areas of their competence. 3. Issue public statements only in an objective and truthful manner. 4. Act for each employer or client as faithful agents or trustees. 5. Avoid deceptive acts. 6. Conduct themselves honorably, responsibly, ethically, and lawfully so as to enhance the honor, reputation, and usefulness of the profession
II. Rules of Practice	1. Engineers shall hold paramount the safety, health, and welfare of the public. 2. Engineers shall perform services only in the areas of their competence. 3. Engineers shall issue public statements only in an objective and truthful manner. 4. Engineers shall act for each employer or client as faithful agents or trustees. 5. Engineers shall avoid deceptive acts.
III. Professional Obligations	1. Engineers shall be guided in all their relations by the highest standards of honesty and integrity. 2. Engineers shall at all times strive to serve the public interest. 3. Engineers shall avoid all conduct or practice that deceives the public. 4. Engineers shall not disclose, without consent, confidential information concerning the business affairs or technical processes of any present or former client or employer, or public body on which they serve. 5. Engineers shall not be influenced in their professional duties by conflicting interests. 6. Engineers shall not attempt to obtain employment or advancement or professional engagements by untruthfully criticizing other engineers, or by other improper or questionable methods. 7. Engineers shall not attempt to injure, maliciously or falsely, directly or indirectly, the professional reputation, prospects, practice, or employment of other engineers. Engineers who believe others are guilty of unethical or illegal practice shall present such information to the proper authority for action. 8. Engineers shall accept personal responsibility for their professional activities, provided, however, that engineers may seek indemnification for services arising out of their practice for other than gross negligence, where the engineer's interests cannot otherwise be protected. 9. Engineers shall give credit for engineering work to those to whom credit is due, and will recognize the proprietary interests of others.

From National Society of Professional Engineers (NSPE). 2022. *Code of Ethics for Engineers.* Copyright NSPE, Used with Permission.

interpreted by religious leaders or humanist philosophers." If an engineer develops good character traits, such as compassion, generosity, or courage, and learns to overcome negative traits such as greed or anger, he or she will make ethical decisions naturally as the result of being a virtuous person.

The Utilitarian Approach

This approach holds that moral behavior produces the greatest good for the greatest number. An engineer or decision-maker is expected to consider the effect of each decision on all parties and select the one that will optimize the benefits for the greatest number of people.

On the surface, it sounds quite admirable; making decisions for the greatest good for the greatest number has a certain appeal to it, and what can be wrong with that? For that reason, many (if not most) large organizations and corporations adopt this approach as their normal ethical guideline. A dividend is awarded equally to each and every shareholder, a nice and simple rule as no decisions need to be made about special circumstances or messy value judgments: just divide the profit by the shares and you're done. However, the utilitarian philosophy may have some serious moral concerns. Utilitarianism is based on a strict meritocracy and equality in all things, irrespective of their contingencies. Persons should naturally receive benefits in direct proportion to their merits. But then, not everyone can be judged on the same comparative scale. Your 75-year-old grandmother with a high school diploma is not able to compete in the workplace equally with a 25-year-old coding expert. The utilitarian ethic tends to not allow nor have room for exceptions to its strict meritocracy judgment. A company following strict utilitarian ethics without substantial exceptions for special cases can quickly become unbalanced and eventually unfair.

Universalist Approach

This framework asserts that all human beings have fundamental rights that cannot be taken away by another individual's decision. Here, an ethically correct decision is one that best maintains the rights of those people affected by it. In this system, these principles are the right to privacy, the right to free consent, and the right to free speech. In a word, everyone would act with an intent based on universal principles.

The Justice Approach

This approach holds that moral standards are based upon the primacy of a single value, which is justice. Moral decisions would be based on standards of equity, fairness, and impartiality. This involves three types of justice. *Distributive Justice* requires that different treatment of people should not be based on arbitrary characteristics such as race or gender. *Procedural Justice* requires that rules be administered fairly. *Compensatory Justice* argues that the party responsible for injury or damage should compensate individuals for the cost of those damages. Conversely, individuals should not be held responsible for matters over which they have no control.

The Pragmatic Approach

The Pragmatic Approach is just as the name implies; it avoids debates about what is right, good, or just and instead bases decisions on the prevailing standards of the engineering profession and larger society, taking the interests of all stakeholders into account. A decision is ethical if it is considered acceptable by the professional or common community. It is one where a manager would not hesitate to publish their actions on a social media site or one that a person would typically feel comfortable explaining to family and friends.

A (somewhat alarming) point of view on the Pragmatic Approach impacting personal ethics comes from Robert Jackall's *Moral Mazes, The World of Corporate Managers*. Here, the overall theme is that engineers constantly adapt to the social environments of their organizations to succeed. In such environments, they have no use for abstract ethical principles but conform to the requirements of the company's bureaucracy, which they will tell you is pragmatism. Several important business implications follow from this belief. As Jackall bluntly states:

> It is hard to see a place for personal ethics in modern organizations at all. The moral ethos of managerial circles emerges directly out of the social context [of the corporation]. It is an ethos most notable for its lack of fixedness. In the … corporate world, morality does not emerge from some set of internally held convictions or principles, but rather from ongoing, albeit changing relationships with some person, some coterie, some social network, some clique that matters to a person. Since these relationships are always multiple, contingent, and in flux, managerial moralities are always situational, always relative.[6]

Whoa … This is an extremely important statement regarding relativism and ethics. Jackall is saying that many, if not most, corporate managers (including engineering managers) will adjust their ethical standards to meet the standards set forth by the organization: a person embedded in an organization will tend to take on at least some of the ethical characteristics of that group. As we just discussed, companies like the utilitarian approach, a structure that avoids those sticky value judgments and calculates its way out of trouble. And many unconsciously apply the Pragmatic Approach simultaneously. If someone will "go along to get along," that means they are now a relativist and therein lies the conflict. Personally, you have a certain belief, but your company requires something different. How do you resolve this disconnect, and how does this play out in day-to-day office life?

This brings us to one of the most important realizations that individuals in engineering or technical fields will come to, and one of the most earnest points I want to share. In the workplace, you will occasionally be presented with an ethical dilemma with absolutely no time to ponder it. You must make a decision, many times a difficult ethical decision, instantaneously. Say your manager (who is known to be vindictive if denied something) may ask you to do something sketchy or even blatantly unethical. He or she will probably want an immediate answer. What will you do?

Over the course of your professional life, you'll collect a number of these experiences. Colleagues have been asked by directors to prepare two versions of the same report, one reporting bad news and the other version deleting that news, with certain individuals receiving one version and others getting the scrubbed edition. Or they've been asked to issue purchase orders without management approval. Or they've been asked to make statements that were false. And this, unfortunately, may happen to you. What is your response?

I suggest two difficult things:

1) You must have your answers to ethical situations worked out ahead of time, and
2) With ethical questions, there should be no moral relativism. Personal ethical positions should not be contingent.

You'll be forced to choose a path, most of the time under pressure, that will be either expedient and easy, or slow and difficult, or right or wrong. You need to know ahead of time which path you are going to take.

CASE EXAMPLE 11.2 THE RISE AND FALL OF FLUIDTECH, INCORPORATED

An excellent example of a conflict between an engineering company's ethics and personal morals has to do with a Research and Development Manager named Ed Krieger.

Ed led R&D for a company named FluidTech. FluidTech was a "Tier 1" (direct) fuel and fluid system supplier to the automotive industry. Their flagship product line were fuel systems made of "SuperTube," a proprietary composite invented and patented by Ed's R&D team, exhibiting unique and superior environmental and safety attributes.

FluidTech was a private company new to the automotive market. Founded in 1989 and owned by two entrepreneurs, it achieved about $40 million sales by 1992. The company was technically unsophisticated but highly motivated to succeed in this new business. The company had a good culture with some truly exceptional people, and FluidTech offered an interesting opportunity for these people to be in on the ground floor of developing new and exciting technologies.

FluidTech's culturally espoused values were, in order of importance:

1) Serve the customer
2) Become the technology leader in automotive fuel systems
3) Grow, grow, grow (they were then doubling sales every 2.5 years).

FluidTech had a culture of egalitarian meritocracy, offering professional opportunities to their employees with growth in responsibility and reward based on performance. The pay was below average, but the culture and fast growth offered their employees an opportunity to progress with substantial freedom, influence within the company, and high levels of authority over their areas of responsibility. In short, it was a fun and dynamic place to work.

In the United States automotive fuel and fluid systems marketplace, FluidTech faced off against three larger, established competitors. These competitors were old-line manufacturing companies with little R&D competence, only offering legacy products.

In automotive fuel systems, there were two major looming technical challenges. The first was Electro-Static Discharge (ESD). The "racetrack" fuel system designs vehicles used at that time supplied more fuel to the engine than was needed and returned the unused fuel back to the tank, as in Figure 11.3. This resulted in high fuel flow velocities that, under conditions of low temperature and humidity, generated dangerous levels of static electricity resulting in "thermal events" (i.e., fires). ESD became a major safety issue when new fuel injection systems moved away from heavier traditional materials that were conductive (continually eliminating static electricity) to lightweight plastic fuel lines that were electrically non-conductive. The solution was to make plastic tubing electrically conductive while maintaining other stringent physical and chemical properties. This was a challenging materials science problem.

The second impending problem was evaporative fuel emissions, soon to be regulated by new United States Clean Air Act standards. Previous emission regulations focused on the tailpipe, but by 1995 more hydrocarbon emissions came from unburned fuel than the exhaust. Permeation through materials was

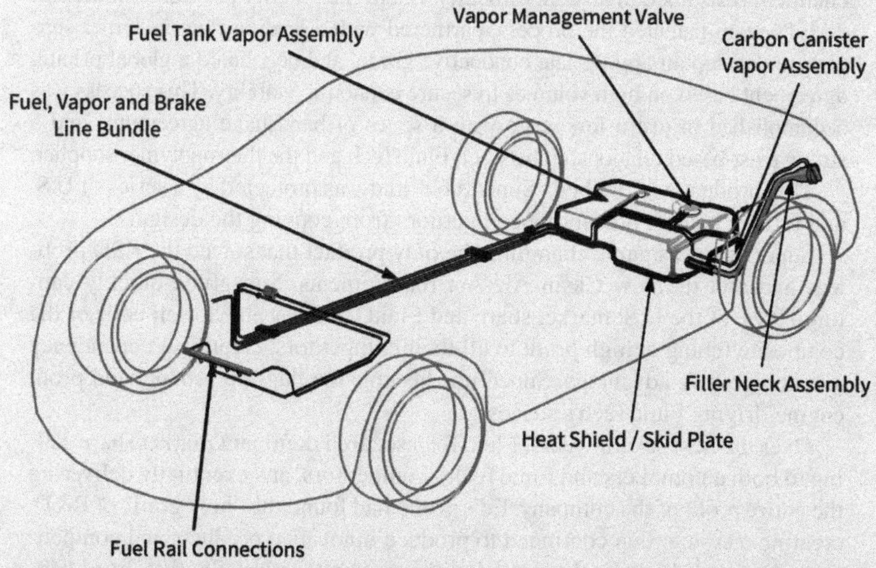

FIGURE 11.3 Racetrack fuel system design.

the issue, and the plastic fuel systems were woefully inadequate to meet the pending standard. This was another significant technical problem.

Ed's team determined what was needed to solve the permeation problem and the increasing temperature and chemical resistance requirements was a fluoropolymer. However, fluoropolymers were prohibitively expensive and failed to meet various physical properties. There were also no acceptable electrically conductive grades available.

Through material science engineering analysis, Ed's group decided the ideal fuel line of the future would be an inner core of a composite combining the permeation, chemical, and thermal properties of a fluoropolymer while utilizing the strength and economics of polyamide as an outer cover. The problem was, fluoropolymers were the original nonstick frying pan material (think Teflon) and there was no known method of bonding the fluoropolymer to other polymers. How do you stick something to a nonstick frying pan?

None of the major plastics companies had a viable fluoropolymer and polyamide combination meeting the ESD, permeation, and other stringent requirements. FluidTech decided on a bold and costly R&D program to staff a high-performance team to figure out how to effectively bond fluoropolymer to polyamide to develop the ideal composite structure.

It worked. Ed's team succeeded in utilizing a mixed gas plasma and other technologies to allow the fluoropolymer to bond to the polyamide. Through a multi-year engineering initiative, they developed the right composite structure to meet the performance goals while bringing superior ESD, permeation, and chemical resistance in an economically viable (i.e., highly profitable) material.

Ed's team patented the process, partnered with a leading fluoropolymer supplier to develop an appropriate conductive grade, and negotiated a global pricing agreement based on high volumes to secure economic viability. This success was accomplished in just a few years with a series of handshake agreements and a strong trust-based relationship between FluidTech and the fluoropolymer supplier.

This product was dubbed "SuperTube" and was protected by a series of U.S. and foreign patents preventing competitors from copying the design.

SuperTube was game changing: the only product that solved the ESD problem and met the new Clean Air Act requirements. SuperTube quickly captured 96% of the U.S. market share and FluidTech was able to sell coils of the composite tubing at high profit to all their competitors, creating a dependency and competitive advantage. SuperTube became the flagship product and profit engine driving FluidTech's success.

Over the next several years, FluidTech secured dominant market share selling to both automakers and FluidTech's competitors, and eventually delivering the entire profit of the company. Ed's group had found the "holy grail" of R&D, creating a system that continued to produce innovative products and competitive advantage, keeping the team and the system together. So, they lived happily ever after, right?

For most startup companies, there comes a time when the initial entrepreneurial owners pass the baton to sophisticated "professional" executive management. In 10 years, FluidTech had gone from $10 to $300 million in annual sales and $1 billion was in sight. While the owners would remain, they realized it was time to bring in someone they believed knew how to manage a large company.

Both owners agreed on the particular individual they wanted as the next Executive Vice President (EVP) who would make the day-to-day business decisions. This incoming EVP (an outside hire) was very well known to Ed. Ed had high respect for the intelligence, market development savvy, and professional polish of the EVP, but he was deeply concerned about the ethical behavior he had witnessed. This EVP had routinely used manipulation, telling people things that were not true to get them to act in a way that he desired. This is counter to accepted business practices in the United States, was strongly counter to the FluidTech culture, and was unacceptable to Ed. The EVP was a "moral relativist," not believing in unchanging right and wrong. He often utilized situational ethics, with a track record of reversing previous ethical positions when business advantage could be obtained. Once Ed discovered who the new EVP would be, he spoke with the major owner about his concerns prior to the announcement. The owner downplayed Ed's concerns but said, "If you don't get along with the new EVP, promise me that you will not leave the company. I will find another boss for you if you can't get along with the new guy." Ed agreed that if things did not go well, he would not leave without talking with the owner to work out a new reporting relationship.

Initially the new EVP came in and generally made good decisions. He expanded Ed's managerial authority substantially. For the first six months, he was the best boss that Ed had ever had. But as Ed recounted, "Bosses are like puppies; they are almost all cute when you first get them, but they don't all remain cute later on." After the initial six months, the EVP began making decisions that almost immediately forced Ed into three important ethical dilemmas.

The first dilemma revolved around performance reviews (PR) and pay increases. At FluidTech, each employee received their annual PR on their date of hire and their salary increase was granted immediately following the review. The EVP instructed all of the department managers to extend the performance review period from 12 months to 18 months while delivering the same percentage increase. This meant that a 6% raise after 18 months was equivalent to a 4% annual raise, so people would really be receiving only two-third of their typical performance increase. Ed said that it should be no problem to extend the review periods but asked what the explanation to the employees was going to be. Were there unexpected expenses the company had to cover, had future business contracts not come through, or was there some other reason? The EVP's response was, "I'm not asking you to explain it to them, I'm telling you to do it." Ed was a bit taken aback and mentioned that in the past the employees had always been willing to endure cost cuts, but that reasons were always provided, and that doing so without an explanation would hurt morale and possibly productivity. The EVP told Ed that there would be no explanation, he just needed to go do it.

Ed disagreed strongly. What were Ed's possible options at that point?

1) Go back to the EVP and try to convince him that an explanation was needed. Ed felt that another meeting would probably be unsuccessful and would likely create a hostile interaction.
2) Obey the EVP and simply implement the policy with no explanation. However, Ed had an understanding with his subordinates that for any decisions that affected them, he would seek their input and explain the context as much as possible. This path would break that bond of trust.
3) Disobey the EVP and put through the employees' performance reviews and raises on the existing 12-month cadence. This would maintain the bond of trust with his employees but be directly insubordinate to his powerful boss, a high-risk option.
4) Go over the EVP's head to the owner and make the case this was a bad decision. Going over a boss's head is a lot more dangerous than it seems. If it doesn't work, and even if it does, you will likely have made a powerful enemy who will not forget. This is an option but a very risky one.
5) Ed could leave the company, but you can't quit your job every time you have a couple of bad days in a row. Leaving is always an option that must remain on the table. During this first ethical conflict, a disagreement at this level would be hard to justify throwing away a very enjoyable and successful career.

Ed chose the third option. He disobeyed the EVP and conducted his performance reviews on the 12-month cycle, and put in pay increase requests for his people that ended up going through. Ed believed that he owed it to his people to give them an explanation if their financial livelihood was going to be adversely affected. Not being allowed to give such an explanation, he believed the integrity of his relationships with his subordinates was more important than the risk of crossing his boss and taking whatever consequences might arise.

The second, more serious dilemma arose a few weeks later, centering on whether your word should be considered a binding contract. Ed was able to report a new, significant success that he hoped might mend his strained relationship with the EVP. The automotive supplier industry is a high volume/low margin industry, and there had been strong efforts to find low volume/high profit, non-automotive market applications for SuperTube. Some very lucrative business had been won in the offshore oil industry, where this outside business was going to bring in profits of about 65%.

There was an understanding in the agreement with the fluoropolymer supplier that Ed and the EVP had negotiated in a handshake agreement. The agreement stipulated that for automotive applications, FluidTech would purchase the fluoropolymer resins at a very attractive cost, about two-thirds of the list price. However, for non-automotive applications, FluidTech would pay the full list price. In this case, the increase in fluoropolymer cost would barely affect the profits. When Ed reported the new offshore oil business deal, and they were going to pay market price for the increased shipment of fluoropolymer resin, the EVP said, "No, we're not." Ed was surprised and reminded him of the handshake deal they had. The EVP looked him in the eye and said with annoyance, "It's none of their damn business what we do with the resin once it hits our door. Buy the resin at the discounted automotive prices and run it."

This ethical dilemma is based on whether your word is a contract. Ed's options here are similar to the first dilemma:

1) Meet one-to-one again with the EVP to try to convince him. While going back to the EVP for a second discussion was a possibility, the first discussion was clear. Going back again would more likely reopen the conflict than result in a solution.
2) Obey the EVP and invoice the fluoropolymer company at the reduced automotive pricing. This would result in FluidTech making slightly more money, could potentially lead to a reconciliation between Ed and the EVP, and the oil industry sales would likely not be detected by the fluoropolymer company. Even if they did find out FluidTech was supplying the offshore oil industry, FluidTech could falsely insist that the resin used came from another company. So, the actual risk in terms of punishment of this option was low: this was the "easy way out." The problem was Ed would be directly breaking his word and lying to the fluoropolymer company.
3) Disobey the EVP and continue the handshake agreement, invoicing the fluoropolymer company at the higher market pricing. Simply paying the list price for the raw material could be done, but Ed could be accused of misallocating tens of thousands of dollars of company money by paying more for the material than the EVP directed. If Ed followed this course and had to explain himself to the owner, there would be no avoiding the fact the EVP was breaking his own word. It is very unlikely that the dispute in front of the owner would end without one of them losing in a very career-damaging way.
4) Go over the EVP's head to the owner. This option has the same risks as before, except here, it would be even more clear to the owner that the EVP was lying. The animosity the EVP already had against Ed would be that much higher.
5) Leave the company. Resigning at this point begins to become a more serious consideration because there are now two important ethical conflicts. Ed and the EVP are in open conflict and there didn't appear to be an easy resolution.

Ed chose the third option again. He directly disobeyed the EVP and directed his accounting person to pay the full list price for the supplier's product used in the offshore application. His word was his bond, and if he had committed to doing something, he was going to do it. While this option certainly carried risk, all the other options were either ethically unacceptable or would be ineffective in getting to an ethical solution.

The third and highest stakes ethical dilemma arrived a few weeks later, centering on a potential safety recall. FluidTech had been purchasing a certain special type of metal tubing from their competitors but had decided to eliminate this dependence by building their own identical product. There were no intellectual property concerns and no patents in force. Ed's R&D organization had developed a new replacement material, and this next-generation FluidTech tubing went into limited production for a single customer. Soon after launch, it was discovered that this new tubing was dimensionally out of specification. Ed's team reported this deviation from specification to the EVP and also stated they had immediately embarked on an investigation, subjecting the new tubing to a complete battery of performance and accelerated durability testing. The product ended up passing all these performance tests with flying colors. Concurrently, service parts were also obtained from the two competitors who also manufactured this type of tubing. These two sets of tubing were also found to dimensionally deviate in the same way from the specification. Ed called a meeting with the EVP and the owner, attended by his R&D Team, to discuss the results. Ed explained that all performance testing met the specification and that both competitors' tubing also exhibited similar dimensional deviations from the specification. Ed recommended that a report be presented to the customer showing the deviation from the specified dimensions but also showing that all the performance testing proved that the tubing was safe and met requirements, also showing that the existing competitor tubing (that they had used for over 30 years) exhibited the same conformation deviation.

By this time, the EVP had begun to see Ed as an opponent because of the two previous disagreements, though he had not confronted Ed about these acts of insubordination. The EVP proposed a theory as to why the tubing deviated and made the case that he thought the tubing constituted an unacceptable safety risk. Ed reviewed all the performance data showing that the tubing was not only safe but actually exceeded the performance specification, undermining the EVP's theory. The EVP then said, "I wouldn't let my kids ride in cars that have this tubing." Ed responded, "Okay, then we need to go to the customer and tell them that we are not confident in the safety of the product and we need to go on the record recommending that they issue a product recall." The EVP said, "We can't do that." Incredulous, Ed looked over at the owner who had made himself busy counting the holes in the ceiling tiles and would not meet Ed's gaze. Ed looked over at his R&D colleagues and they were wide eyed. Ed turned back to the EVP and said

> You can't have it both ways. Either the tubing is safe, and we go to the carmaker and explain the situation and recommend continued production while we investigate whether the dimensional deviation can be fixed, or we think the tubing is dangerous and advocate for a recall.

The EVP glared at Ed and said, "We can't." Confused, Ed looked at the EVP again and said, "If it's your judgment that the tubing is unsafe, then we have to advocate for recall." The EVP looked at Ed and said, "We can't do that. If we did, we would go out of business." Ed looked again at the major owner who continued to study the ceiling. Ed then made a brief note of the conversation, looked once more at the EVP and the willfully distracted major owner, got up, and walked out of the meeting.

While the testing Ed had ordered had clearly shown that the product was passing all performance and durability tests, FluidTech's contract with the customer required that FluidTech report any deviations from specification, and this deviation fell under this requirement. The EVP had further raised the stakes by claiming that he believed that the tubing being produced was in fact unsafe, so he had chosen to make this a safety issue despite the test data. Depending upon how things played out, this could result in legal action with possible penalties affecting both FluidTech and its employees. This could also result in a rupture of the relationship with an important customer which could threaten FluidTech's survival.

This ethical dilemma is based on honesty to the customer and potentially safety. Again, Ed's options were similar to the choices after the first two dilemmas but not identical:

1) Meet one-to-one again with the EVP to try to convince him. This option was no longer viable as the owner would need to be involved since he had witnessed the acrimonious disagreement. Given the open hostility between the EVP and Ed, and the owner avoiding the conflict, Ed perceived the chance of a second meeting resolving the issue to be negligible.
2) Obey the EVP and not tell the customer about the conformity issue which would mean that Ed would be knowingly participating in a coverup, with all of the ethical and career risks this choice entails.
3) Disobey the EVP and tell the customer, meaning Ed is acting as a "whistleblower" by exposing the lie. If Ed believed the product was truly unsafe, there would be no dilemma: he had an obligation to go to the customer and expose the issue. However, the tube had easily passed all relevant testing, so Ed and his experts were convinced that the product was indeed safe, especially given the high engineering "safety factor" for this tubing built into the design and specification. Ed now believed that EVP had chosen to classify this as a safety issue, despite the conclusive contrary test data, as a tactic of bringing the

conflict with Ed to a decisive conclusion by blaming him for a safety problem and an existential business risk. That result? Ed would be fired.

4) Go over the EVP's head to the owner, which was unlikely to yield success because the owner refused to take any leadership responsibility during the crisis.
5) The fifth option, leaving the company, was clearly on the table and could no longer be denied as a serious likelihood. Whether the machinations of the EVP could get Ed fired or whether Ed would need to leave of his own free will was not clear, but it was now unavoidably apparent that his ethical framework was antithetical to the direction of the EVP.

Ed's worst fears about the EVP had come to pass. He drove home and that evening called the leader of his R&D team to discuss the meeting. Ed's R&D experts were in shock from what had happened. The next day, Ed had a private meeting with the owner.

Ed opened the meeting reminding the owner of the discussion they had had when Ed had warned against hiring the new EVP, in which the owner had made Ed promise not to leave the company and committed to finding him a different boss if things didn't work out. The owner agreed that he did remember the conversation and what he had said. For a brief moment, Ed held out hope that he could salvage his fast-rising but now very tenuous career at FluidTech. Ed told the owner about all three ethical dilemmas, making clear in all three cases the EVP's direction was unethical, and Ed was unwilling to execute the direction that the EVP had given. Ed told the owner that there was an impasse: the EVP was unwilling to budge, and Ed was unwilling to follow directions he deemed unethical.

Then the surprise. The owner said, "I've placed my bet on the new EVP to run this company. You guys need to find a way to work things out. I'm not going to give you another boss. You two have to work things out." At that moment, Ed realized there was no way out; he felt almost physically sick realizing that the bright future he had at FluidTech was now ashes.

He looked at the owner, and with effort, kept his voice level, said "Then I'm going to give myself another boss. This is my two weeks' notice. Thank you for the opportunities you have given me over the last nine years." The owner was taken aback. He said that he didn't want that to happen, and he would meet with Ed to talk further the next morning. But Ed knew it was over.

The next morning, the EVP showed up in Ed's office. He told Ed that he had spoken with the owner, there was nothing more to say, and he wanted Ed gone right away. After a difficult goodbye with his shocked team, Ed left FluidTech, deeply concerned about the future of the company, his many friends there, and not knowing what the future would hold for himself personally.

Ed may ask: "What did you do about the third dilemma?" Ed's answer was that he was completely convinced that there was no safety issue based upon all the test data and the competitive products that had been in the field for decades with the same dimensional deviation.

Once he left FluidTech, he never found out what they did about either talking to their customer or attempting to hide the deviation.

Now let's turn our attention to the fallout from these ethical dilemmas. First, what happened to FluidTech over the next several years?

The hiring of the EVP dramatically changed the values, culture, and environment of FluidTech. Under the EVP's leadership, the relationship with the supplier fluoropolymer company deteriorated rapidly. The fluoropolymer company ended the handshake agreement they had with Ed on working exclusively with FluidTech and began working to develop a competitive product to SuperTube, which they licensed to FluidTech's competitors, eliminating FluidTech's most profitable product: selling bulk SuperTube to the competition. It also introduced price competition eroding FluidTech's profit margins when they sold to the automotive companies. FluidTech's most powerful profit engine was rapidly reduced to a small fraction in a two-year period.

Relationships with the most important automotive customers also deteriorated rapidly under the EVP. The previous "serve the customer" cultural value attenuated under the EVP, and FluidTech increasingly failed to win new business, stopping their decade-plus run of doubling in size every 2.5 years.

Most important was the effect the EVP's leadership had on the employees of FluidTech. While Ed was not the first senior manager to depart, his resignation was quickly followed by a wave of other key management departures driven by the EVP's changes to company culture and values. Shortly after the flight of key managers, a large-scale exodus of the leading technical people took place.

A few months later, the EVP was demoted, and shortly thereafter demoted a second time and then fired. He was replaced by a second EVP who reached out to Ed and some of the other key managers who had left, trying to convince them to return. An interview made it clear to Ed that the second EVP and his team had no real understanding of the company and that they would be unable to turn things around. Shortly thereafter, the company went into bankruptcy. FluidTech was bought out of bankruptcy by a notorious turnaround firm, experienced a grueling season of investment and employee cuts and, with no significant new business on the books, the remnants of the company were peddled to an unsophisticated old-line manufacturing company for a tidy profit. The original owner, who had been worth about $100 million, ended up working for a salary at the turnaround firm and then retired.

Most importantly, what happened to Ed? Ed was 33 years old when he left FluidTech. He ended up at a major automotive company with global responsibility for upstream research partnerships. The job was exceptionally rewarding, and also allowed him the flexibility he desired.

Now here's something you may not yet have thought of.

Ed had a young son, a second child on the way, and his spouse worked half time. When he returned home after resigning, rather than complaining, criticizing, or worrying, his wife responded, "Good! I'm glad you're gone from that place." Ed was inclined to immediately search for another job, but his wife (whom he always described as being "the wiser of the two of us") urged him to join her during her pending maternity leave to take some time and really figure out what he wanted to do next and give himself a chance to detoxify from the recent past. The job market was strong, and Ed turned away several serious opportunities, took his wife's advice, and took maternity leave with her. They strictly controlled their expenses, dipping cautiously into savings.

As a result, Ed learned to be a more purposeful husband and father as he determined what he would like to do during the next phase of his career. Looking back, Ed had the same reaction to FluidTech that many military veterans have to their time at war: he would never do it again, but he wouldn't trade the experience for anything. As unpleasant as it was at the end, Ed was able to leave the company with his ethics intact, knowing he had done what he believed to be right.

So, what are the takeaways from our journey with FluidTech? There are many. First, there is an absolute certainty that over the course of your working career, your personal ethics will be tested. You cannot avoid this fact by saying you'll never do anything that would put yourself at ethical risk, because in the majority of cases you will not have any control over a dilemma being thrust upon you. To paraphrase Leon Trotsky's famous quote regarding war:

> "You may not be interested in ethics, but ethics is interested in you."

Next, it is imperative that you determine your ethics before a dilemma arises. It's radically more difficult to figure out (and have the courage to do) the right thing in the midst of an immediate, high-stress situation with angry executives and your career on the line. From the military and athletics we know that, in times of extreme stress, you revert to your training. In the middle of a dilemma, you really don't have the time nor necessarily the mental focus to think carefully about ethics. You need to be able to instantly revert to an ironclad ethical code that you determined long before the crisis.

Another learning is if you ever truly need a job, you may be in deep trouble because you will do whatever it takes to keep that job rather than stay on your ethical path.

Importantly, most ethical choices in technology and engineering are not simple nor obvious: notice the many pages and level of detail required to tell the story of FluidTech. Ethical problems can exist deep in your organization or work: it may be hard to even describe the issue, let alone deal with it effectively.

It bears repeating: many times, ethical dilemmas require an immediate decision, response, or action on your part, be it a difficult question from an authority figure, taking an "innocent" action, or agreeing to do something sketchy. The common factor here is that *you will not have any time to ponder your answer*; you will be forced to choose a path immediately. Your ethics must be firmly in place before a crisis arrives.

The Fluid Tech story highlights something else that may not be apparent: an organizational or corporate ethical issue becoming entangled with personal ethical choices. Rarely is a personal ethical dilemma neatly separated from the organizational problem. For Ed, the choice was to do what the organization said and remain employed or take a personal stand and likely suffer punishment. In the FluidTech case, the company's unethical behavior led directly to the loss of the highly valued employee.

Here's probably the most important point regarding ethics: personal ethics is the one area of this book that *does not* follow contingency theory. "Moral relativism" and "situational ethics" are a trap; it's easy to apply them as a rationalization to justify wrong or unethical behavior. "Everyone else was doing it" is not an ethical justification; it's merely an excuse.

Finally, you will make ethical mistakes. Fix them and remain committed to your path.

A personal note. Remember, Ed and his wife had a small child and one on the way. Ed's wife was working part time, and soon to be on extended leave. Savings were slim. It's one thing to take a stand on principle when there is money in the bank. It's quite another to come home, look your spouse in the eye and say: "I quit my job over an ethical issue." That's where the strength of your relationships outside of your workplace becomes critical. As mentioned, workplace ethical issues can easily flow into your home and may cause additional stress and heartache,

Let's change gears for a moment. What about corporate ethics and its impact on you in the workplace? Before the arrival of ubiquitous digital information, many corporations (and their employees) would attempt to hide embarrassing or even illegal information from public view. With the onset of digital data, it is effectively impossible to hide corporate or personal information for even a short amount of time: anyone who thinks so is sadly, sadly mistaken. In today's ethical reality, this means whatever information you have, be it good or bad, embarrassing or positive, people will eventually know about it; the only question will be how quickly they obtain it.

It is amazing how even today, certain organizations have the hubris to attempt to hide deleterious information from investigators or the general public. A recent example has to do with the Volkswagen diesel emissions scandal uncovered in 2014. The company purposely installed an emissions-altering software system in Volkswagen diesel engines to defeat European government pollution requirements. Apparently, Volkswagen management thought this software could be hidden forever. It wasn't. As of this writing, this mistake of arrogance cost Volkswagen and its related brands a total of some $34.7 billion to date in fines and untold costs in reputational damage.[7]

We've spent a lot of time on personal morals, so what about professional ethics? We should take a moment and touch on the ethics of the engineering profession as opposed to strictly personal or corporate ethics.

With the rise of digital information is a corresponding rise in the public's expectation of transparency in all we do professionally. Every professional act has the potential to be examined, including your own. Simply put, conduct your professional life with the certain expectation that all your work will come to light, if not to the public, then certainly to your boss. In short, it's foolish to attempt to hide major mistakes or missteps.

One last point. Professional ethics are unique in that they fall squarely between personal and corporate morals: both organizational and personal points are mentioned as shared in Table 11.2. This is good, but not totally sufficient. While professional ethics guidelines are a good place to start when beginning to develop your initial values framework, it is only a starting point. You will have to eventually move on to the personal moral constructs to "flesh out" a workable, strong point of view.

Obviously, ethics are overwhelmingly personal, but I think you would agree there are some ethical positions you may prefer over others. Whatever the case, the important point is to actively begin to address your moral or ethical positions now. You'll need it soon.

11.5 REALIZATIONS

We've spent substantial time discussing two important ideas you need to absorb before you can finally create your personal roadmap: connecting the components of the core into a strong, dependable, holistic model of mutual understanding and dependency, and then applying a capstone of personal ethical considerations to your model. These final steps will complete the preparations for you to develop your personal guide, the navigation tool you will use each day transitioning into the professional ranks. The good news is that neither of these tasks is particularly difficult. Establishing the obvious linkages between components is straightforward, and developing your personal ethical understanding is fundamentally simple once you understand the main points of ethical belief and behavior. The only challenge is applying some mindfulness to both these activities.

Combining the individual components of the Essential Engineering Framework requires understanding the purpose, characteristics, and limitations of each component well enough to naturally and quickly see the dependencies between them. And not just the obvious ones, but the hidden connections as well. That's what melding the components means.

Note that not everyone has the same definition of connecting the components: your definition may be different from that of your colleagues or boss. Ensuring consistent definitions is your first task with any assignment involving joining components together. This means that no matter how quick the evaluation deadline is, you need to pause and think about it. Sensemaking may help. It is an attempt to reduce multiple meanings and handle the complex information used by other people in your organization. This means strategies, plans, rules, and goals are not things that exist objectively, but within a group member's way of thinking. In sensemaking, whether someone's view of the world is correct is not relevant: the correctness of a position is contingent on the personal filters used for evaluation. The basic idea of sensemaking is that reality is an ongoing realization that emerges from individual's efforts, meaning organizing is a continuous, evolutionary process with the goal of reducing confusion by placing limitations upon the organization as they make sense of themselves and their environment. Communication becomes doubly important because of its role in this sensemaking process.

Our second idea, personal ethics, is one of the most important subjects you will wrestle with. Personal ethics establishes a strong boundary between acceptable behavior and actions that would be damaging to yourself and others. While corporate and professional ethics also impact your work life, it is your personal ethics that will have the greatest impact on most aspects of your work life.

As our examples show, ethics can be quite complex, demanding a response that you may not have time to develop. Personal ethics can be buried deep within technical decision-making. And ethics can literally change your life if you do not understand them.

Most importantly, ethics are the one topic that does not follow contingency theory. They are constant and unchanging in a contingent world; relativistic ethics is a siren song of behavior. Do not be fooled by its lure of easy solutions to whatever situation you are experiencing.

Personal ethics will cause you to experience some uncertainty that making your way in a technical organization brings. This is normal. After all, uncertainty results from change, and change is what we're after. It's simple: if you don't know, then you need to educate yourself on what understanding you are missing and fill in the missing knowledge

Enough moralizing. Let's get on with creating your roadmap.

NOTES

1. Weick, Karl. 1995. *Sensemaking in Organizations.* Thousand Oaks, CA: Sage Publications.
2. Neill, Stern, McKee, Daryl and Rose, Gregory. 2007. Organizational Sensemaking. http://psychology.iresearchnet.com/industrial-organizational-psychology/organizational-development/organizational-sensemaking/.
3. Argyris, Chris. 1999. *On Organizational Learning.* Hoboken, NJ: Wiley-Blackwell.
4. National Society of Professional Engineers (NSPE). 2022. *Code of Ethics for Engineers.* Copyright NSPE, Used with Permission. www.nspe.org/resources/ethics/code-ethics.
5. Daft, Richard. 2010. *Organizational Theory and Design.* Mason, OH: Cengage Learning.
6. Jackall, Robert. 2009. *Moral Mazes: The World of Corporate Managers.* Oxford: Oxford University Press.
7. Taylor, Edward. 2020. Volkswagen Says Diesel Scandal Has Cost It 31.3 Billion Euros. *Reuters News Service.* https://www.reuters.com/article/us-volkswagen-results-diesel-idUSKBN2141JB.

12 Crafting the Roadmap
Creating Your Personal Guide

12.1 CONSTRUCTING YOUR ROADMAP

At last, it's time to take some action: gathering what we've discussed and creating your personal roadmap, designed to get you over the Great Divide and navigate through the organization you've just joined. But this leaves us with just one question: How do we actually create that roadmap for you and your individual company? How in the world can you develop a workable roadmap incorporating who you are and your own personal understanding of your engineering organization? Where is the checklist, the process document, the instructions you can follow, or the boxes to check off as you complete each step? In short, what do you do next?

No worries, we'll get there.

Let's first define what a personal roadmap is. Most personal guidance plans are a variant of the business roadmap, which is a strategic document defining a corporate goal and the major steps necessary to achieve it over time. Much like its corporate cousin, personal roadmaps are touted as "personal action plans" featuring highly prescribed steps to achieving individual goals. Filled with checklists, milestones, and your own "mission statements," these highly codified plans are reassuring and comforting. They are also goal based, sterile, and relentlessly rational. And that's not what we're talking about here.

For us, a *roadmap* is a framework of you and your organization, broken down into components to which are applied your current and desired characteristics in the future, and identifying the actions needed to achieve them. Our approach to the individual roadmap is about gaining organizational skills, not objectives or goals. It's about personal development as a professional engineer. And it's simple. We take the five components from the Essential Engineering Framework, add to it the common functions and skills needed to successfully operate within each component, and assess how each skill stacks up against your current and future capabilities. This identifies the action you need to take within each component. Ultimately, it's about insight: insight into your current engineering organization, and most importantly, awareness of yourself as a new engineer.

Let's break this down. We already know what the five components are. Functions and skills are quite easy to identify: they are the topics we've already discussed in Chapters 5 through 10. We then add to those skills your contingency topics; those

DOI: 10.1201/9781003214397-16

abilities that are unique to you and your situation that are not specifically covered in this book. Add to the roadmap your ethical positions, and your guide is complete.

What we haven't really discussed yet is this roadmap is time-based, containing a starting point (your situation today) and an ending point (a time somewhere in your future). Your knowledge today and your desired knowledge tomorrow; your capabilities now and your needed capabilities in the future. Your roadmap will include both these points in time.

You see, a roadmap is no good if it is stuck at a single moment. You change, the organization changes, your colleagues change, and so does the environment. Everything shifts, sometimes rapidly, and your roadmap must be flexible enough to provide for that change. Otherwise, your map would be valid for about three months and then quickly become more and more irrelevant. So, you will be assessing yourself at two points in time: now and then.

Just like the difference between structures and frameworks discussed in Chapter 4, we'll take the same approach to developing your plan. Your roadmap is a framework, not a structure. Remember, a framework is able to absorb the ongoing contingencies and unique boundaries that a structure cannot, which is essential for what you are about to do.

So, let's begin. Here's the framework:

Section 1

1) Schedule at least one hour of uninterrupted time in a comfortable location.
2) If you have not yet done so, select and complete at least three of the six self-assessments we've covered: Myers-Briggs, Five Factors, Thomas-Kilmann, Gardner Multiple Intelligences, Demographics, and Bolling 4-Step Exploration, found in Chapters 2, 3, and 6. This is not optional. A self-assessment is an important first step in determining the time and energy investment you are about to make. It can drastically reduce the work you need to accomplish in creating your map, as you can ignore certain irrelevant personal characteristics and focus on others. It can also identify aspects of yourself you may have yet to realize. Make a summary of each self-assessment and keep it close by.
3) Also have a copy of the Essential Engineering Framework handy. You may need to refer to it during your evaluation.
4) Create a modifiable copy of Table 12.1. You may download one at the Roadmap Creation Template" within the Ancillaries section of the Routledge website at https://www.routledge.com/9781032102511. This will be the start of your plan.
5) Select the time frame you intend your map to be valid. Obviously, your starting point is today; your ending point can be a moment in time, attaining a certain professional position (becoming a respected worker, senior engineer, supervisor, or technical specialist) or any other meaningful signpost of your future. Add this to your table.
6) Each of the five components is divided into two types: the *standard skills* derived from our prior component discussions and *unique skills* supporting

TABLE 12.1
Roadmap Creation Template

		Current Location in Organization		Future Location in Organization	
		Add Current Position Here		Add Future Position Here	
Component	**Skills and Capabilities**	**Knowledge/ Skill Required Today?**	**Level of Knowledge/ Skill Required Today**	**Knowledge/ Skill Required in Future?**	**Level of Knowledge/ Skill Required in Future**
Technology System					
	Decision-Making	Yes/No	0 to 5	Yes/No	0 to 5
	Testing	Yes/No	0 to 5	Yes/No	0 to 5
	Data Analysis	Yes/No	0 to 5	Yes/No	0 to 5
	R&D	Yes/No	0 to 5	Yes/No	0 to 5
	Design	Yes/No	0 to 5	Yes/No	0 to 5
	Manufacturing	Yes/No	0 to 5	Yes/No	0 to 5
	Service Systems	Yes/No	0 to 5	Yes/No	0 to 5
	Process	Yes/No	0 to 5	Yes/No	0 to 5
	Success Measures	Yes/No	0 to 5	Yes/No	0 to 5
	Operations	Yes/No	0 to 5	Yes/No	0 to 5
	Add Unique Skill/ Capability Here	Yes/No	0 to 5	Yes/No	0 to 5
	Add Unique Skill/ Capability Here	Yes/No	0 to 5	Yes/No	0 to 5
	Add Unique Skill/ Capability Here	Yes/No	0 to 5	Yes/No	0 to 5
	Add Unique Skill/ Capability Here	Yes/No	0 to 5	Yes/No	0 to 5
Formal Organization					
	Organizational Structures	Yes/No	0 to 5	Yes/No	0 to 5
	The Nature of Timing	Yes/No	0 to 5	Yes/No	0 to 5
	Power	Yes/No	0 to 5	Yes/No	0 to 5
	Conflict	Yes/No	0 to 5	Yes/No	0 to 5
	Politics	Yes/No	0 to 5	Yes/No	0 to 5
	Management Characteristics	Yes/No	0 to 5	Yes/No	0 to 5
	Standards	Yes/No	0 to 5	Yes/No	0 to 5
	Standard Operating Procedures	Yes/No	0 to 5	Yes/No	0 to 5
	Boundaries	Yes/No	0 to 5	Yes/No	0 to 5

(*Continued*)

TABLE 12.1 (CONTINUED)
Roadmap Creation Template

		Current Location in Organization		Future Location in Organization	
		Add Current Position Here		Add Future Position Here	
Component	**Skills and Capabilities**	**Knowledge/ Skill Required Today?**	**Level of Knowledge/ Skill Required Today**	**Knowledge/ Skill Required in Future?**	**Level of Knowledge/ Skill Required in Future**
	Policy Deployment	Yes/No	0 to 5	Yes/No	0 to 5
	Add Unique Skill/ Capability Here	Yes/No	0 to 5	Yes/No	0 to 5
	Add Unique Skill/ Capability Here	Yes/No	0 to 5	Yes/No	0 to 5
	Add Unique Skill/ Capability Here	Yes/No	0 to 5	Yes/No	0 to 5
	Add Unique Skill/ Capability Here	Yes/No	0 to 5	Yes/No	0 to 5
Human and Hard Resources					
	HR Office Functions	Yes/No	0 to 5	Yes/No	0 to 5
	Performance Reviews	Yes/No	0 to 5	Yes/No	0 to 5
	Reward & Recognition	Yes/No	0 to 5	Yes/No	0 to 5
	Dual Ladder	Yes/No	0 to 5	Yes/No	0 to 5
	Teams	Yes/No	0 to 5	Yes/No	0 to 5
	Mentorship	Yes/No	0 to 5	Yes/No	0 to 5
	Budgeting	Yes/No	0 to 5	Yes/No	0 to 5
	Project Funding	Yes/No	0 to 5	Yes/No	0 to 5
	Economy	Yes/No	0 to 5	Yes/No	0 to 5
	Hidden Resources	Yes/No	0 to 5	Yes/No	0 to 5
	Add Unique Skill/ Capability Here	Yes/No	0 to 5	Yes/No	0 to 5
	Add Unique Skill/ Capability Here	Yes/No	0 to 5	Yes/No	0 to 5
	Add Unique Skill/ Capability Here	Yes/No	0 to 5	Yes/No	0 to 5
	Add Unique Skill/ Capability Here	Yes/No	0 to 5	Yes/No	0 to 5

(Continued)

TABLE 12.1 (CONTINUED)
Roadmap Creation Template

		Current Location in Organization		Future Location in Organization	
		Add Current Position Here		Add Future Position Here	
Component	Skills and Capabilities	Knowledge/ Skill Required Today?	Level of Knowledge/ Skill Required Today	Knowledge/ Skill Required in Future?	Level of Knowledge/ Skill Required in Future
Hidden Organization					
	Being Visible Training	Yes/No	0 to 5	Yes/No	0 to 5
	Models of culture	Yes/No	0 to 5	Yes/No	0 to 5
	Social Rites	Yes/No	0 to 5	Yes/No	0 to 5
	Dynamics	Yes/No	0 to 5	Yes/No	0 to 5
	Subcultures	Yes/No	0 to 5	Yes/No	0 to 5
	Ownership	Yes/No	0 to 5	Yes/No	0 to 5
	Ethics (Type)	Yes/No	0 to 5	Yes/No	0 to 5
	Add Unique Skill/ Capability Here	Yes/No	0 to 5	Yes/No	0 to 5
	Add Unique Skill/ Capability Here	Yes/No	0 to 5	Yes/No	0 to 5
	Add Unique Skill/ Capability Here	Yes/No	0 to 5	Yes/No	0 to 5
	Add Unique Skill/ Capability Here	Yes/No	0 to 5	Yes/No	0 to 5
Communication					
	Framework	Yes/No	0 to 5	Yes/No	0 to 5
	Opportunities and Situations	Yes/No	0 to 5	Yes/No	0 to 5
	Control Issues	Yes/No	0 to 5	Yes/No	0 to 5
	Standardizing	Yes/No	0 to 5	Yes/No	0 to 5
	Special Cases	Yes/No	0 to 5	Yes/No	0 to 5
	Add Unique Skill/ Capability Here	Yes/No	0 to 5	Yes/No	0 to 5
	Add Unique Skill/ Capability Here	Yes/No	0 to 5	Yes/No	0 to 5
	Add Unique Skill/ Capability Here	Yes/No	0 to 5	Yes/No	0 to 5
	Add Unique Skill/ Capability Here	Yes/No	0 to 5	Yes/No	0 to 5

(Continued)

TABLE 12.1 (CONTINUED)
Roadmap Creation Template

		Current Location in Organization		Future Location in Organization	
		Add Current Position Here		Add Future Position Here	
Component	Skills and Capabilities	Knowledge/ Skill Required Today?	Level of Knowledge/ Skill Required Today	Knowledge/ Skill Required in Future?	Level of Knowledge/ Skill Required in Future
Ethical System					
	Confident/ Comfortable in Current Structure?	Yes/No	0 to 5	Yes/No	0 to 5

your current and future positions stemming from your current role and what role you wish to attain. Say part of your current position requires you to evaluate end-customers' use of a new keyboard design. The required skill is to probe behind the customer's façade and get at their true feelings about using the product. In the future, you will still need this talent. Here, it's appropriate to place "verbal evaluation skills" as a topic in the unique portion of the map.

7) If a unique skill is needed today but not in the future, list it. Conversely, if you anticipate a capability will be needed in the future but not today, still list it. The key is to list as many unique skills as appropriate, now and for the future.
8) Dependent on the time spent so far in your engineering organization, there may not be many entries in the unique area. This is fine, as this chart will become a living document and many more unique capabilities will be added over time.
9) Set the roadmap aside for a few days.

Section 2

10) As before, reserve at least one hour of quiet time in a calm location. Do not allow interruptions.
11) On your map, focus only on the columns "Current Location in Organization." Do not consider the "Future Position" columns at this time.
12) The next steps are quite easy. For each component skill and capability shown, honestly and truthfully assess if that specific skill is needed in your current position and if so, what is your current mastery of that topic. If no

knowledge or skill, place a "0"; if you are a master of that domain, place a "5." Do this realistically. Be sure to place your mark where you are currently at, not where you hope to be. Don't be surprised if most of your self-evaluation scores are very low. Remember, you are assessing if that skill is needed now, and if so, what is your current capability. Do this for all standard and unique skills listed.

13) Now return to the top and repeat his procedure for the "Future" columns. This should be a much more uncertain assessment; some may be indeterminate. This is normal; just evaluate with the same honesty as before.
14) *Do not* add any of these ratings together to create some sort of "summary" grade or overall score. This is not a contest.
15) This last step is even easier and more obvious: select the current vs. future self-evaluations ratings, search for the greatest differences between the two, and then decide what actions you must take to close the gap. Make a note of these and place them in an easy-to-access location.
16) And *voilà,* a map of the path through the engineering organization, customized for you and your reality. It is a summary of organizational skills, focused on the five components of the inner core, estimating the type and amount of personal skill needed.

Well, actually that's not totally it. The challenge now is to consider exactly how you close the gaps you've identified. Dependent on the particular gap and the level of skill required to master the topic, it may be as simple as taking an approved company training class (such as extending your coding skills) all the way to completing a one or two-year rotational assignment (say, a sensor design and implementation position). No matter what the gap actually is, you need to become educated on the topic, which means adopting a "curious" frame of mind. And like that investigative reporter we discussed earlier, you should take the initiative to accomplish this. No one will come and tell you to do this. You are on your own.

One final note about the roadmap. As mentioned earlier, this is not a "one and done" process. We all change, some quickly and some slowly. Sometimes our rate of change begins slowly and accelerates, other times we begin with a sprint and end with a crawl. Whatever the case, you should revisit your map periodically and update it to who you now are at that moment. Your Myers-Briggs assessment will change, so will your Five Factor assessment. And your map will (and should) change as well. You pick the time interval or goal, but the key is actually revisiting the work. Return to this exercise periodically as your experiences grow and you continue to gain personal insights.

CASE EXAMPLE 12.1 CONTAINERIZED SHIPPING

We need to run through an example. Imad Makki is a new hire, working for the past nine months as a bachelors-level electrical engineer. He is part of an effort to develop a breakthrough logistics system for containerized ocean-going vessels, to address the continuing supply chain issues brought on by the COVID-19 pandemic.

As of 2021, the number of cargo container ships operating worldwide stood at a combined capacity of about 24.6 million 20 ft containers. Additionally, since 2000 the average vessel size of container ships has more than doubled, until today the largest container ships can carry about 24,000 of those 20 ft. containers each.[1] If successful, this new logistics system would expand the number of sensors placed on each standard 20-foot container to over 400, or some 9.6 million sensors per ship. These sensors would do everything, from controlling the interior temperature and humidity environment of each container (reducing spoilage due to excessive heat during long-distance voyages), to providing pin-point positioning of every parcel in a container and its condition. A game-changing innovation but also a substantial challenge due to the very large amount of data to be continuously monitored.

Imad has a typical engineering background, completing his bachelor's degree in 4.5 years, focused on electrical engineering with a GPA of 3.56. His classes were overwhelmingly technical: only two composition or writing classes, one presentation class, and two unrestricted electives in which he selected technology-related subjects. While he was a member of many classroom teams, the average duration of these teams was three months, composed of group members he already knew. His single capstone course in product design was of mixed success due to two group members electing to not contribute to the group, leaving the other members to cover the extra assigned roles.

While a student, Imad spent one summer as an intern in the main office of a national civil engineering and construction firm. His 10-week internship was a disappointment: he converted stress simulation data into bar charts for management reviews. While he was able to attend these management meetings, his field experience was limited to "tours" of construction sites with other interns.

Imad wishes to spend his career in this logistics engineering specialty, but being a new hire, he is uncertain at this moment what his personal path forward should be. He does know he would like to eventually lead a group of analysts in logistics systems improvement, but at an uncertain time in the future. With this as a starting point, Imad creates the following map:

The Technology System

In the standard categories, Imad has introductory knowledge in design, operations, and data analysis. In the unique functions, he has good skills in digital communication and low capabilities in C++ coding. He anticipates in his next position he will need knowledge in sensors, instrumentation, electronic design, and improved data analytics. Case Example Table 12.1a shows the resulting Technology System chart for Imad at this point.

Imad discovers two things: First, his existing knowledge and capabilities are substandard in all but one category. He needs education and experience in the remaining 11 to attain at least a minimum competency rating, giving

CASE EXAMPLE TABLE 12.1A
Example Assessment for the Technology System

		Current Location in Organization		Future Location in Organization	
		Logistics Electrical Engineer		Lead Engineer for Logistic Analytics Group	
Component	**Skills and Capabilities Categories**	**Knowledge/ Skill Required Today?**	**Your Knowledge/ Skill Level Today**	**Knowledge/ Skill Required in Future?**	**Level of Knowledge/ Skill Required in Future**
Technology System					
Rating Definitions:	Decision-Making	No	N/A	Yes	4
0 = No Knowledge	Testing	No	N/A	Yes	1
1 = Awareness	Data Analysis	Yes	1	Yes	2
2 = Surface Knowledge	R&D	No	N/A	Yes	1
3 = Minimum Competency	Design	Yes	2	Yes	4
4 = Journeyman Capability	Manufacturing	No	N/A	No	N/A
5 = Expert Knowledge	Service Systems	No	N/A	No	N/A
	Processes	Yes	2	No	N/A
	Success Measures	Yes	0	Yes	3
	Operations	Yes	1	Yes	5
	C++ Coding	Yes	2	No	N/A
	Digital Communication	Yes	3	Yes	4
	Sensor Usage	Yes	1	Yes	2
	Instrumentation and Test Methods	Yes	1	Yes	2
	Electrical Design	Yes	2	Yes	3
	Data Analytics	Yes	1	Yes	2

him the insights to prioritize the improvements he needs to accomplish. Imad needs to prioritize the 11 categories in rank order to systematically improve his technical skills standing.

The future lead engineer position assessment is more interesting. Twelve categories need attention, requiring enhanced skills to attain an appropriate future skill set. Note that some categories need competence, other skills merely an awareness. The Operations category needs expert knowledge but the C++ category, interestingly, becomes irrelevant.

With the number of technical skills needing improvement, Imad (in association with his supervisor) needs to create a more formal development and education plan with appropriate time reserved to improve these ratings.

The Formal Organization

Knowledge and capability in understanding the formal organization are obvious. Let's take a look at Imad's case.

At his starting point, Imad essentially has no knowledge of the formal organization. This is to be expected, as his engineering school did not teach organization theory and his internship omitted it in the short time he was present. Here, he has a larger deficit to overcome than with the Technical System. Imad has very limited knowledge of organizational structures (only the hierarchy) and some industry standards. Overall, he has essentially no unique skills in this component. He anticipates his next position will require at least a basic knowledge in the standard categories, but again he has no idea of the skills required at that future time. Case Example Table 12.1b represents his current and future state of affairs.

CASE EXAMPLE TABLE 12.1B
Example Assessment for the Formal Organization

		Current Location in Organization		Future Location in Organization	
		Logistics Electrical Engineer		Lead Engineer for Logistic Analytics Group	
Component	**Skills and Capabilities Categories**	**Knowledge/ Skill Required Today?**	**Your Knowledge/ Skill Level Today**	**Knowledge/ Skill Required in Future?**	**Level of Knowledge/ Skill Required in Future**
Formal Organization					
Rating Definitions: 0 = No Knowledge 1 = Awareness 2 = Surface Knowledge 3 = Minimum Competency 4 = Journeyman Capability 5 = Expert Knowledge	Organizational Structures	Yes	1	Yes	3
	The Nature of Timing	Yes	0	Yes	3
	Power	No		Yes	1
	Conflict	Yes	0	Yes	2
	Politics	No	N/A	No	N/A
	Management Characteristics	Yes	0	Yes	3
	Standards	Yes	1	Yes	4
	Standard Operating Procedures	Yes	0	Yes	3
	Boundaries	Yes	0	Yes	3
	Policy Deployment	Yes	0	Yes	2

On examination of his roadmap, Imad became alarmed: his existing knowledge and capabilities in the formal organization were nonexistent, and he was very uncertain as to what unique skills he needed to concentrate on for the future, other than "everything." He needs education, quickly. He begins to "interview" a wide range of individuals, similar to the self-assessment discussed in Chapter 3. These interviews are not formal: they are informal conversations with peers, a range of engineers, experienced supervisors, engineers in different groups, and especially any administrative assistants who may exist somewhere in the organizations. It means being curious. Now is the time for Imad to be that engineering anthropologist.

Human and Hard Resources

Things are faring better for Imad in the Human and Hard Resources zone, as he had some exposure to the hard resources area while at his internship but nothing in the human category other than a six-month performance review. One of Imad's hard resources tasks was to reconcile actual invoices for equipment and services rendered against their budget to determine if any funds were being underspent, creating some discretionary money for the department. He understood basic accounting principles from his one engineering economy course and could calculate net present value and hurdle rates but not much more. Any understanding of human resources subjects was essentially zero. He knows for the future he needs substantial skill in understanding and practicing engineering group management. Case Example Table 12.1c defines his path through this important area.

Thankfully, Imad's current skill level in the human and hard resources area is average for a person with only nine months of company experience. At this point, the Human Resources department is still a mystery, and hard resources are still controlled by the supervisor or manager. In these areas, hands-on, "learning by doing" is the only realistic way of gaining the necessary skills to improve his capability over time. This means Imad should volunteer for simple budgeting tasks to relieve his supervisor of this normally onerous work.

For a future group leader position, the necessary human resources experience is harder to obtain. To pursue the group leader position involves the paradigm of promotion discussed in Chapter 3: Imad must acquire the human behavior skills to run a group before he can actually run it. To break this cycle, Imad should ask for minor human resources-related tasks, nothing too important or sensitive yet, and practice this craft. He needs to ask for advice within his department and elsewhere. Once again, Imad should be that anthropologist and uncover the essential skills he needs in this soft subject.

The Hidden Organization

This category is really hard to assess, not only in absorbing the basic skills to work within it but to even see it. Luckily, no one expects a new engineer to operate successfully in the hidden organization until a year or two after

CASE EXAMPLE TABLE 12.1C

Example Assessment for Human and Hard Resources

		Current Location in Organization		Future Location in Organization	
		Logistics Electrical Engineer		Lead Engineer for Logistic Analytics Group	
Component	**Skills and Capabilities Categories**	**Knowledge/ Skill Required Today?**	**Your Knowledge/ Skill Level Today**	**Knowledge/ Skill Required in Future?**	**Level of Knowledge/ Skill Required in Future**
Human and Hard Resources					
Rating Definitions: 0 = No Knowledge 1 = Awareness 2 = Surface Knowledge 3 = Minimum Competency 4 = Journeyman Capability 5 = Expert Knowledge	HR Office Functions	No	N/A	Yes	3
	Performance Reviews	Yes	1	Yes	3
	Reward & Recognition	No	N/A	Yes	3
	Dual Ladder	No	N/A	Yes	1
	Teams and Groups	Yes	1	Yes	4
	Mentorship	No	N/A	Yes	1
	Budgeting	Yes	1	Yes	3
	Project Funding	Yes	2	Yes	3
	System Economy	No	N/A	Yes	1
	Hidden Resources	No	N/A	Yes	2
	Tracking Funding Burn Rate	Yes	2	Yes	4
	Adjusting Funding Requirements	Yes	2	Yes	4
	Responding to Budget Reductions	No	N/A	Yes	4

onboarding, as obtaining these skills can't be taught: they must be learned through daily experience.

Here, Imad's assessment of his current knowledge is easy to map: he has none. In the standard categories shown in Case Example Table 12.1d, he needs to begin to even understand what the categories are. In the unique function area, his goal is to just get a handle on the special cultural forces in play.

In his next position, he anticipates he will need knowledge of the basic cultural cues, subcultures, models of culture, ownership and social rites, just enough to know how to avoid the basic landmines.

CASE EXAMPLE TABLE 12.1D
Example Assessment for the Hidden Organization

		Current Location in Organization		Future Location in Organization	
		Logistics Electrical Engineer		Lead Engineer for Logistic Analytics Group	
Component	**Skills and Capabilities Categories**	**Knowledge/ Skill Required Today?**	**Your Knowledge/ Skill Level Today**	**Knowledge/ Skill Required in Future?**	**Level of Knowledge/ Skill Required in Future**
Hidden Organization					
Rating Definitions: 0 = No Knowledge 1 = Awareness 2 = Surface Knowledge 3 = Minimum Competency 4 = Journeyman Capability 5 = Expert Knowledge	Being Visible	Yes	0	Yes	3
	Models of culture	No	N/A	Yes	2
	Social Rites	No	N/A	Yes	3
	Dynamics	No	N/A	Yes	3
	Subcultures	No	N/A	Yes	3
	Ownership	No	N/A	No	N/A
	Understand Broad Categories in Play	Yes	0	Yes	4

The good news is that absorbing the corporate culture and group subculture is not on a strict timetable. Unlike a SMART objective, a new hire can't declare "By December 31st, I will have a 3rd level capability in identifying specific cultural types." Instead, a novice like Imad adopts a mindset to continuously look for examples of cultural impacts and, more importantly, what do those markers mean. Imad begins to see the benefit of observation and reconciliation of actions and meanings. As suggested previously, he resolves to be more curious.

Ubiquitous Communications

This brings Imad to his next category: communication. Fortunately for him, Imad already had some natural advantages in his area. In school, he had no reluctance to speak up in class; at university, he was vice president of a student service organization which required him to make short, verbal presentations to groups of strangers. In these areas at least, Imad was not a strict novice. For the other areas of communication, including identifying unique skills,

CASE EXAMPLE TABLE 12.1E
Example Assessment for Communications

		Current Location in Organization		Future Location in Organization	
		Logistics Electrical Engineer		Lead Engineer for Logistic Analytics Group	
Component	**Skills and Capabilities Categories**	**Knowledge/ Skill Required Today?**	**Your Knowledge/ Skill Level Today**	**Knowledge/ Skill Required in Future?**	**Level of Knowledge/ Skill Required in Future**
Communication					
Rating Definitions: 0 = No Knowledge 1 = Awareness 2 = Surface Knowledge 3 = Minimum Competency 4 = Journeyman Capability 5 = Expert Knowledge	Framework	Yes	1	Yes	4
	Opportunities and situations	Yes	3	Yes	4
	Control Issues	No	N/A	Yes	3
	Standardizing	No	N/A	Yes	3
	Special Cases	No	N/A	Yes	2
	Investigate and gain general communication experience	Yes	1	Yes	4

Imad lacked experience. For future communication skills, Imad only had a fuzzy idea that he needed to know "all matters great and small." The resulting assessment looked like Case Example Table 12.1e.

On examining his roadmap, Imad discovers that his communication skills are not as good as he thought they might be. While he was familiar with a certain subsegment of communication (ad hoc presentations), he didn't know that certain communication categories even existed. He needs education and experience in the remaining skills to attain at least a minimum competency rating. Again, this insight allows him to prioritize the work he needs to do. In consultation with his supervisor, Imad needs to take specific communications training, including presentation techniques, and request every opportunity to speak and communicate in public and in a variety of inter-organizational situations.

Ethics

Upon arrival, a new employee needs to have at least a rudimentary understanding of their individual ethical makeup. Both an individual's current and future

CASE EXAMPLE TABLE 12.1F
Example Assessment for Ethics

		Current Location in Organization		Future Location in Organization	
		Add Current Position Here		Add Future Position Here	
Component	Skills and Capabilities	Knowledge/ Skill Required Today?	Your Knowledge/ Skill Level Today	Knowledge/ Skill Required in Future?	Level of Knowledge/ Skill Required in Future
Ethical System					
	Confident/ Comfortable in Current Structure?	Yes	3	Yes	4

"values of self" inform and bind all the assessments they make within the five components. Both consciously and unconsciously, an individual's beliefs condition their roadmap. In a nutshell, knowledge of self and a person's ethical framework are critical boundaries and ground rules for creating a valid roadmap.

Imad is no exception. Fortunately for him, Imad attended a secondary school that offered two courses in philosophy and ethics. Imad took both, coming away with a decent understanding of the role both subjects can play in corporate life. As a result, Imad believes his ethical beliefs are appropriate for him at this time, and his future knowledge should build upon his existing foundational belief. While he knows he might be tested, he has reasonable confidence in his response, demonstrated in Case Example Table 12.1f.

So, there it is: a completed road map of where Imad believes he is today and where he wants to be tomorrow. Systematically, Imad deconstructed the necessary organizational skills and capabilities, assessed himself today, and began to consider the skills he needs to achieve operational competency within his organization at a certain future point in time.

A few notes are in order. First, obviously this is not meant to be a precise exercise revealing an accurate "answer." This is about gaining a skill in mapping your current abilities and assessing your way forward, and practice like this hones that skill. Second, as mentioned earlier, this roadmap should be revisited periodically. Each iteration will produce ever-better and higher-confidence results. Remember, this process is like a physician: always "practicing" the art of medicine to learn more about how to best improve a patient's health.

Finally, this process should make you a bit uncomfortable. After all, this is what it's designed to do. Forcing new professionals to earnestly face their areas of skill (or non-skill) gets them closer to having a quality roadmap to handle the many situations coming their way.

12.2 CROSSING THE DIVIDE: TURNING INSIGHT INTO ACTION

There is a very old phrase that is appropriate here: "Talk is cheap, but action is priceless": it is easier to talk about doing something than actually doing it. Many say they will do something but never do. Others say that something should be done, but they never envision themselves as the individual to do it. This roadmap is a prime example.

At this point, we have a defined framework, forged links between components, and developed a list of do's and don'ts for each component. We have shared a process to develop a personal map to find a way across the Divide. But all this is of limited value unless these insights can be put into action, to do something with this knowledge. Something that can be applied, especially on a moment's notice, a solid path forward that addresses the barriers and contingencies within each situation and how to plow a way through it.

The famous aviator Charles Lindbergh, in his memoir *The Spirit of St. Louis* tells the story of his decision to make the takeoff to start his nonstop trip across the Atlantic. He wrote of his great uncertainty in making the attempt, especially the many others who had crashed and died over the weeks before his own attempt. He wrote:

> THIRTY REVOLUTIONS Low! The engine's vibrating roar throbs back through the fuselage and drums heavily on taut fabric skin. I close the throttle and look out at tense faces beside my plane. Life and death lies mirrored in them – rigid, silent, waiting for my word. Thirty revolutions low – a soft runway, a tail wind, an overload … The wind changed at daybreak, changed after the Spirit of St. Louis was in take-off position on the west side of the field, changed after all those barrels of gasoline were filtered into the tanks, changed from head to tail - five miles an hour tail! Taking off from west to east with a tail wind is dangerous enough - there are only telephone wires and a road at the far end of the field, but …
>
> A missing cylinder and – "Hit a house. Crashed. Burned." – I can hear the pilots saying it – the end of another transatlantic flight.
>
> I lean against the side of the cockpit and look ahead, through the idling blades of the propeller, over the runway's wet and glistening surface … . A curtain of mist shuts off all trace of the horizon. Wind, weather, power, load – a balancing act.
>
> Wind, weather, power, load – gradually these elements stop churning in my mind. It's less a decision of logic than of feeling, the kind of feeling that comes when you gauge the distance to be jumped between two stones across a brook. Something within you disengages itself from your body and travels ahead with your vision to make the test. You can feel it try the jump as you stand looking. Then uncertainty gives way to the conviction that it can or can't be done. Sitting in the cockpit, in seconds, minutes long, the conviction surges through me that the wheels will leave the ground, that the wings will rise above the wires, that it is time to start the flight. I buckle my safety belt, pull

goggles down over my eyes, turn to the men at the blocks, and nod. Frozen figures leap to action. A yank on the ropes – the wheels are free. I brace myself against the left side of the cockpit, sight along the edge of the runway, and ease the throttle wide open. Now, in seconds, we'll have the answer. *Action brings confidence and relief.*[2]

Action does bring confidence and relief. Taking action is hard. It takes practice, lots of practice. As I just said, it's easy to read and talk; it's hard to actually *do*. You are going to make mistakes before you get it right, and that may take some time. And while it may be frustrating to do this work but not get an immediate benefit, learning how to take action will pay increasing dividends from the day you begin until you become a master at handling different professional situations. What you can expect in applying these guidelines and taking action include:

1) Certain early failure
2) Slow progress
3) Frequent forgetting to apply underlying principles
4) Struggle for patience
5) Relief when a single action finally works out
6) Intense satisfaction when beginning to avoid a mistake or dodge a bullet
7) Remembering those failed experiences for future use

Some may become frustrated with this roadmap. Today, there tends to be a high expectation from digital natives: that of instant answers and more importantly, instant truth. Ask a question on Google: you get hundreds of immediate answers (even if some are wrong). Doing your banking now requires a single click, taking 2.3 seconds to execute. Dinner magically arrives at your door in 30 minutes or less. But these are conveniences that unfortunately lead to an expectation that meaningful personal work is equally instantaneous. This personal work should have a different kind of expectation: one requiring continuous, sometimes uncomfortable, serious and meaningful effort. It is an investment in yourself that may not come easily.

The key point here is to not give yourself an excuse to avoid taking any action except reading this book. Taking action is where the benefit of this work is truly found.

It's time to move on, but before we do, I'm going to leave you with a simple question:

What are you going to do?

12.3 REALIZATIONS

As you have seen, developing your roadmap is a fundamentally simple process. It's a process of "start here (today) and go there (the future)," identifying your skills and capabilities in each component, bounded by an ethical boundary. Remember, the entire point in creating your map is in actually performing the analysis. There is no correct or incorrect answer, only an answer representing a point in time that will be readdressed, reconsidered, and weighed against ongoing organizational experience. The value is in defining what the "today" is, and what the future means, and untangling the interconnecting pathways between the components through the lens of self and your ethical framework.

To do this, you must practice the difficult and uncomfortable skill of moving from the strictly quantitative to the qualitative. Develop your roadmap in sections or pieces: remember, this is not an exercise in efficiency but an investigation into identity. Attempt to sensitize yourself to a more holistic view of your work and by extension, yourself. As you've heard before, embrace the uncertainty, apply mindfulness to this roadmap, and know that you are conducting an investigative process.

That said, let's now look at what's ahead and see some "previews of coming attractions".

NOTES

1. Costamare. 2022. *Container Facts*. Monaco: Costamare, Inc. https://www.costamare.com/industry_containerisation.
2. Lindbergh, Charles. 1952. *The Spirit of St. Louis*. New York: Charles Scribner's Sons.

13 Coming Attractions

13.1 PREVIEWS OF COMING ATTRACTIONS

We'll next briefly discuss what's to come. Like previews at a movie theater, several "attractions" are coming your way: one concerning the immediate future and the other addressing the longer-term time horizon, as much as such a thing can be realistically done. The point of this is to reinforce the notion that your roadmap, once created, will soon need to change. You'll notice this initially as small updates arrive (say, an improvement to an existing measurement technology that requires new training) to a major change to an HR conformance policy (say, requiring Darwinian "survival of the fittest" competition between department colleagues. Later, there will be even more substantial changes triggered by the external environment (think Tesla's entry into the global automotive market). At some point your roadmap will need a solid update reflecting the new reality at that moment. With this in mind, let's talk about the near-term, as well as long-term factors that may be approaching.

13.2 FACING THE IMMEDIATE

As mentioned above, the immediate future will begin to impact your roadmap soon after you finish it. These short-term changes are primarily caused by new information arriving from the enterprise-level organization to your engineering group or department. This can be good: every day means learning more, seeing new connections between components that will enrich your roadmap. Each day will produce new insights and understandings.

Let's consider some examples.

13.2.1 Continuing Sources of Information

At this point you may be sick of reading. After something like 16+ years of absorbing textbooks, websites, screen content, and other sources of information, anyone would be worn out. But one thing should be considered: throughout every professional career is the continuing need to absorb new information from a wide variety of sources. We're not just talking about journals specific to your engineering specialty but also a much wider variety of information topics and types. At this point in any career, there comes a need to broaden your viewpoint in addition to diving deeper within a technical specialty. Fortunately, as you know there are literally hundreds of adjacent technical and non-technical information sources (blogs, YouTube briefings, TED talks, and all the rest) to cull. Perhaps generate a list by category: some technical, some social, some balanced political, some economic (offshore sources are very good in this area), and some for pure enjoyment. Your purpose is not to make a list

DOI: 10.1201/9781003214397-17

of the "best" sources, checking each one off in turn, but instead looking for thought starters to broaden where to learn more. Just base your personal selections on two factors: the reliability of the information presented and the excellence in writing or speaking. Both can pay dividends in the long term.

Finally, please consider making this a weekly habit. Once established, you may find you do not have to carve out special time for reading, listening, or watching; it's already been done.

13.2.2 You May Not Be in Control

Many new engineers express surprise that disrupters appear soon after their job begins: they really have no "control" over their work. Think about this. At university many (sometimes even a majority) of assignments are "team-based," meaning not only is the work done in groups, but each member may also be graded as a group, not as an individual. If the assignment earns a "B," then everyone does, including the group superstar and the two slackers who don't bother to show up. The biggest frustration for the committed group members is not having true control over their grades.

Like at university, work assignments are also expected to be accomplished despite your lack of control. So many potential derailers: interruptions, other deadlines, uncooperative group members, no supporting data, inadequate budgets, the boss's wishes, changes in direction, and cancellations halfway through. The list can be long and difficult. Consider the common control issues between university and professional assignments in Table 13.1.

The solution to this is to plan your work, but not as you might think. You have to actually plan two things: the technical work itself and also plan your *work system*. The work system is not complicated. This is merely setting in advance the work process boundaries and ground rules to enable the technical work to proceed without an

TABLE 13.1
University vs. Professional Assignment Control Factors

Characteristic	Assignment Control at University	Assignment Control in Professional Settings
Assignment due date	Unchanging	High chance of moving in either direction
Choose group members	Generally, Yes	No
Data available to complete assignment	Always available	Low availability, many times none
Method to accomplish assignment	Well defined, provided by Professor	Ill defined, many times none. Requires invention
Interruptions	Can install barriers relatively easily	Constant. Expected availability to boss makes this difficult
Iterations	None. Rework of assignment is rare	Many iterations common; additional reworks with surprise due dates

TABLE 13.2
Phases of an Engineering Group Project

Phase	Outcome
Phase One	Enthusiasm for the assignment
Phase Two	Development and execution of the plan
Phase Three	Initial bad results
Phase Four	Upset management
Phase Five	Search for the guilty
Phase Six	Persecution of the innocent
Phase Seven	Replacement of assignment lead engineer
Phase Eight	Eventual project completion with everyone relieved
Phase Nine	Disband the group with a promise never to do that again

outside force stopping your progress. You've already done this in the past. Preparing for a final exam may have sent you to the library stacks, closing the door to your room, or working in the middle of the night to avoid pop-ins. In other words, establish privacy when required. This seems trivial, yet the open floor plans so common today have one major complaint: lack of privacy and excessive interruption. Just plan for your work before you work.

Finally, let's leave this section with an old joke (summarized in Table 13.2) about the "alternative" phases of an engineering group assignment or project every engineer eventually experiences.

13.2.3 Management Fad or Management Insight?

The book *The Second Curve: Thoughts on Reinventing Society* touched on the idea of identifying reoccurring corporate "fads" versus truly new and effective corporate management techniques.[1] The examples are many: *The One Minute Manager, Re-engineering, Blue Ocean Strategy* and the like. An important (and famous) example was the emphasis from 1970 onwards on an idea called *shareholder value*. This notion, most prominently expressed by the economist Milton Friedman, argued that the singular social responsibility of a business is to increase its profits, nothing more or less.[2]

Friedman argued that a firm is nothing more than a group of contractual relationships and the company management should strictly align themselves with the shareholders. This approach emphasizes attention on the company's immediate or short-term profits at the expense of the medium-and long-term outcomes. This became the popular view to hold, meaning companies should use every technique possible to make money for shareholders, not necessarily creating world-class products. In this philosophy, profit was a managed end goal, not a measuring stick tracking the overall success of the firm. Financiers replaced engineers at the head of manufacturing companies and corporate cultures changed.

Luckily, the shareholder value fad for the most part disappeared in the second decade of the 21st century. According to *The Second Curve*, "society has not benefited, indeed it almost collapsed in 2008 when banks and businesses overreached themselves."[3]

The lesson here for new engineers and other recent hires is this: companies tend to go through eras (or sometimes fads) of new management thought. Each era is believed to provide a new, "progressive" way of doing business. If and when in the future your company adopts one of these ideas, you will be obligated and compelled to follow it. Just bear in mind that this new way may not be the panacea advertised, but instead a well-intentioned but flawed attempt at modernization or change. The good news is that true fads will slowly disappear as management sees the shortcomings of the new way.

13.3 ANTICIPATING THE LONG TERM

It's easy to see that a long-term assessment will result in a radical change to a roadmap. The immediate future will quickly fall away, replaced by a future with major unknowns as significant societal, technological, economic, environmental, and political factors emerge and morph into a kaleidoscope of change. These longer-term changes are primarily caused by new information, techniques, inventions, and innovations from outside the firm, industry, or even current society. In this view, each day can produce major change, risks, and opportunities.

Let's consider some examples.

13.3.1 Impact of the Changing External Environment

The external environment is a pesky and annoying problem. It just won't stay still long enough for us to truly get a handle on it. And in the future, this problem will only get worse. Of all the major external forces you must deal with, each one is accelerating, and so its impact on you: seen as an increasing force of change and the emergence and disappearance of relevant factors. Many engineering planners are surprised when their six-month analysis of the external environment is rendered obsolete by the time they present their work to upper management. It's a frustrating and embarrassing situation when, in a meeting, management contradicts your work on the external environment.

Take for instance our friend Qianyan Shea, whom we met in Chapter 10. As you remember, she was her organization's technology "futurist." In the early 2000s, Qianyan identified nanotechnology, the use of molecular configurations such as carbon to construct physical structures that are strong, cheap, and easy to be made. At that time, it was a significant emerging technology that could drastically change the company's manufacturing methods and product performance. She correctly identified the emerging potential of this game changer.

It took about nine months of researching this technology to develop her analysis to where it could be briefed to senior management. Unfortunately, just a week before she was scheduled to present, nano took a significant turn from scanning tunneling microscope-based carbon manufacturing (STM) to a more biologically based growth engine using a building block approach. This change in direction was announced in the technology press. This turn of events quickly made the original carbon-based STM nanotechnology obsolete for her company's product engineering process. Yet Qianyan was told by her management she was still obligated to present even though her material was now out-of-date. Their justification was it would become a "background only" briefing. What happened? Soon after she began, she was stopped by the vice president of product development and was told that this original technology was already obsolete, as the vice president had just read the announcement and was not now interested in a background-only talk. Needless to say, Qianyan was embarrassed, and her management lost some confidence in her when she was stopped from presenting even though she was directed to give the presentation.

Or take the example of fossil fuels for the automotive industry. All battery-electric, plug-in electric, conventional hybrid, diesel, and internal combustion engines continuously change their cost-benefit ratios due to commodity oil prices fluctuating on a daily basis, plus in the longer-term movement in electricity prices, especially when considering external foreign markets such as China, Middle East, Russia, Europe, and in North and South America. This instability in the energy external environment requires continuous analysis to anticipate the energy environment three to six years in the future to support the selection of an appropriate power plant. Ask any strategic planner and they will give you chapter and verse on the headaches caused by this extreme environmental fluctuation.

13.3.2 Understanding the Organization's Lifecycle

Just like any organism, a technical organization is continuously moving through a lifecycle, a growth and decline dynamic where the organization is born, grows, plateaus, and finally declines unless it experiences a rebirth. This is a macro trend exhibited by any organization, and we need to understand what that lifecycle is, where a given organization currently is within that cycle and where it might be headed.

Figure 13.1 shows a generalized chart of a typical company lifecycle. This is an average; a particular engineering organization may or may not follow this exact trend line. But the important point is that there is a dynamic growth and decline aspect to every organization.

Briefly, there are five general stages of a firm's lifecycle:

Entrepreneurial Stage

This stage is identified by the initial marshaling of resources, investigating many ideas, and performing niche entrepreneurial activities with little planning while attempting to be a first mover within that niche. Creativity is the goal, while the company's size is very small.

Collectivity Stage

Collectivity is denoted by informal communication, structure, and a sense of group cohesion, resulting in a high commitment (i.e., spending long hours at work) originating from a strong sense of mission. Innovation continues at a high level.

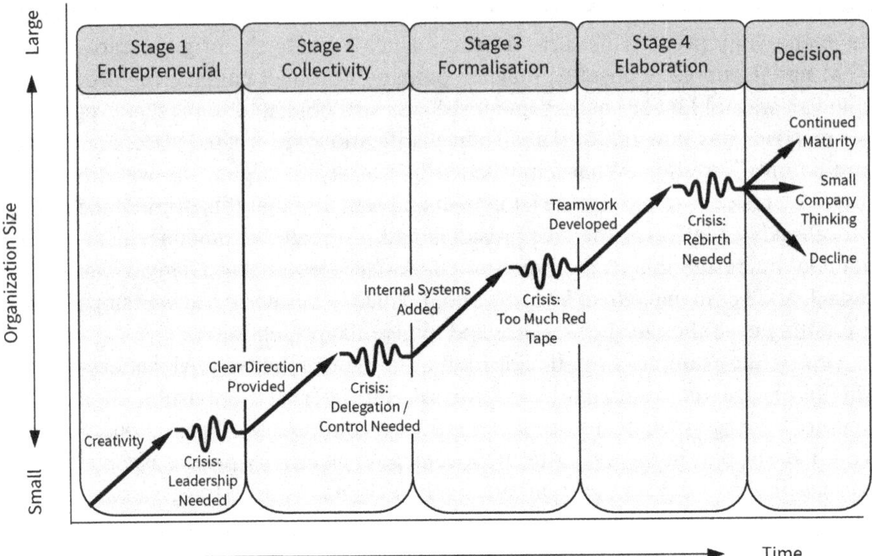

FIGURE 13.1 Generalized organizational lifecycle.

Formalization Stage

Here, standard organizational rules are developed and adopted, communication becomes more structured, and a stable structure emerges with an emphasis on coordination and control systems while expending a minimum of the firm's resources. Bureaucracy is now embedded in the firm.

Elaboration Stage

Expansion of the firm with increased structural complexity, decentralization into local profit centers, and expansion of markets while working in a post-bureaucratic manner, including associated widespread adoption of new methods. Expressed as a general sense of company maturity.

Decision Stage

Completion of the elaboration stage forces the firm into a strategic decision of what direction the company should now take to remain vital. This decision can include continued maturity, adopting small company thinking, breaking up into separate units, or doing nothing (which greatly increases the chances of decline).[4]

Table 13.3 provides additional characteristics of these four stages to pair with Figure 13.1.[5]

13.3.3 Reconciling Expectations

Let's now change our focus. Some time has passed since your arrival. You are comfortable in your position, and many of the work methods, assignments, and tasks are

TABLE 13.3
Characteristics of the Organizational Lifecycle

	Stage One – Entrepreneurial	Stage Two – Collectivity	Stage Three – Formalization	Stage Four – Elaboration	Stage Five – Decision
Characteristic	Non-bureaucratic	Pre-bureaucratic	Bureaucratic	Very Bureaucratic	
Structure	Informal, "one-person" show	Mostly informal, some procedures	Formal procedures, division of labor, specialties added	Very Rigid	Variable depending on path decision
Products or Services	Single product or service	Major product or service with variations	Line of products or services	Multiple product or services lines	May be legacy or innovative
Reward and Control Systems	Personal, paternalistic	Personal, contribution to success	Impersonal, formalized systems	Extensive, tailored to product and department	On spectrum from legacy to innovative
Innovation	By owner-manager	By employees and managers	By separate innovation group	By institutionalized R&D	On spectrum from evolutionary to revolutionary
Goal	Survival	Growth	Internal stability, market expansion	Reputation, complete organization	Secure future viability
Top Management Style	Individualistic, entrepreneurial	Charismatic, direction-giving	Delegation with control	Team approach, attack bureaucracy	Agents of change

Source: Adapted from Greiner, Larry E. 1972. Evolution and Revolution as Organizations Grow. *Harvard Business Review* 50: 37–46.

now familiar and beginning to become enjoyable. You seem to fit well with your colleagues, and your reputation is positive and growing. Assuming that you wish to stay, let's say your future career expectation is a choice of either spending some additional time in your present position before rotating to that design position you've always wanted or being promoted into an advanced technology position in the R&D organization. You've shared this with your supervisor who is understanding and supportive. Maybe you've already had several interesting, short-term rotations in other departments. And whenever you asked your supervisor if you could rotate there, the answer is always an immediate "yes."

Then one day that same supervisor sides up to you and says: "Have you ever thought about getting your master's in business administration?" You state you haven't. The supervisor replies: "Maybe you should think about it. Let's talk later."

You now know what's coming: the battle over expectations.

Every new engineer has a series of personal expectations guiding their work path, normally a series of steps leading to a satisfying goal. These are important to you; after all, fulfilling these expectations is the main reason you joined the company in the first place. Yet at some point, at a time and place unknown to you, you will be confronted with another set of expectations: those of your organization.

The overall hypothesis is this. As a new hire, your company and professional knowledge are very limited. Essentially any position within the company will gain you experience. Through your management's lens, wherever a new hire wants to go is granted. It's a low-risk play: the engineer gets much-needed experience while seeing themselves as in control of their career path. This may go on for several years.

As more time passes, the supervisor or manager begins to exert a bit more control over the engineer's job choices. This is evidenced by those friendly questions, and those suggestions that come out of the blue. There's an agenda here. This may become apparent when, during the weekly 1:1 meeting, the supervisor asks if you would like to take on a certain role, either inside or outside the department. Of course, it's totally your choice.

The firm's expectations of you are normally well hidden, and many times couched in tempting terms. "You've done so well; we think you're the right person to do this job." Or "This is a great growth opportunity for your long-range development." Or even "This is part of the great plans we have for you." This may all be true. But there is another possibility: They need someone, now, to take over a position or do an onerous task which is risk-filled, unpleasant, or seen as a backwater. You are picked because they need their expectations filled, not yours.

You can tell if this is happening by the manner this is communicated: you are informed, not asked.

If this sounds a bit familiar, it should. This results in something commonly referred to as the *golden handcuffs*, where the past and current rewards and investments you've made in your current company prevents you from moving to a new position without risk. Over time, you find it harder and harder to leave. Ask anyone who has worked for a firm for over five years and they can tell you all about it.

The purpose of bringing this up has to do with awareness. Be aware this situation will be coming to you. And don't be surprised, as nearly everyone in any organization experiences it. Just be thinking about this question: what are the company expectations that you are able to live with and those you are not? What if you are asked to transfer to a different department in your building? To a different building? A different city, or state, or country? Like ethics, this type of company expectation questions needs to be pondered in advance before you are asked.

13.4 SMALL MATTERS TO BE ATTENDED TO

This is a section I initially resisted writing, but in a book of this type, we can't escape those pithy, one-line nuggets of advice that convey the wisdom of the ages in a dozen words or less. What follows is a list of these statements that you will collect in both the near and far term future. It's easy to come up with a general list of "do this" or "be sure not to do that." And it's simple to receive advice but also equally easy to ignore. Yet in some instances we just can't do that. So here are a few points that you will probably see or hear again: just remember you got them here first:[6]

1) Be careful with the commitments you make. If you make a commitment, you must fulfill it.
2) Learn from others: set up and operate an efficient and effective information gathering system.
3) Work on the right projects, meaning learn how to say no.
4) Plan your work, then work your plan, efficiently.
5) Allow reserve time. Don't schedule 100% of your work hours. Leave slack time to ensure your work can be finished despite surprises and self-inflicted wounds.
6) Value yourself in the job, and value yourself personally.
7) Always keep your sense of humor.
8) The grapevine probably knew something about you before you arrived on the job. Find out what that something was.
9) Choose the style you are comfortable with and be yourself.
10) A common rule: "There is no sure thing as a stupid question." Its corollary: "As long as it's asked only once."
11) Fitting in can be very difficult at first. Everyone knows everyone and you are new, so you are the pariah.
12) Of course, it's easier to follow someone who has been fired.
13) Identify who are the friends of your local organization.
14) Talk to people, but more importantly, listen to people.
15) Be seen.
16) Get the skills you need quickly: this means training.
17) Understand the status quo.
18) Be right the first time on the first assignment. Be right the first time on the second assignment. You should be predominately right on the third assignment.
19) Once you are ranked as a person with low potential, that corporate opinion will rarely change. If it does change, it takes a long, long time.
20) If you don't have a network, you must build one, as current information helps you prepare for change.
21) Join into the social system as you are allowed.
22) Congratulate others' accomplishments and successes as soon as possible.
23) The process and systems in your new organization will always be in flux. Be prepared for this.
24) Be cheerful and have fun.
25) Stop worrying.
26) Succeed.

13.5 REALIZATIONS

Anticipating the immediate and far-field future is obviously an uncertain business, yet there are a few actions available to better the chances of getting it right. The carefully constructed roadmap you've created will need to be updated periodically; at what frequency is up to you. Like a single loop control system, roadmaps drift away from their normal value over time and do not return until brought back into relevancy by revisiting the original work. In the meantime, one of the actions to slow down this normal drift between revisits is to absorb fresh information from wide-ranging sources, not just technology-related fields. The idea is to round out your personal knowledge over time. "Lifelong education" is not just a slogan; it's a way of life.

One of the most frustrating facts about working within engineering is the inherent lack of control in your assignments and daily activities. The hoped-for straight line between start and finish rarely (if ever) happens, and the actual circuitous route has many sharp turns that can easily put you into the ditch. A mindset of expecting the sharp curves and meandering route may be the best way to prevent some unwanted frustration and anxiety.

A large contributor to this lack of control is management's continuous desire to improve their operations. Major frustration occurs when everyone is forced to absorb another change to the general management philosophy. While this change is less prevalent now than in the recent past, pleas to "think nimbly" and "be entrepreneurial" can actually hurt productivity and effectiveness if presented without worker education, an implementation plan, and a serious mindset by management. Overwise "management by slogan" results.

Approaching the longer-term time horizon can be fraught with danger or at least has a high potential for embarrassment. Just knowing (and viscerally understanding) that the external environment will radically change in the farther-term time frame goes a long way to condition your mindset to anticipate the changes in your path. Being proactive rather than reactive is always a good bet, especially when considering your own engineering company's lifecycle.

For instance, know where the firm is at this moment in time and attempt to judge where it might be headed. Is the firm's path moving in a good direction? Is it addressing the normal issues at its current life stage? Significantly, does the company's life stage match with your own personal stage of life? This can be telling. For example, say the firm is at the collectivity stage, with its informal structure, high commitment and required long hours at work. Are these long hours consistent with your personal life stage, such as caring for a significant person, or raising young children, or climbing El Capitan? Consider the other non-engineering factors in your home and work. Comparing your company's life stage with your own personal journey can really help assess your current path and plan for your next steps.

NOTES

1. Handy, C. 2015. *The Second Curve: Thoughts on Reinventing Society.* London: Random House.
2. Posner, Eric. 2019. Milton Friedman Was Wrong. *The Atlantic.* https://www.theatlantic.com/ideas/archive/2019/08/milton-friedman-shareholder-wrong/596545/.
3. Ibid., Handy.

4. Daft, Richard. 2010. *Organizational Theory and Design*. Mason, OH: Cengage Learning.
5. Quinn, Robert and Cameron, Kim. 1983. Organizational Life Cycles and Shifting Criteria of Effectiveness: Some Preliminary Evidence. *Management Science* 29(1983): 33–51; et al., and Daft, 340–344.
6. Bolling, G. Fredric. Professor, University of Michigan, Dearborn. Interview by Robert M. Santer, August 3, 2000. Transcript.

14 Some Final Thoughts

14.1 THE MEANING OF THE GREAT DIVIDE

We've reached the final part of our journey together, the "reflect and conclude" phase of making our way across the Great Divide. In our time together, we've covered a lot of territory and met a wide range of people; all different yet sharing the common challenge of working successfully within various technical organizations.

When reflecting on the major messages in this book, two immediately come to mind. First, the environment the organization operates in will continue to change at an exponential rate, meaning any insights presented today will evolve significantly in ten years' time. Your future understandings will be informed by the foundation presented here, but it is your responsibility to keep the roadmap fresh and relevant. In short, contingency theory influences all that you do today, as well as hope to do for the foreseeable future.

Secondly, the important role of personal ethics and values needs to be developed and settled upon very early in a new engineer's professional life. As repeatedly stated, ethics is the one subject that does not follow contingency theory. Values are absolute, not relativistic, and set a constant and fundamental context for all professional work. Making decisions within a values context significantly improves the quality and success of the work surrounding that decision.

That said, let's consider some final thoughts to conclude our journey across the Great Divide.

14.2 DEMONSTRATING TECHNICAL MASTERY

This idea of technical mastery, that is, knowing your technical "stuff" at the expert level, depends on the technology involved and the organization you work in. Some organizations that value strong technical knowledge will articulate and express that value often and without hesitation. Other organizations tend to assume technology is a prior expectation merely in service of the greater goal of bringing an awesome product to market. Products addressing customer safety or important public policy tend to follow the former. Organizations that provide more convenience-based consumer products will not emphasize technology as much, as convenience products do not normally impact public safety. They have the luxury of being less concerned about important functionality than companies such as autos, aircraft, pharmaceuticals, and nuclear energy.

Yet no matter the case, being an expert in a technical specialty leads to respect and value even though management may not express it often. Consider when an engineering problem emerges. Management is interested in solving the technical problem as quickly as possible at the minimum cost, either in dollars, reputation,

DOI: 10.1201/9781003214397-18

or both. As part of the task force solving the problem, you as an engineer will bring to bear your engineering knowledge in a "nuts and bolts" approach to solving the issue. Management assumes you know how to do this and will not necessarily realize the depth of technical skill and knowledge needed to successfully accomplish this task. Yet without your technical knowledge, the problem may not be able to be solved at all.

So, at some point, what happens after spending a number of years within the technical core? Assuming you stay in the firm, do you stay in the core and continue a career in a technical area with the mastery that comes with it, or do you elect to move on to a different part of the enterprise? This question touches on the dual ladder concept we discussed in Chapter 7.

It's also a very interesting question, and I will not venture individual advice on this complex and highly personal puzzle. But most people know that technical prowess and strong reputation is an indicator of future success. As the old saying goes, "The best indication of future success is past success." Just know the question is out there and will arrive someday.

A word about new tools. New, emerging tools are an exciting and critical part of any dynamic organization. Luckily for new employees, it's also a place for enhanced job satisfaction, positive exposure, and a place to grow an excellent reputation.

Sadly, a surprising number of experienced engineers will tend to shy away from new tools, either from lack of confidence in learning them, resistance to new ways of doing business, or just plain laziness. New hires are excellent candidates to be "first movers" in adopting, learning, and applying these new tools. This opens up an excellent opportunity to provide a unique skill to the organization. You would become part of a small group possessing that specific technical skill, making you more valuable to your department and especially your boss. There is very little downside to quickly adopting and learning these new tools, whatever they may be.

A great example of adopting a new organizational "tool" had to do with a new organization at a U.S. mobility firm being established in Palo Alto, California, in the mid-2010s. This company decided that the "Silicon Valley Way" of quickly creating and marketing products needed to be investigated and perhaps adopted for their own use. This firm decided to establish a large satellite campus adjacent to Stanford University. Here, engineers and coders would discover and develop breakthrough mobility products while demonstrating the superiority of new methods to the firm's legacy engineers. It was also meant to be a "listening post" for emerging technologies as well as being a critical contributor to the company's methods of working. Basically, this new facility was to become a huge "technical tool" for new product development.

One of the key approaches to this new way of working was the adoption of the "skunk works" methodology. Here, the Palo Alto facility was purposely placed some thousands of miles away from the headquarters "mothership." This allowed the engineering team to work in relative anonymity, fostering those fresh methods of working. For a new engineer this can be heaven, but only if management provides meaningful support to let these new tools and processes work themselves out before being applied to the organization as a whole.

The story of the development of this Palo Alto facility is a long and interesting one. The conflicts, fault paths, and uncertainties convincing management this is a good idea and resulting employee benefits are just a few of the many challenges that need to be overcome. In fact, it took more than one attempt to get the facility up and running. Yet today, after almost a decade, the Palo Alto site is finally being accepted as a legitimate and forward-looking component of the overall organization, and adoption of this new tool is finally proceeding. And those located there are pursuing mastery of a new technology: a new way of working.

CASE EXAMPLE 14.1 THE IMPACT OF DR. HAREN GANDHI

This brings us to an excellent example of how a single individual's technical mastery literally changed an entire planet.

We begin with a gentleman named Dr. Haren Gandhi, an automotive research chemist in a global mobility firm specializing in the nascent field of atmospheric pollution. In the mid to late 1960s, global tailpipe emissions of carbon monoxide, hydrocarbons, and nitric oxide were reaching crisis proportions, and governments worldwide began to realize that regulatory action was required to reverse this alarming trend. In the United States and elsewhere, extremely aggressive new laws were passed, requiring automotive companies to substantially reduce tailpipe emissions, and quickly.

Starting in 1967, Haren focused his work on emissions mitigation. One promising area was three-way catalysts (TWCs) which perform three functions: convert carbon monoxide into carbon dioxide, hydrocarbons into carbon dioxide and water, and nitric and nitrogen oxides into nitrogen and water. What prevented TWCs from being used was the very high cost of three critical elements used in the catalyst: platinum, palladium, and rhodium. Individually, no one could afford these precious metals. If TWCs were to work, increased availability and reduced cost of these base elements across the industry were the challenge.

Haren's solution was to establish a global infrastructure system through which all automotive companies worldwide would share their TWC technologies and together tackle precious metal utilization and recycling, making this catalyst feasible and affordable.

It worked. By sharing the best of global TWC designs, and banding manufacturers together to completely change the precious metals industry, pollution from hydrocarbons and carbon monoxide was cut by 90% from 1970 to 1975, and nitric oxide pollution was reduced by 90% from 1970 to 1976.

The short story is this: Haren invented and led the implementation of modern catalytic convertor system, the basic method for reducing global tailpipe pollution used to this day. Haren's leadership brought the automotive industry through the crises.

In short, Haren literally changed the world.

Today, Haren's legacy crosses all borders, having served as an automotive emissions panel expert appointee of the United Nations and advisor to the Indian government. His four-decade-long research led him to be regarded by the National Academy of Engineering as "one of the world's foremost authorities in the area of automotive emissions control."

A postscript.

One day in 2003 while working at his office, Haren received a phone call from a strange number.

Haren: "Hello?"
Stranger: "Hello, is this Dr. Haren Gandhi?"
Haren: "Yes, it is"
Stranger: "Dr. Gandhi, this is the White House calling. The President of the United States has awarded you the National Medal of Technology and Innovation. He was wondering if you would join him in the East Room where he will present the award to you. Are you available?"
Haren: "I think I am."

Haren won the nation's highest honor for technology for his global leadership in reducing atmospheric pollution. It was the first time ever that an auto industry researcher had been awarded the Medal of Technology, joining such past winners as Bill Gates, Steve Wozniak, Gordon Moore, Ray Kurzweil, and many others.

An important point emerges here that deserves mention. Mainstream thinking tells us that to make a global impact on society (and get rich in the process), a person should create a startup, fight for venture capital funding, and through superior ideation and plain old "smarts," implement a game-changing technology or innovation. The world changes as a result.

This has certainly worked. But there is another way that may be a higher percentage play.

Large, established companies (especially mass-market firms) have power that startups don't. Internal funding, name recognition, hard resources, and existing infrastructure: all make an engineer's life infinitely easier. Large firm engineers can book meetings and present ideas that startups will never be able to achieve. Yes, large firms can install internal barriers startups don't need to worry about. But the point is this: joining larger, established companies can leverage power and resources that startups generally lack. And as Haren proves to us, there is more than one way to impact our world.

14.3 WHAT MAKES YOU SPECIAL?

In the early days of globalization, North American and European companies struggled with the disruption caused by digital communications and the rapid movement of goods and services offshore, especially to Asia and India. Relatively comfortable organizations suddenly found themselves attacked on cost performance due to so many services moving to low-wage, higher-education countries. Tom Peters, the influential management writer and author, came up with an effective counter to the overwhelming uncertainty of the time. It came in the form of a simple question: "So, what exactly makes you special?"[1]

Simply put, what product, service, knowledge, or talent do you as an individual (or collectively as a firm) possess that cannot be replicated digitally by anyone, overseas or local to your home market?

Peter's thinking was that the digital revolution took the elements of geography and distance and removed them from the marketplace. With the sudden realization that any knowledge work can instantaneously move anywhere on the globe forced a rapid and radical rethinking of what is an individual's value to the engineering organization. In other words, what is your personal competitive advantage you bring to any technical group? Mark Karbow, an IT engineer with about 25 years' of experience, witnessed this firsthand. Very early in his career, Mark learned a specialized 1980s computer language called Erlang. He found it easy and satisfying to work with. Unfortunately, Erlang was quickly surpassed by more sophisticated and robust languages in the 1990s such as Haskell and Java, and Mark naturally learned those languages and continued his career. Some three decades later, the original Erlang language had a bit of a renaissance as original programming done in Erlang needed to be updated to support current and emerging hardware and software systems. Suddenly, there was a strong demand for anyone with experience in Erlang, and Mark fit the bill perfectly. Soon, he had all the work he could deal with converting that legacy language into 21st-century uses.

That language made Mark special. It made him unique, and there was obviously a high demand for that special skill he brought to the workplace.

So, what makes *you* special?

14.4 WOULD YOU RATHER BE LIKED OR RESPECTED?

This simple phrase incorporates several of the points we've been discussing throughout this book. A decision is looming: in the workplace, do you wish to be *liked* or *respected*? To be either liked or respected is a very personal "Y" in the road, a choice that can have a life-long impact on your work and personal life. The answer to this question can help direct you in so many situations and guide the decisions you eventually will be faced with.

Being liked involves several beliefs and actions you might consider. If you remember our conversation about conflict strategies, you'll find being liked tends to fall into the passive category. It involves loss of control where you give up your original position, sacrificing it to be accepted by another person or entity.

To be respected is a different animal. Respect means that you pursue fair outcomes to conflict, that people trust your actions, words, and beliefs, and will do the right thing even if it causes additional effort or loss. Respect means integrity. It does not mean the avoidance of conflict but the pursuit of fairness.

Making the decision to be liked or respected needs to be made early and reinforced repeatedly, so it becomes second nature; to naturally follow that philosophy without consciously thinking about it.

I was introduced to this idea by my wife Beth. Beth is the smartest person I know, an amazing combination of first-class thinking combined with the special gift of empathy and the ability to understand people and their interrelationships. She is a wonderful combination of giving and insight.

Whenever I face an important question and am unsure how to proceed, Beth has a simple question for me: "Bob, would you rather be liked or respected?" The simplicity of this question belies its importance.

It's great if you can be both liked and respected. Unfortunately, it's normally not that easy. Being respected is harder, yet no matter how uncomfortable you are in sharing unpleasant news with others or speaking truth to power, it's a good bet that you will come out of the interaction with the respect of others. Respect makes for long-term benefits for you and the people around you. It's nice to be liked, but it's even nicer to be respected.

14.5 COMMITTING ENGINEERING

You are about to commit engineering. While a cute quip, there is a nugget of importance in the meaning of that phrase. The act of committing engineering is a deliberate, important, and impactful action, where careers can be advanced or broken and lives won or lost. While we like to talk about the fun stuff in engineering: developing a new app or developing a cool and exciting new product that makes you eager to share it with your friends, there is a certain core importance we need to realize and commit to at the start of any professional career.

Soon after I began my work at Boeing, a terrible tragedy occurred on the F-15 Eagle fighter program. The Eagle had been in production for many years, and each finished aircraft was routinely flown on an acceptance test flight to ensure all systems were functioning correctly and the plane was ready to be delivered to the customer.

During an acceptance test for a new F-15, the aircraft suffered a major structural failure. A life raft inadvertently deployed from under the pilot's seat, jamming the control stick forward. The test pilot immediately lost control and crashed into the woods of western Missouri. As is normal in any crash investigation, the parts of the ill-fated aircraft were recovered and sent to a Boeing facility, where each part was reassembled (as best as possible) to help determine the root cause of the tragedy.

Soon after my arrival at Boeing, I was summoned to a meeting with a number of my newly hired colleagues. Not sure why we had been called, our group entered a large hangar where spread out on the floor and on tables throughout the room were the remains of the crashed fighter. Parts were everywhere: crumpled, broken, and covered in hydraulic fluid and oil. The scene was like something from a disaster movie. Our guide told us to take the next hour and examine, touch, and try to put together the remains in a way we could learn something.

As we gathered back together, our guide held up a shattered piece of the clear plastic canopy that once protected the pilot. On the inside surface were three long scratches. Our guide asked: "Does anybody know where these scratches came from?" Naturally, no one did. The guide then held up the test pilot's helmet. On the left side of the helmet were three screws protruding from its surface. Our guide placed the helmet against the three long scratches, showing a perfect match to the three screws on the helmet. The conclusion was clear. During the accident, the pilot exited the vehicle through the top of the canopy. We all became very quiet.

It then occurred to me why we were there. This is not about reconstructing an accident by putting bits and pieces back together. This was about learning our responsibilities as aerospace engineers; that our actions, our designs, our checking and double checking our work all had a direct and literal impact on the lives of the pilots who flew these aircraft. We had a serious responsibility as engineers to ensure no one needed to visit a hangar in the future. A group of solemn young engineers left the building that day, better not only for the oil on our hands but for the visceral experience we just had.

A reminder of that experience occurred a few years later. The F-18A fighter had moved from design into the manufacturing phase with the building of a dozen pre-production prototypes. As a liaison engineer in the factory, I was called on to solve a problem inside the pilot's cockpit. An avionics electronics box would not slide into its mounting rack located by the pilot's right elbow. The avionics box was about a 1/16th inch too high for its rack, and a rework of the rack would be needed. As I climbed into the pilot seat, I thought I could wiggle the box and force it into its space on the rack. I began to push and shove, trying to get some leverage to force the box in. Out of the corner of my eye, I saw a stranger approaching the cockpit.

Stranger: Hi, what 'cha doing?
Santer (huffing): "Trying to get this avionics box into this support rack ..."
Stranger: "Looks like it's kind of tough, huh?"
Santer (puffing): "Yeah, but I think I can bend it enough to get it in. It's only a sixteenth interference. If I can just twist it, I can avoid having to rework the entire rack structure."
Stranger (icily): "Well, I suggest you do it right. Because when this plane takes off for the first time, my ass is going to be sitting where yours is right now."

It was the test pilot.

I immediately knew what I had done wrong. In my hope of avoiding a bit of work and "letting it slide," I was willing to trade my inconvenience for the risk caused by not doing things right, and it took the actual test pilot to wake me up. The fact that I was looking at the face of the real person who would fly my product was sobering. The point here is not the severity of the cheat but the fact that I was willing to do it. It told me I was beginning to let standards relax. That pilot performed an excellent service for me that day. He reminded me of my responsibility; to watch for my tendency to take a shortcut. The fact that I clearly remember this incident after so many years tells me its importance to this day.

14.6 THE LEGACY OF YOUR WORK

At this point it may feel premature to be thinking about the legacy of your work. After all, just hitting your marks each day, learning as fast as possible about each new situation; these thoughts are certainly going to push out of your mind the legacy aspect of your work. No one would blame you.

Yet as I gained time in various jobs and positions in my career, I began (slowly at first) to ponder two ideas: first, the "legacy" of my work and, secondly, its overall importance. What I mean by that is, how much longevity does my work have? What is the "mean time" of the relevancy of what I do each day as a professional? What is the relative importance of my efforts? No one wants their work to quickly become obsolete or insignificant. For many engineers, career satisfaction is directly sourced from both the impact and longevity of their work.

This is something that you don't need to consider right now (after all, getting a job and then keeping it in your early days is the main event), but you might want to put these questions in your memory and use them to guide your experiences toward that future job that fulfills your definition of high impact and lasting usefulness.

CASE EXAMPLE 14.2 THE JOURNEY OF YUAN LONGPING

We live in both possible and probable times. As engineers, we want to see what is possible: a new material, the new bioengineered limb, a new data analysis tool, the new vaccine delivery device. Yes, all these are certainly possible. But the next question is harder. Are these things probable? Are we smart enough and savvy enough to move these desires from the possible to the probable? Is there a limit to the size of the immense problems we as engineers can overcome, either as an individual or in groups, large or small?

I like to look at the story of Yuan Longping, an agronomist who died in 2021 at age 90. A resident of Hunan Providence in China, Yuan spent his life on one thing: rice. Amazing, special, life-changing little bits of nutrition.

From early on, Yuan was fascinated with all things rice, growing it, tending it, and improving it; constantly striving to increase its size and nutrition. After a lifetime of work, "Super Rice" was born: a hybrid seed producing an amazing 20% to 30% higher yield, meaning some 60 million more people could be fed globally. With his help, China's rice crop rose from just 57 million tons in 1950 to about 195 million tons in 2017, moving the entire country from food scarcity to security. His higher ratios allowed growers to use the now-extra land for other crops such as fruit, vegetables, and fish. Diets became more balanced: today, 20% of all rice in the world comes from his hybrids.[2]

So what can we learn from Yuan? Like Haren Gandhi, Yuan is a simple story of how a single passionate man from a Chinese province changed the world in a massive and positive way. Belief, knowledge, and action made this possible.

14.7 ACROSS THE GREAT DIVIDE

There is an old proverb attributed to Geoffrey Chaucer: "All good things must come to an end."

Nothing lasts forever, and all things and situations are temporary, including enjoyable ones. For me, it also applies to this book. The people we've met are remarkable for their many experiences. We've met Bob Tyler, the aircraft factory foreman who taught lessons about bias. We've met Laura Kendric, the high-flying marketing executive who ignored the key facet of culture in organizations. We've witnessed Ed Krieger struggle with the ethical conflict between himself and his company, and so many more people with insights to share. Tom Siligato, Greg Scribner, Qianyan Shea: all contributed their experiences and learnings to our time together.

Your new organization wants you to succeed. The organization has already invested time, talent, and money (and for some, their reputation) in your recruitment. By hiring you they are placing a bet that you will perform to their expectation and most importantly, fit into their existing community and society. That is not to say that you don't have to prove yourself, but it does provide you some cushion, as they will expect mistakes and missteps as you enter into your new professional world. And while they are interested that you succeed, that does not mean that your transition will be made easy. It will be your professional responsibility to use this opportunity to secure your ongoing place in your chosen organization.

So here we are. You're now facing the impending problem you knew was out there, but now it's knocking at the front door. That Great Divide is here. The transition has started, no matter your thoughts on it. It may be happening not at the time of your choosing and its duration is indeterminate. Some transitions are mild and short, others are somewhat painful and long, some begin easily and suddenly hit like a punch to the gut. All are totally at the discretion of you and your new professional organization, and you must be prepared for all these situations.

14.8 AU REVOIR

Finally, thank you for coming this far. I hope you have enjoyed this little journey across the Great Divide and have learned how to better operate in your future technical and engineering home. I prefer to think of this moment not as a goodbye but instead consider as the French would say: *au revoir*; until we meet again.

NOTES

1. Peters, Thomas. 1998. *The Circle of Innovation.* Interview with Charlie Rose. First aired on The Charlie Rose Show, PBS in 1998.
2. Economist. 2021. To Feed the World. *The Economist.* May 29, 2021.

Index

Printed in the United States
by Baker & Taylor Publisher Services